Neuropeptide Systems as Targets for Parasite and Pest Control

ADVANCES IN EXPERIMENTAL MEDICINE AND BIOLOGY

Recent Volumes in this Series

Volume 684
MEMORY T CELLS
Edited by Maurizio Zanetti and Stephen P. Schoenberger

Volume 685
DISEASES OF DNA REPAIR
Edited by Shamim I. Ahmad

Volume 686
RARE DISEASES EPIDEMIOLOGY
Edited by Manuel Posada de la Paz and Stephen C. Groft

Volume 687
BCL-2 PROTEIN FAMILY: ESSENTIAL REGULATORS
OF CELL DEATH
Edited by Claudio Hetz

Volume 688
SPHINGOLIPIDS AS SIGNALING AND REGULATORY MOLECULES
Edited by Charles Chalfant and Maurizio Del Poeta

Volume 689
HOX GENES: STUDIES FROM THE 20TH TO THE 21ST CENTURY
Edited by Jean S. Deutsch

Volume 690
THE RENIN-ANGIOTENSIN SYSTEM: CURRENT RESEARCH PROGRESS
IN THE PANCREAS
Edited by Po Sing Leung

Volume 691
ADVANCES IN TNF FAMILY RESEARCH
Edited by David Wallach

Volume 692
NEUROPEPTIDE SYSTEMS AS TARGETS FOR PARASITE AND PEST CONTROL
Edited by Timothy G. Geary and Aaron G. Maule

Neuropeptide Systems as Targets for Parasite and Pest Control

Edited by

Timothy G. Geary, PhD
Institute of Parasitology, McGill University, Montréal, Québec, Canada

Aaron G. Maule, PhD
Parasitology, School of Biological Sciences, Medical Biology Centre, Queen's University Belfast, Belfast, UK

Springer Science+Business Media, LLC

Landes Bioscience

Springer Science+Business Media, LLC
Landes Bioscience

Printed in the USA.

Springer Science+Business Media, LLC, 233 Spring Street, New York, New York 10013, USA
http://www.springer.com

Please address all inquiries to the publishers:
Landes Bioscience, 1002 West Avenue, Austin, Texas 78701, USA
Phone: 512/ 637 6050; FAX: 512/ 637 6079
http://www.landesbioscience.com

The chapters in this book are available in the Madame Curie Bioscience Database.
http://www.landesbioscience.com/curie

Neuropeptide Systems as Targets for Parasite and Pest Control, edited by Timothy G. Geary and Aaron G. Maule. Landes Bioscience / Springer Science+Business Media, LLC dual imprint / Springer series: Advances in Experimental Medicine and Biology.

ISBN: 978-1-4419-6901-9

Library of Congress Cataloging-in-Publication Data

Neuropeptide systems as targets for parasite and pest control / edited by Timothy G. Geary and Aaron G. Maule.
p. ; cm. -- (Advances in experimental medicine and biology ; v. 692)
Includes bibliographical references and index.
ISBN 978-1-4419-6901-9
1. Neuropeptides. 2. Insect hormones. 3. Pesticides. I. Geary, Timothy G., 1952- II. Maule, Aaron G. III. Series: Advances in experimental medicine and biology ; v. 692. 0065-2598
[DNLM: 1. Neuropeptides. 2. Antiparasitic Agents--pharmacology. 3. Arthropods--drug effects. 4. Helminths--drug effects. 5. Receptors, Neuropeptide. W1 AD559 v.692 2010 / WL 104 N4942077 2010]
QP552.N39N47455 2010
612'.015756--dc22

2010021837

PREFACE

The need to continually discover new agents for the control or treatment of invertebrate pests and pathogens is undeniable. Agriculture, both animal and plant, succeeds only to the extent that arthropod and helminth consumers, vectors and pathogens can be kept at bay. Humans and their companion animals are also plagued by invertebrate parasites. The deployment of chemical agents for these purposes inevitably elicits the selection of resistant populations of the targets of control, necessitating a regular introduction of new kinds of molecules.

Experience in other areas of chemotherapy has shown that a thorough understanding of the biology of disease is an essential platform upon which to build a discovery program. Unfortunately, investment of research resources into understanding the basic physiology of invertebrates as a strategy to illuminate new molecular targets for pesticide and parasiticide discovery has been scarce, and the pace of introduction of new molecules for these indications has been slowed as a result. An exciting and so far unexploited area to explore in this regard is invertebrate neuropeptide physiology. This book was assembled to focus attention on this promising field by compiling a comprehensive review of recent research on neuropeptides in arthropods and helminths, with contributions from many of the leading laboratories working on these systems.

The Editors have been involved in neuropeptide studies in helminths for more than 20 years, including a productive period of intense collaboration on the identification and physiological characterization of nematode neuropeptides at The Upjohn Company (now part of Pfizer, Inc.) in Kalamazoo, Michigan. Aaron Maule continued in the basic research realm at Queen's University Belfast, while Timothy Geary remained focused on the discovery of non-peptide ligands for neuropeptide receptors as candidate antiparasitic drugs in Kalamazoo. The Editors would like to thank the many people involved in those efforts in both our laboratories; the number is too large to permit an exhaustive list, but the literature citations in the chapters provide the evidence. However, credit must be given in particular to Prof. David Halton, now an Emeritus Professor at Queen's, who first brought us together. His wisdom and foresight in encouraging us to pursue this field of research (in which he was a pioneer) has been rewarded (or at least so we hope).

The contributing authors have made notable and far-ranging contributions to the understanding of neuropeptide physiology and pharmacology in invertebrates. The organization of the chapters is intended to provide an overview of the organism-level biology of neuropeptidergic function in insects and helminths, progressing to an understanding of the molecular biology of the genes that encode their precursors and receptors in these organisms. We include for perspective consideration of the state of the art in discovery of insecticides and anthelmintics, a review of drugs that affect similar systems in nematodes, and a summary of drug discovery efforts that target mammalian neuropeptide receptors for therapy of non-infectious diseases in humans.

The authors have done a remarkable job of producing a coherent and highly valuable book. We hope it will stimulate new work in this exciting area. The burgeoning accumulation of genomic data will offer an unprecedented view into the genes that underlie neuropeptide physiology in invertebrates; this book should remind us that sequence data are of limited value unless interpreted through functional studies in organisms. We thank them for their insight and thoughtful reflections as documented in this volume. We also thank the many people at Landes Bioscience who facilitated its publication.

Timothy G. Geary, PhD
Aaron G. Maule, PhD

ABOUT THE EDITORS...

TIMOTHY G. GEARY is a Tier I Canada Research Chair and is Professor and Director of the Institute of Parasitology at McGill University in Montréal, Québec, Canada. Geary received a BSc degree from the University of Notre Dame in South Bend, IN (USA) and a PhD in Pharmacology from the University of Michigan in Ann Arbor, MI (USA, 1980). He worked on malaria chemotherapy at Michigan State University in East Lansing, MI (USA) until 1985, when he joined The Upjohn Company in Kalamazoo, MI (USA) as a staff scientist working on discovery of antiparasitic drugs. Geary joined the Faculty at McGill in 2005. His research interests include proteomics and genomics analyses of the host-parasite interface, discovery of anthelmintics and the pharmacology of antiparasitic drugs.

ABOUT THE EDITORS...

AARON G. MAULE is a Director of Research for Molecular Biosciences and Professor of Molecular Parasitology at Queen's University Belfast, Northern Ireland. Maule was awarded a BSc(Hons) in Biology (1986) and a PhD in Experimental Parasitology (1989) at Queen's where he began working on parasite neuromuscular systems. He focused on molecular aspects of parasite neurobiology during postdoctoral positions at Queen's and at The Upjohn Company in Kalamazoo, MI (USA). Maule returned to Queen's as a member of academic staff in 1995 to pursue the basic biology of helminth signaling systems. His research interests include parasite neurobiology and the development and exploitation of gene silencing platforms for target validation and parasite control.

PARTICIPANTS

Miriam Altstein
Department of Entomology
The Volcani Center
Bet Dagan
Israel

Liesbeth Badisco
Research Unit Functional Genomics and Proteomics
Zoological Institute
Katholieke Universitet Leuven
Leuven
Belgium

Geert Baggerman
Research Unit Functional Genomics and Proteomics
Katholieke Universitet Leuven
Leuven
Belgium

William G. Bendena
Department of Biology
Queen's University
Kingston, Ontario
Canada

Cheryl Butler
Pfizer Global Research & Development, Pfizer Ltd.
Sandwich, Kent
UK

Elke Clynen
Research Unit Functional Genomics and Proteomics
Katholieke Universitet Leuven
Leuven
Belgium

Tim A. Day
Neuroscience Program and Department of Biomedical Sciences
Iowa State University
Ames, Iowa
USA

Timothy G. Geary
Institute of Parasitology
McGill University
Montréal, Québec
Canada

Karen Greenwood
Pfizer Animal Health
Kalamazoo, Michigan
USA

Kyuhyung Kim
Department of Biology
Brandeis University
Waltham, Massachusetts
USA

Michael J. Kimber
Neuroscience Program and Department of Biomedical Sciences
Iowa State University
Ames, Iowa
USA

Chris Li
Department of Biology
City College of the City University of New York
New York, New York
USA

Nikki J. Marks
Parasitology, School of Biological Sciences
Medical Biology Centre
Queen's University Belfast
Belfast
UK

Richard J. Martin
Department of Biomedical Sciences
Iowa State University
Ames, Iowa
USA

Aaron G. Maule
Parasitology, School of Biological Sciences
Medical Biology Centre
Queen's University Belfast
Belfast
UK

Inge Mertens
Research Unit Functional Genomics and Proteomics
Katholieke Universitet Leuven
Leuven
Belgium

Angela Mousley
Neuroscience Program and Department of Biomedical Sciences
Iowa State University
Ames, Iowa
USA

Ronald J. Nachman
Pest Management Research Unit
Southern Plains Agricultural Research Center
US Department of Agriculture
College Station, Texas
USA

Dick R. Nässel
Department of Zoology
Stockholm University
Stockholm
Sweden

Ekaterina Novozhilova
Neuroscience Program and Department of Biomedical Sciences
Iowa State University
Ames, Iowa
USA

Patricia V. Pietrantonio
Department of Entomology
Texas A&M University
College Station, Texas
USA

Jeroen Poels
Research Unit Functional Genomics and Proteomics
Zoological Institute
Katholieke Universitet Leuven
Leuven
Belgium

Ank Reumer
Research Unit Functional Genomics and Proteomics
Katholieke Universitet Leuven
Leuven
Belgium

Alan P. Robertson
Department of Biomedical Sciences
Iowa State University
Ames, Iowa
USA

Liliane Schoofs
Research Unit Functional Genomics and Proteomics
Katholieke Universitet Leuven
Leuven
Belgium

Jozef Vanden Broeck
Research Unit Functional Genomics and Proteomics
Zoological Institute
Katholieke Universitet Leuven
Leuven
Belgium

Hans Peter Vandersmissen
Research Unit Functional Genomics and Proteomics
Zoological Institute
Katholieke Universitet Leuven
Leuven
Belgium

Matthias B. Van Hiel
Research Unit Functional Genomics and Proteomics
Zoological Institute
Katholieke Universitet Leuven
Leuven
Belgium

Tom Van Loy
Research Unit Functional Genomics and Proteomics
Zoological Institute
Katholieke Universitet Leuven
Leuven
Belgium

Heleen Verlinden
Research Unit Functional Genomics and Proteomics
Zoological Institute
Katholieke Universitet Leuven
Leuven
Belgium

Tracey Williams
Pfizer Animal Health
Kalamazoo, Michigan
USA

Debra Woods
Pfizer Animal Health
Kalamazoo, Michigan
USA

CONTENTS

Chapter 1

Receptor-Based Discovery Strategies for Insecticides and Parasiticides:

A Review

Debra Woods,* Cheryl Butler, Tracey Williams and Karen Greenwood

Abstract

Drug discovery is an iterative process with high risks and low chance of success. New genomics technologies allow veterinary medicine and agrochemical companies to validate and functionally screen new receptor-based targets, including neuropeptide G-protein coupled receptors, which were previously not amenable to high throughput screening. However this is just the first step in a long process to translate a mechanistic assay hit into a drug on the market. In addition to effectively eradicating pests on crops and parasites on their host, the molecules must also be safe, cheap to synthesise, formulatable and patentable. This is a costly process in which early attrition of unsuitable molecules is key to any successful program. Although first principle discovery is risky the ultimate benefits are considerable and future genomics resources will help to generate higher quality hits to strengthen the discovery pipeline.

Introduction

Identification of valid targets for anti-parasitic screens is considerably more difficult than the flood of patents and papers claiming to have found new chemical targets would suggest (Table 1). In fact the number of patents has remained at a constant level over the last 6 years with total anti-parasitic patents published averaging 5460 per year among which 370 patents describe novel molecules. Despite this, only a few new classes of molecule reach the market per decade.

Historically, new anti-parasitic agents were discovered by screening against intact invertebrate organisms; preferably the parasites themselves but also model organisms such as *Drosophila melanogaster* or *Caenorhabditis elegans*. Even then there are issues of translation, especially for veterinary parasites. Screening using animal models of parasitic disease is ultimately required for product approval, but is not an appropriate tool for high throughput research. The pharmacokinetic parameters of the agent in the target species is usually unknown, as again, this requires work in the patient species which will not happen until there is considerable confidence in the effectiveness of the molecule. Many more filtering assays are used before in life work to maximise the chance of pursuing an active series whilst increasing confidence in safety. This is less of a hurdle in agrochemical research, but laboratory or limited field trial conditions may still raise issues, for example around parasite and host strain selection. For parasite-in-a-dish screening the method is limited by the available endpoints. Agents that affect the ability of a parasite to interact with the host or undermine the host defensive systems would not be detected, nor would those that have a subtle effect on the viability of the parasite. Furthermore most veterinary

*Corresponding Author: Debra Woods—Pfizer Animal Health, Pfizer Inc, 7000 Portage Road, Kalamazoo, MI, 49009, United States. Email: debra.j.woods@pfizer.com

Neuropeptide Systems as Targets for Parasite and Pest Control, edited by Timothy G. Geary and Aaron G. Maule.

Table 1. Anti-parasitic patents

Publication Year	Total Anti-Parasitic Patents	Anti-Parasitic Patents Containing New Compounds
2000	4495	344
2001	5269	365
2002	5485	370
2003	5803	404
2004	5945	336
2005	5904	399

Table 1 lists the total number of anti-parasitic patents published each year since 2000, including the subset of patents containing new compound classes. Each patent refers to a patent family. Data were obtained from the Derwent World Patents Index.

parasites are difficult to culture so only parts of the life cycle can be used in laboratory trials. There can be issues with delivery of test agents, which may require different properties in order to access both the host and each parasite species targeted. Moving to model organisms such as *D. melanogaster* or *C. elegans* that offer genetic and biochemical tools is yet another hurdle as there is no guarantee that the model organisms mimic the behaviour or biochemistry of the parasites, nor do model organisms reflect the parasitic aspects of the target organisms.

A valid target must be genuinely deleterious to the parasite in the 'real world', preferably causing death of the parasite, but otherwise ensuring that it cannot harm the host or propagate itself. The target can be specific to one organism only, although to cover the cost of development of a new pharmaceutical or agrochemical agent the single organism must provide a valid market. It must also work against any strain of that organism and in any host, which in itself is a challenge. To meet the economic realities of the industry it should be valid across parasite species and preferably parasite phyla. Such agents are difficult to identify!

As well as agrochemical and veterinary medicine markets, many neglected tropical parasitic diseases lack safe, effective medicines and contribute to sustaining poverty in many regions. Although the majority of anti-parasitic discovery is carried out by the veterinary medicine and agrochemical industries, with drugs being historically leveraged for tropical diseases; the scarcity of new chemical entities, as well as an increase in the availability of funds from the World Health Organisation, philanthropic foundations and a number of nongovernmental organizations means that tropical parasitic drug discovery research is being stepped up. A recent paper by Nwaka and Hudson[1] from WHO-TDR describes the new paradigm of 'integrated drug discovery' whereby coordinated, integrated partnerships and networks of academic institutions and industry are working together towards the discovery of new agents to treat tropical parasitic diseases. Recognition of the hurdles associated with discovering novel anti-parasitics has driven this significant culture shift, partnering the drug discovery expertise of industry with humanitarian efforts to eliminate the economic and social burden of parasitic disease in the poorest regions of the world.

As the number of potential new targets increases simple screening strategies are required to identify potential compound leads. A good example of this is the recent work on invertebrate neuropeptides. These offer an exciting new set of targets that would potentially act across phyla and are amenable to high throughput mechanistic screens. Receptor-based screening however is a long way from delivering an active, safe and cost-effective therapy and the chances of success are therefore low.

Mechanistic Receptor-Based Screens

Following identification of a neuropeptide receptor as a valid target for the pharmaceutical treatment of parasitic infections, it is essential to establish a suitable assay for testing potential nonpeptide agonists or antagonists.

Historically, exposure of intact invertebrates to peptide ligands or synthetic mimics thereof has been shown to impact viability or cause phenotypic changes. Although these assays are a fundamental tool for target validation they are often low throughput and require relatively large quantities of the test molecule. Furthermore, inactivation of the peptide or compound by endogenous enzymes can disguise its intrinsic efficacy.[3] However, the direct testing of small molecules on in vivo systems continues to be the primary screening method applied within the agrochemical research industry and has generated lead candidates at a steady rate for over 30 years.[4]

Tissue preparations can also be used to demonstrate that they have specific pharmacological effects. For example the modulatory action of FMRFamide-related peptides or FaRPs on nematode muscle systems has been studied extensively using isometric tension tests of somatic body wall muscle strips isolated from *Ascaris suum*.[2] Due to the complexity of these models it is not possible to definitively identify which receptor or other target the molecules are active against. The low throughput nature of the models means that they are not often successful for identification of lead compounds.

The subsequent movement towards mechanism-based assays has enabled the identification of synthetic compounds which act upon specific invertebrate targets. The pharmacological characterisation of neuropeptide G-protein coupled receptors (GPCRs) has been aided considerably by the development of ligand binding assays using parasitic membrane fractions. An example is identification of thiadiazole analogs by displacement of [^{125}I] AF2 (KHEYLRF-NH_2) from binding sites in *A. suum* muscle membranes.[5] The isolation and preparation of tissue from parasite relevant species can be labour intensive and, from a drug discovery perspective, it is difficult to prepare the quantities required for screening large numbers of compounds.

Expression of invertebrate neuropeptide receptors in recombinant cell systems allows considerably increased throughput of compound screening. The completion of the genome project, EST databases and characterisation of the molecular genetics of the free-living *C. elegans* and fruit fly *D. melanogaster* has elucidated 54 nematode and potentially 50 *D. melanogaster* neuropeptide GPCRs, many of which occur in differentially spliced forms.[6,7] Subsequent cloning and transfection of *C. elegans* and *D. melanogaster* cDNA into mammalian expression systems has facilitated the pharmacological evaluation of a number of invertebrate receptors. These include the Dm Type-A Allatostatin Receptors DAR-1 and DAR-2, which have been transiently transfected into Chinese Hamster Ovary (CHO) cells for [^{35}S] guanosine 5′-*O*-(thiotriphosphate) (GTPγS) binding studies.[8] The successful expression of nematode GPCRs in systems developed for mammalian genes and proteins has however proved problematic. Reports in the literature suggest modifications to cell culture conditions to include a temperature reduction 24 hours post-transfection are required to achieve detectable functional expression of selected *C. elegans* neuropeptide GPCRs in CHO cells.[9,10]

Another option is use of recombinant micro-organisms, such as bacteria or yeast, for the expression of parasitic GPCRs. The yeast *Saccharomyces cerevisiae* has been utilised extensively in the investigation of receptor pharmacology due to its simple, inexpensive maintenance and ease of genetic manipulation. Furthermore, yeasts possess an endogenous eukaryotic G-protein transduction system, the pheromone response pathway.[11]

A mechanistic assay designed for the identification of molecules which act as agonists or antagonists of target receptors should be robust (i.e., with a high signal-to-noise ratio) and reproducible. In addition, within a drug discovery screening programme, the assay must be amenable to miniaturisation to a microtitre plate format to increase screening throughput.

A recent review[12] describes several of the key assay methodologies applicable to the development of high throughput functional GPCR screens. At the outset it is of critical importance that the expression system chosen replicates the ability of the invertebrate neuropeptide receptor to respond to the test molecule through a transduction system which produces a measurable quantitative endpoint e.g., viability, chemiluminescence or fluorescence. The emergence of highly specific radiolabelled ligands, fluorescent dyes and reporter gene systems has enabled the monitoring of intracellular calcium levels,[13] cyclic AMP levels[14] and guanidine nucleotide exchange[8] in mammalian cells expressing *D. melanogaster* and *C. elegans* neuropeptide receptors. Recent advances in genetic modification capabilities have aided the exploitation of the endogenous pheromone response pathway in *S. cerevisiae* for rapid and cost-effective in vitro screening of test compounds. Mammalian GPCRs expressed in *S. cerevisiae* have been successfully coupled to the endogenous transduction system via the native Gα (Gpa1) or chimeric mammalian Gα subunits.[15,16] Subsequent transcription of pheromone-inducible promoter sequences fused with reporter cDNA, including *lacZ* or *His3*, is therefore directly correlated to ligand binding at the heterologous receptor. This GPCR bioassay has recently been applied extensively in the high-throughput screening of more than 30 *C. elegans* and *D. melanogaster* neuropeptide receptors, measuring cell viability on histidine-deficient growth media as an indication of *His3* sequence induction (Williams, Cox and Greenwood, unpublished data).

The process of testing thousands of compounds for pharmaceutical activity against a biological target in a minimal time period is termed high throughput screening (HTS). This method of blind, random high throughput (HT) testing is employed routinely within the drug discovery industry to detect new templates which form the primary basis for rational drug design. Advances in recombinant expression technology have enabled the development of HTSs to identify non-peptide ligands of invertebrate GPCRs. The successful cloning and transformation of the target receptor into mammalian cell, yeast or bacteria systems has facilitated the miniaturisation of in vitro functional assays into 96, 384 and 1536 well microtitre plates. In addition to increasing throughput, miniaturisation of such mechanism based assays minimises compound and reagent requirements and importantly cost.[17]

Modification of an assay into a total reaction volume ≤50 μl, as required for a standard 384-well microtitre plate, necessitates several additional considerations surrounding liquid handling, automation, quality control and data management issues. Liquid handling robots that are capable of rapid, multiple dispensing in the nanolitre range are utilised as stand alone workstations or integrated into fully automated robotic screening platforms. Such platforms routinely incorporate a robotic arm which can transfer a microtitre plate between all assay instrumentation, including incubators, shakers and endpoint or kinetic plate readers. Assay protocols should ideally be homogeneous, without separation steps such as centrifugation or filtration.[18] Application of automated systems can permit further elevated throughput, particularly if operating 24 hr/day. The vast number of data points generated from automated HT screening requires the development of complex data management processes for tracking, analysis, storage and visualization. Each stage requires specific software applications which are linked to a core relational database system containing information including assay parameters, plate co-ordinates and compound structures. The resulting user interface is able to cross-reference between applications and therefore assist the implementation of flexible queries, for example relating primary screening data to compound plate maps.[19]

Quality control (QC) procedures allow the monitoring of intrinsic biological and instrumentation variability and are an essential element of any assay. They are of particular importance during the HTS validation process due to the risks of false positive or negative results inherent when testing such large quantities of substrate. The QC parameters established for any HTS can incorporate considerations around assay sensitivity, physicochemical properties of the compounds, reagent and signal stability and the accuracy of the liquid handling equipment utilised. A recent review[20] describes several of the frequently used statistical methods for evaluating the quality of HTS assay data. Traditionally, QC has focused primarily on the analysis of data obtained from a series of control wells incorporated into each screening plate and subsequent calculations of signal-to-noise ratios, including the Z' factor. Limited control well data however is not always indicative of data from sample areas therefore whole screen and individual well level QC analysis methods are now being introduced to identify underlying trends in biological screening.[21]

Selection of Compounds for Screening

Chemical leads for anti-parasitic and pesticide drug discovery come from a wide range of sources. Scientific and patent literature is always a rich resource for new molecules, giving a starting point for chemists to optimise potency at the target receptor; as was described in the introduction, in 2005 there were 5904 new anti-parasitic patent applications with 399 exemplifying new molecules (Table 1). Traditionally, medicinal chemists worked on single molecules investigating Structure Activity Relationships (SAR) in a stepwise manner. In more recent years parallel chemistry technologies such as combinatorial chemistry and high speed analoging (in which chemistry is carried out using similar reaction conditions either in the same reaction vessel or individually in parallel using semi-automated synthesis)[22] have developed to such a level that, where structures are amenable, tens, hundreds and even thousands of compounds can be synthesized in a relatively short space of time. This allows scientists to investigate chemical space which would previously have been impossible and in so doing increasing the probability of identifying potent active leads.

Despite complications due to isolation and production of the active components, natural products are an attractive source of new molecules. Feher & Schmidt[23] showed that in terms of chemical diversity space combinatorial compounds densely populate a small area whereas natural products are more diverse, which may partially explain the failure of combinatorial chemistry to produce de novo synthetic small molecule drug candidates to date. Natural products have historically been a rich source of pesticides and natural products. Macrocyclic lactones such as Ivermectin, Doramectin and Selamectin are ground-breaking endectocidal anti-parasitics which were originally isolated from the soil organism *Streptomyces avermitilis*; while Spinosad, a nicotinic agonist insecticide, is produced by fermentation of the soil actinomycete *Saccharopolyspora spinosa*. Discovery scientists must however be realistic when embarking on a natural product discovery program, understanding the timescales and risks associated with isolating active constituents and producing medicines commercially. Unless the active agent can be produced at a commercial price synthetically (and a large proportion of natural products are extremely large and complex), production will require scale up of a fermentation process. Micro-organisms are more amenable to scale up, as plant and marine derived actives would require cloning and expression in a host organism to allow commercial production.

If no chemical leads are available it can be tempting to use native ligands as a starting point. In general such molecules are not amenable to delivery as drugs, having poor ADME (Absorption, Delivery, Metabolism, Excretion) properties. Native molecules are as a rule too polar for absorption and are highly metabolically unstable, hence requiring these properties to be altered to make the molecules 'drug-like'.

HTS is utilised when no tractable chemical lead is available or when discovery scientists are seeking a new chemical class to address the target. As has already been discussed, this requires a miniaturised assay with the accompanying automation and data management infrastructure. If screens are compatible and resources available whole file screening is ideally carried out. Many company files are now however so large (1-3 million compounds) that screening single compounds requires 1536 well plate formatted screens to complete file testing in a realistic timescale. Some companies may therefore choose to screen using 'compressed' compound plates, in which each well contains a mixture of several compounds. The core relational database system, combined with decompression where necessary, may then be used to identify the active compound. The risk is that compound interactions may prevent identification of a hit molecule, but this is balanced by the benefits of screening the whole file. Strategically, it could be more practical to screen smaller subsets of the file (10-50,000 compounds). These subsets can be prepared based on either biological or chemical criteria and may be a more rapid method of filtering through the multitude of compounds to find the hits.

In silico screening is another attractive approach to drug discovery, with the potential for screening millions of compounds from a virtual library, allowing chemists to focus synthetic resource on molecules most likely to be active and potent at the target site. This is still a considerable challenge when using structure-based methods, as the X-ray structure of many target proteins is unknown, especially for membrane-bound receptors such as GPCRs. Recent research describes approaches using homology modelling of the transmembrane domain of GPCRs based on the crystal structure of bovine rhodopsin and bacteriorhodopsin, reviewed by Becker et al.[24] The same authors have also reported a de novo modelling approach (using PREDICT algorithm) for GPCRs, which does not rely on the rhodopsin X-ray structure and which they claim can be applied to a range of GPCRs resulting in generation of promising hit molecules.[25] In addition to in silico HTS medicinal chemists are also increasingly successfully utilising computational chemistry to design molecules based on preliminary SAR data.

Finally it is worth considering why to date no pesticide or anti-parasitic small molecule discovered by mechanistic HTS has been brought to market. In fact every anti-parasitic and pesticide drug currently available was derived from whole organism screening, either using intact pests/parasites or infected animal models. The quality and content of the file is one factor, which is dependent on the historical background of the company and the therapeutic areas they have focused on. This biases the compounds towards activity against certain targets and screens. Most pharmaceutical company compound files are 'Rule of 5' compliant. 'Rule of 5' compliant molecules are amenable to oral dosing in humans; whereas agrochemical and anti-parasitic therapies are often dosed by other routes, such as crop spraying or topical dosing of animals. 'Rule of 5' derived its name from the cut-off values for each of the four parameters that define the "drug-likeness" of the potential drug candidates, the values of which are all close to five or a multiple of five.[26] The 'Rule of 5' states that poor absorption or permeation is more likely when:

- There are more than 5 H-bond donors (expressed as the sum of OHs and NHs).
- The molecular weight is over 500.
- The c$\text{Log}P$ is over 5.
- There are more than 10 H-bond acceptors (expressed as the sum of Ns and Os).

Natural products and compound classes that are substrates for biological transporters are exceptions to the rule. Therefore to increase chances of success if the file is 'Rule of 5′ compliant, the targets selected for HTS should also be amenable to binding 'Rule of 5′ compliant small molecules, a property termed 'druggability'.[27]

Whether screening the whole file or a subset of the file, a HTS hit must have a realistic chance of directing chemistry towards generation of a potent lead molecule. As HTSs are run with test compounds at relatively high concentrations (1-10 μM), hit molecules are expected to be effective at >70-80% of the control well effect. As already discussed, the hit rate and quality of hits will be driven by the file substrate, but ideally scientists are seeking hit rates of <1%, to achieve both a manageable hit rate and to select the most potent molecules.

'Hit-to-Lead': Converting a Mechanistic Screen Active into a Lead Molecule

It is a big leap from efficacy in a mechanistic screen to discovery of a chemical lead with anti-parasitic or pesticide activity. The first stage is to identify the most potent mechanistic hit for testing against the pest/parasite. By searching the large chemical files of agrochemical and pharmaceutical companies, discovery scientists can rapidly determine whether more potent molecules are available and start to determine whether there is a structure-activity relationship (SAR) for these compounds in their efficacy at the target receptor.

Ideally, the mechanistic screen will filter a large number of hits into a more manageable quantity for progression to in vitro pest and parasite screens. The definition of a lead anti-parasitic or agrochemical molecule is one which has the desired activity against an enzyme or agonist/antagonist activity against a receptor in vitro with additional activity against a living target organism in vivo, which in itself is not an easy thing to identify.[28] Furthermore a lead nonpeptide ligand identified from a HTS is unlikely to be the molecule which is progressed to market for a particular target or disease. Synthesis of analogues around the chemical structure is used to address shortcomings such as insufficient potency and/or selectivity. The newly synthesised molecules are then resubmitted for testing in the biological assays and SAR evaluated by comparing assay data and structural changes to the molecule, leading to further modification.

Potency in the mechanism-based screen is only one factor required for achieving efficacy in whole organism screens. Solubility, penetration and chemical/metabolic stability are all parameters important in delivering an active molecule to its site of action. For anti-parasitic discovery scientists there is the additional hurdle of delivering the molecules to the host, so even if the molecules are active against the target parasite in vitro, they may fail to control the disease in vivo. This is described in greater detail in the next section.

Selecting a spectrum of screens with both systemic and contact administration of compound; incorporating different parasites/pests and lifecycle stages; plus running screens in the presence of metabolic inhibitor are all factors which increase the chances of success when progressing a HTS hit to discovery of an active lead molecule. Investigation of properties of the molecules, such as in vitro metabolism in microsomes (to determine susceptibility to Phase I oxidative metabolism) and hepatocytes (Phase I and II metabolism); and physicochemical parameters such as solubility, log D, polar surface area and plasma protein binding also aid in determining which factors must be addressed to advance the series. Efficacy against the parasite/pest in vitro guides selection of the best molecules to either crop field testing or small animal models, however the limitations of investigating parasites in the absence of their natural host must be recognised.

Lead-to-Candidate: Identification of Molecules for Progression to Market

The translation from in vitro to in vivo activity is another difficult problem encountered within anti-parasitic drug discovery. Safety, both human food safety and target animal safety; parasite/pest spectrum; lifecycle stage and route of administration (such as topical, oral or injection delivery for anti-parasitics) all determine the potential of a molecule to fit desired product profiles. It may take several iterations to find a compound that has the desired spectrum of properties required for a successful drug.

Successful progression of molecules to the market is dependant on several key factors including safety, efficacy and pharmacokinetics. Understanding the pharmacokinetics (PK) of a molecule at an early time point is particularly important in the veterinary medicine pharmaceutical industry, where the molecule is administered to livestock or companion animals. In addition to determining the potential of a molecule to deliver efficacy in line with the product profile; for livestock products this will also give an early indication of issues which might hinder progression of the molecule, such as human food safety and withhold or withdrawal periods. Efficacy of a molecule can be demonstrated in small field trials, either in the target animal (livestock or companion animal) for Animal Health applications or small plot studies for agrochemical utility (described by Mazumdar in http://www.ficci.com/media-room/speeches-presentations/2006/jan/agrochem/SessionIII/MrMazumdar.ppt). These studies allow the investigation of dose-activity relationships and duration of effect using the relevant route of administration. Once a promising compound has been identified extensive safety testing is required. Evaluation of the potential toxicity of molecules can be assessed using in silico structural alert software. This may then be followed by in vitro genetic toxicity assays such as bioluminescence Ames and in vitro micronucleus assays. Submission of the molecule to a broad cross screening panel of receptors will also give an indication of receptor selectivity and whether the molecule is likely to have any side effects in mammals. Compounds would then be progressed to a range of safety screens depending on the eventual use. Within the agrochemical area, toxicological studies must be conducted in mammals and other organisms (fish, aquatic organisms, birds, beneficials and micro-organisms) and environmental safety, metabolism and residue studies carried out in plants, animals, soil, water and air to ensure the molecules are safe to be used. This is also a consideration for the veterinary industry, where the impact of excretion of compound or metabolite must be evaluated in the field.

In the pharmaceutical industry clinical trials include further studies to investigate acute, sub-chronic and chronic toxicity together with any mutagenic, carcinogenic and teratogenic effects of the molecule in mammals. For both livestock and agrochemical products, residue and human food safety studies must also be completed to ensure residue levels are below the FDA defined safe concentration for the treatment of the molecule or its metabolites in meat, milk or plants destined for consumption.

The discovery and development process for new compounds is a costly business, with development of a livestock anti-parasitic compound costing up to $100 million. The chance of success, particularly for first principle discovery such as the neuropeptide research described in this review, is very low. The cost of discovery and of compounds that fail in development, particularly at late stages where considerable money has been invested, all has to be funded by the compounds that do make it to the market.

Within both the pharmaceutical and agrochemical industries obtaining intellectual property (IP) around candidate molecules of interest is essential. 'Freedom to operate', which verifies whether a company is free to commercialise a product without infringing existing patents, must be established and composition of matter filed for the candidate. This will protect the molecule and associated chemistries from competitor infringement. The patent protects the compound for 20 years from the priority date of the patent, although patents are often filed many years before the compounds finally reach the market, reducing the period of market exclusivity.

The progression of a candidate molecule to market is fraught with disaster and can fail at any of the stages described above. Within the pharmaceutical industry, around 90% of programmes within discovery do not identify lead molecules and of the successful 10% which progress to exploratory development, 66% of programmes do not reach the market. Any candidates moving from discovery into development need to be of a very high quality. They must be differentiated from other products already on the market; have an acceptable therapeutic index; work at a cost-effective efficacious dosage, with a straightforward synthetic route; be formulatable and deliverable to the animal or crop by the desired route; and finally intellectual property surrounding the molecule must be owned or licensed by the company.

Conclusion and Discussion

With the advent of new genomics tools,[29] the potential for both novel target validation and generation of high throughput parasite and model organism screens have increased significantly. In fact it is relatively easy to develop a mechanism-based screen and identify nanomolar hits from the growing compound files of pharmaceutical and agrochemical companies. What is much more difficult is translating this 'hit' into an anti-parasitic drug which can be delivered by a suitable route to control the required spectrum of parasites safely. However the benefits of delivering a new class of anti-parasitic or agrochemical to the market, with no cross-resistance to current commercial agents, outweigh the risks. Key for the discovery process is to balance precedented and novel targets; and then identify the issues with any new chemical class, especially those which are insurmountable, as early in the discovery process as possible.

Despite these considerations, as these '-omic' technologies advance and more tools become available, drug discovery scientists will have access to an increasing number of validated targets and functional screens; for example a growing number of parasite genome sequencing projects are being funded, including those with veterinary relevance such as *Haemonchus contortus* and *Ixodes scapularis*.[29] Access to screening tools for progressively more validated novel, druggable targets will increase the number and quality of hits identified; expand the chemical substrate and elevate the chances of success in identifying the next blockbuster.

Acknowledgements

The authors wish to acknowledge the help of Clare Challenger who produced the anti-parasitic patent literature review. They would also like to thank Pfizer colleagues Denis Billen, Christelle Lauret and Liz Littlewood for reviewing the manuscript.

References

1. Nwaka S, Hudson A. Innovative lead discovery strategies for tropical diseases. Nat Rev Drug Discov 2006; 5:941-955.
2. Maule A, Geary T, Bowman J et al. Inhibitory effects of nematode fmrfamide-related peptides (farps) on muscle strips from Ascaris suum. Inv Neurosci 1995; 1(3):255-265.

3. Greenwood K, Williams T, Geary T. Nematode neuropeptide receptors and their development as anthelmintic screens. Parasitology 2005; 131:S169-S177.
4. Ward E, Bernasconi P. Target-based discovery of crop protection chemicals. Nat Biotech 1999; 7:618-619.
5. Lee B, Dutton F, Clothier M et al. Synthesis and biological activity of anthelmintic thiadiazoles using an AF-2 receptor binding assay. Bioorg Med Chem Lett 1999; 9:1727-1732.
6. Hauser F, Williamson M, Cazzamali G et al. Identifying neuropeptide and protein hormone receptors in Drosophila melanogaster by exploiting genomic data. Brief Funct Genomic Proteomic 2006; 4(4):321-330.
7. Lowery D, Geary T, Kubiak T et al. G-protein coupled receptors and modulators thereof. International Patent 2003; WO 01/38533 A2.
8. Larsen M, Burton K, Zantello M et al. Type A allatostatins from Drosophila melanogaster and Diplotera puncata activate two Drosophila allatostatin receptors, DAR-1 and DAR-2, expressed in CHO cells. Biochem Biophys Res Comm 2001; 286: 895-901.
9. Kubiak T, Larsen M, Nulf S et al. Differential activation of "social" and "solitary" variants of the Caenorhabditis elegans G-protein coupled receptor NPR-1 by its cognate ligand AF9. J Biol Chem 2003; 278(36): 33724-33729.
10. Kubiak T, Larsen M, Zantell M et al. Functional annotation of the putative orphan Caenorhabditis elegans G-protein coupled receptor C10C6.2 as a FLP15 peptide receptor. J Biol Chem 2003; 278(43); 42115-42120.
11. Minic J, Sautel M, Salesse R et al. Yeast system as a screening tool for pharmacological assessment of G-protein coupled receptors. Curr Med Chem 2005; 12: 961-969.
12. Thomsen W, Frazer J, Unett D. Functional assays for screening GPCR targets. Curr Opin Biotech 2005; 16:655-665.
13. Mertens I, Vandingeen A, Meeuen T et al. Functional characterization of the putative orphan neuropeptide G-protein coupled receptor C26F1.6 in Caenorhabditis elegans. FEBS Lett 2004; 573:55-60.
14. Birse R, Johnson E, Taghert P et al. Widely distributed Drosophila G-protein coupled receptor (CG7887) is activated by endogenous tachykinin-related peptides. J Neurobiol 2005; 66:33-46.
15. Cismowski M, Takesono A, Ma C et al. Genetic screens in yeast to identify mammalian nonreceptor modulators of G-protein signalling. Nat Biotech 1999; 17(9):878-883.
16. Wang Z, Broach J, Peiper S. Functional expression of CXCR4 in Saccharomyces cerevisiae in the development of powerful tools for the pharmacological characterization of CXCR4. Meth Mol Biol 2006; 332:115-127.
17. Wolcke J, Ullmann D. Miniaturized HTS technologies—uHTS. Drug Discov Today 2001; 6(12):637-646.
18. Sittampalam G, Kahl S, Janzen W. High-throughput screening: advances in assay technologies. Curr Opin Chem Biol 1997; 1:384-391.
19. Kenny B, Bushfield M, Parry-Smith D et al. The application of high-throughput screening to novel lead discovery. Prog Drug Res 1998; 51:245-269.
20. Macarron R, Hertzberg R. Design and implementation of high throughput screening assays. Meth Mol Biol 2002; 190:1-29.
21. Gribbon P, Lyons R, Laflin P et al. Evaluating real-life high-throughput screening data. J Biomol Screen 2005; 10(2):99-107.
22. Terrett NK, Gardner M, Gordon DW et al. Combinatorial synthesis—the design of compound libraries and their application to drug discovery. Tetrahedron 1995; 51(30):8135-8173.
23. Feher M, Schmidt JM. Property distributions: differences between drugs, natural products and molecules from combinatorial chemistry. J Chem Inf Comput Sci 2003; 43:218-22.
24. Becker OM, Shacham S, Marantz Y et al. Modeling the 3D structure of GPCRs: advances and applications to drug discovery. Curr Opin Drug Discov Devel 2003; 6(3):353-361.
25. Becker OM, Marantz Y, Shacham S et al. G-protein coupled receptors: in silico drug discovery in 3D. PNAS 2004; 101(31) :11304–11309.
26. Lipinski CA, Lombardo F, Dominy BW et al. Experimental and computational approaches to estimate solubility and permeability in drug discovery and development settings. Adv Drug Delivery Rev 1997; 23(1-3):3-25.
27. Hopkins AL, Groom CR. The druggable genome. Nat Rev Drug Discov 2002; 1(9):727-730.
28. Walter M. From serendipity to design—making agrochemicals to order. Pesticide Outlook 2003; 14:27-31.
29. Gilleard JS, Woods DJ, Dow JAT. Model organism genomics in veterinary parasite drug discovery. Trends Parasitol 2005; 21(7):302-305.

Chapter 2

Nonpeptide Ligands for Peptidergic G Protein-Coupled Receptors

Timothy G. Geary*

Abstract

Neuropeptides play essential roles in many physiological systems in vertebrates and invertebrates. Peptides per se are difficult to use as therapeutic agents, as they are generally very unstable in biological fluid environments and cross biological membranes poorly. Recognition that nonpeptide ligands for peptide receptors have clinical utility came from the discovery that opiates (such as morphine) act by binding to G protein-coupled receptors (GPCRs) for which the endogenous ligands are a family of neuropeptides (enkephalins and endorphins). Basic research has revealed a very large number of distinct neuropeptides that influence virtually every aspect of mammalian physiology and considerable effort has been expended in the pursuit of new drugs that act through peptidergic signaling systems. Although useful drugs have been found to affect various aspects of neuropeptide biology, most work has been devoted to the discovery of nonpeptide ligands that act as agonists or antagonists at peptidergic GPCRs. Similar opportunities are apparent for the discovery of nonpeptide ligands that act on invertebrate GPCRs. A consideration of the knowledge gained from the process as conducted for mammalian peptidergic systems can inform and illuminate promising strategies for the discovery of new drugs for the treatment and control of pests and parasites.

Introduction

Neuropeptides are short chains of amino acids (< ~50 residues) that are released from nerve cells and influence the activity or function of other tissues, including other neurons, muscles or glands. The origin of the concept that peptides released from neurons can affect neuromuscular or glandular systems in animals is difficult to date with precision, but certainly occurred >80 years ago with observations on an insect pupation hormone. The discovery of substance P in 1931, a peptide derived from brain (and gut) that contracts ileum tissue, may be taken as the first conclusive demonstration of a bioactive peptide in mammals.[1,2] Progress in understanding the pharmacology of neuropeptides was slowed at least in part by the lack of methods for defining amino acid sequences and for purifying or synthesizing sufficient amounts for detailed investigation (first accomplished in the early 1950s for oxytocin). In vivo studies were hampered by the very low oral bioavailability of peptides and their typically very short half-lives in the circulation. However, by the 1960s, the importance of neuropeptides in both invertebrate and vertebrate neuromuscular systems was generally recognized, as was their involvement in essentially every aspect of animal physiology. To date, some 100 distinct neuropeptides have been purified from mammalian CNS tissue.[2,3] Much remains to be learned about how these peptides contribute to the function of neuromuscular systems in

*Timothy G. Geary—Institute of Parasitology, McGill University, 21111 Lakeshore Road, Ste-Anne-de-Bellevue QC Canada H9X 3V9. Email: timothy.g.geary@mcgill.ca

Neuropeptide Systems as Targets for Parasite and Pest Control, edited by Timothy G. Geary and Aaron G. Maule. ©2010 Landes Bioscience and Springer Science+Business Media.

health and disease. Nonetheless, neuropeptidergic systems have become important, though still mostly potential, targets for therapeutic intervention in humans.[2,4,5]

Considerable research on neuropeptide biology has also been conducted in invertebrates. Neuropeptides profoundly influence the behavior and physiology of arthropods[6] and helminths,[7-9] the metazoan organisms that as parasites and pests affect plant, human and animal health and productivity. The broad and essential roles of invertebrate neuropeptides make them theoretical targets for chemotherapy and pest control.[10-14] The focus of subsequent chapters in this book is the potential of neuropeptidergic systems in those organisms as targets for compounds that can be used to control them. To provide perspective for this topic, it is useful to review work on the discovery of therapeutic agents that act on neuropeptide systems in mammals, in which by far the most work has been done.

Druggable Targets in Neuropeptidergic Signaling

Ligands that act through neuropeptide systems fall into two general pharmacological categories: agonists, which mimic the effects of the endogenous peptide ligand and antagonists, which act by preventing the response to the endogenous peptide; antagonists will have physiologically detectable effects only when the endogenous peptide contributes actively or constitutively to the behavior of the system. A number of points of potential value for chemical interdiction are found within the biology of peptidergic signaling (see Fig. 1). Neuropeptides are typically encoded as distinct amino acid sequences in genes; translated from mRNA, the gene product is linear (A). The biology of neuropeptide gene expression is not currently targeted by drugs and presents no apparent opportunities for selective pharmacology. The posttranslational processing of neuropeptide gene products by proteolysis and in some cases by additional modification, especially amidation at the carboxyl terminus, does constitute a druggable opportunity (B). Postprocessing, the controlled release of neuropeptides from storage vesicles at the synapse following arrival of an action potential and synaptic depolarization is required for biological effects (C); as for mRNA generation, neuropeptide storage and release is governed by conserved pathways and is not an obvious target for selective intervention. Termination of the peptide effect is commonly due to proteolytic digestion, either at the synapse or in biological fluids, which is also a druggable target (D; and see below).

Neuropeptides almost always act as agonists of G protein-coupled receptors (GPCRs), which exhibit a conserved 7 transmembrane domain structure (E). Binding of a cognate peptide ligand to its receptor liberates G protein subunits (α, $\beta + \gamma$) from an inactive trimeric complex. Released G protein subunits then effect downstream changes in a variety of second messenger systems (F), such as adenylyl cyclase, phosphoinositol kinases, etc. Specificity in signaling is achieved by cell-specific receptor expression and by the use of selective G_α subunits which target specific second messenger systems. A limited number of signaling mechanisms or pathways is available to service a much larger number of receptor-ligand pairs. This common infrastructure means that it has been difficult to exploit postreceptor signaling mechanisms for selective therapeutic purposes. Instead, the vast majority of drugs which target peptidergic systems act at the level of the GPCR and the vast majority of discovery efforts are similarly targeted. This bias reflects the fact that mammalian physiological homeostasis relies on the regulated actions of dozens of neuropeptide systems which share a common set of proteolytic enzymes and downstream signaling pathways. Thus, the specificity required for therapeutic use is most often attained by using drugs which interact selectively with a targeted GPCR.

The situation is complicated by the fact that all small molecule neurotransmitters (such as the monoamines) and almost all neuropeptides act on families of related GPCRs, known as receptor subtypes. Receptor subtypes are characterized by a fairly moderate but wide range of amino acid sequence identity, even though they recognize the same ligand and presumably were derived in evolution from a series of gene duplication events, as opposed to convergent evolution from distinct precursors. Receptor subtypes typically mediate distinct physiological effects and usually exhibit distinct patterns of expression at the tissue level. It is critical that drugs targeted to these GPCRs discriminate among subtypes in order to minimize unwanted side effects. As noted below, this factor adds considerable difficulty to the search for nonpeptide ligands for peptide receptors.

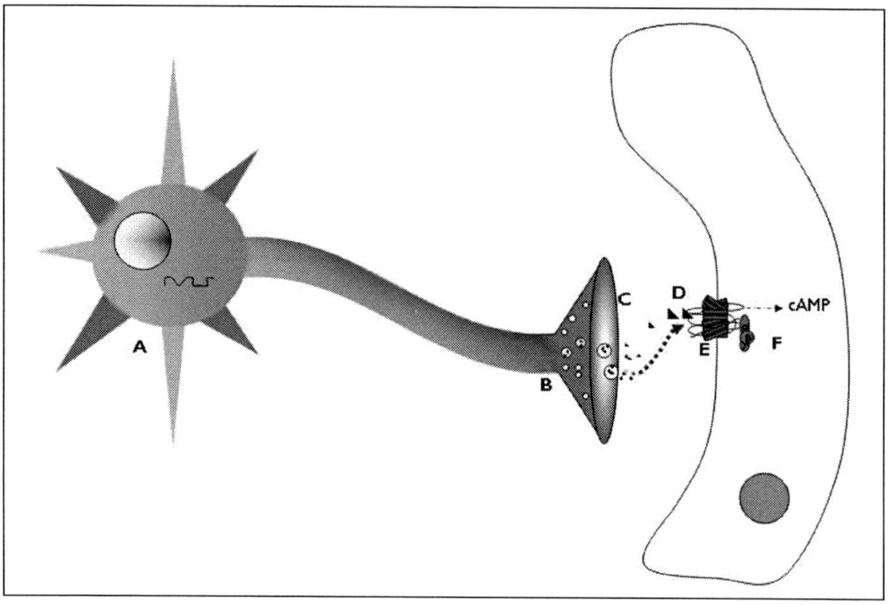

Figure 1. Possible sites of therapeutic intervention in peptide signaling pathways. A) In the neuron, transcription of mRNA from neuropeptide-ending genes and its translation into a linear propeptide protein provide few opportunities for selective interdiction. B) Processing of the propeptide into functional neuropeptides involves proteases and peptide modifying enzymes (e.g., amidation); these are potential but little exploited sites for drug action. C) Storage of peptides in synaptic vesicles and their release in response to the arrival of an action potential are not currently targeted by drugs. D) Termination of peptide signaling in the synapse is achieved by proteolysis; inhibition of peptide destruction is a validated drug site. E) Interaction of peptides with a GPCR at the postsynaptic membrane is the primary site for drug discovery in these systems. F) Downstream signaling pathways (e.g., an increase in cAMP generated by adenylyl cyclase) stimulated by ligand-induced receptor activation are initiated by the liberation of G protein subunits from an inactive complex; these pathways, though of considerable interest, have been elusive targets for selective pharmacological manipulation.

Neuropeptide-Targeted Drugs: Basic Features

Drugs that act on peptidergic signal transmission fall into two general structural classes: peptides or nonpeptides; the former includes molecules with modified, nonnatural amino acids but which retain the basic chemical features of peptides. As noted, drugs may act on peptidergic systems in a few basic ways. One is by interfering with production of peptides by cleavage from precursor proteins or their modification (e.g., by amidation). Prevention of peptide production or processing has the same consequences as administration of an antagonist. Alternatively, inhibition of proteases that terminate the action of peptides would prolong their presence at the receptor; such inhibitors would generate the same action as an agonist. A third option for drugs that act through peptide signaling pathways is at the receptor itself. With very few exceptions, neuropeptide receptors fall into the class termed G protein-coupled receptors, or GPCRs. Drugs that are ligands for these receptors can stimulate the receptor, like the peptide itself (agonists), or block of the response of the receptor to the native peptide (antagonists).

Peptides as Drugs

Despite their obvious potential for therapeutic intervention in various areas of human physiology, peptides per se are relatively unimportant therapeutic agents. Peptides as chemicals poorly survive the acidic and proteolytic environment of the gastrointestinal tract and are readily degraded by enzymes in the bloodstream and tissues, making delivery by continuous intravenous infusion (or synthetic modification to enhance metabolic stability and/or bioavailability) typically required for therapeutic utility. In the same context, their limited stability, relatively large size (effective molecular diameter) and generally hydrophilic nature preclude or limit their diffusion across the gut and across the external surfaces and internal membranes of parasitic arthropods and helminths. This property, along with their cost and extreme environmental fragility, makes them poor candidates for use as agricultural insecticides.

There are instructive exceptions in human medicine. Oxytocin is used to induce labor in late-term pregnancies; a continuous IV drip is acceptable in this case. Removal of the drip causes an almost immediate cessation of the peptide effect, a desirable feature. Conversely, peptides derived from oxytocin by amino acid modification have been introduced as antagonists for the cessation of preterm labor,[15] administered parenterally. Even so, considerable interest in nonpeptide ligands for oxytocin receptors is apparent,[16] especially for prolonged therapy. Modified versions of vasopressin also have therapeutic indications. Desmopressin, modified by removal of the terminal amine and substitution of D-arginine for the L-isomer at position 8, is used to reduce the frequency of bedwetting.[17] This molecule is dosed orally despite having very low (<1%) bioavailability by this route. Desmopressin is an agonist, like the parent peptide, but antagonists are also of interest for a variety of therapeutic goals. In this case, nonpeptide ligands for vasopressin receptors are the drugs of choice.[18] Vasopressin receptors can be classified into several different types based on sequence and pharmacology. Nonpeptide antagonists at the V1a receptor subtype show promise for conditions such as dysmenorrhea and Raynaud's Syndrome (peripheral vasoconstriction), while V1b receptor antagonists may be useful in some psychiatric disorders and V2 receptor antagonists can cause a salt-sparing diuresis, which is useful in management of certain cardiovascular conditions.[18]

Interference with Peptide Production or Stability

Chemical intervention in with small molecule enzyme inhibitors can be deployed to block the proteolytic cleavage of peptide precursors or to prevent their processing, mimicking a receptor antagonist. Notable in this regard are angiotensin converting enzyme (ACE) inhibitors such as captopril, which prevent the proteolytic processing of inactive angiotensin I to the vasoconstricting derivative angiotensin II. The therapeutic potential of ACE inhibitors for the treatment of hypertension was demonstrated by the discovery in a snake venom of a peptide that caused vasodilation by inhibiting this enzyme, which led to the discovery of captopril; their clinical utility is well recognized.[19] ACE also inactivates the vasodilating peptide bradykinin, a process that may contribute to efficacy but also may contribute to a higher incidence of cough in treated patients. This side effect has led to the development of nonpeptide ligands that are antagonists at angiotensin II receptors.[20] Angiotensin receptor antagonists such as losartan have the same therapeutic benefits as ACE inhibitors but seem to have fewer side effects[19,20] and have become widely used drugs for the management of hypertension.

Alternatively, inhibition of proteases that degrade neuropeptides can be employed to induce an agonist effect. Glucagon-like peptide 1 (GLP-1) is an incretin. It stimulates insulin release and reduces glucagon release, leading to reduced appetite and food intake. Agents which promote GLP-1 effects have definite therapeutic potential for the treatment of Type 2 diabetes mellitus. Dipeptidyl peptidase IV (DPP-4) is a serum protease that degrades many peptides, including GLP-1; inhibition of DPP-4 leads to enhanced GLP-I levels and a DDP-4 inhibitor, sitagliptin phosphate, is licensed for the treatment of diabetes[21] based on this action. Interestingly, stable peptide GLP-I agonists have also been introduced for this purpose;[22,23] the first licensed product, a 39-amino acid peptide called exenatide, is the synthetic version of a peptide called exendin-4, first isolated from Gila monster saliva. Given by sc injection, this and related peptides provide stable GLP-1 agonist effects and have been well received for the treatment of Type 2 diabetes.

Drug-Receptor Interactions

As noted, compounds that act directly on peptide receptors are highly sought as therapeutic agents because specificity can be difficult to achieve with protease inhibitors. In that context,

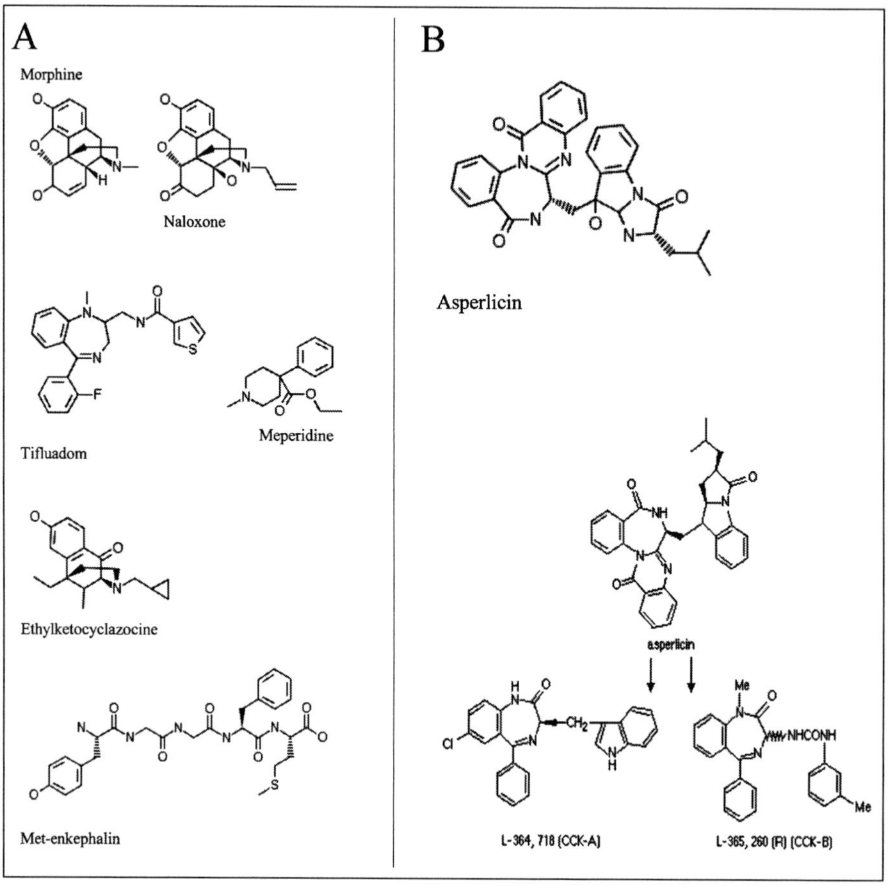

Figure 2. Examples of nonpeptide ligands for mammalian receptors. A) The family of nonpeptidic opiate ligands includes compounds derived from many different templates and with different receptor subtype preferences; ethylketocyclazocine is selective for the κ subtype while morphine and meperidine are preferential ligands for the μ subtype. Naloxone is a pure antagonist. Met-enkephalin is one of the endogenous peptides that act at opiate receptors in the central nervous system. B) Asperlicin is the first nonpeptide ligand for a peptide receptor derived from high-throughput screening. Although a poor lead compound, subsequent medicinal chemistry efforts focused on the benzodiazepine moiety generated derivatives with very high potency.

nonpeptide ligands for peptide receptors are typically more desirable than peptidic agents based on pharmaceutical flexibility, especially in terms of pharmacokinetics and bioavailability. Consequently, therapeutic interventions targeted to peptidergic systems have primarily focused on the discovery of pharmaceutically tractable nonpeptide ligands for neuropeptide receptors that act either as agonists or antagonists. Although a considerable amount of work has been done on the rational design of peptide mimetics, intended to create peptide-like molecules with acceptable pharmaceutical properties,[24] the major thrust in small molecule discovery for peptide receptors has come from screening of large chemical libraries in high-throughput, mechanism-based assays.

Although GPCRs comprise the predominant drug target class of interest for neuropeptides, two invertebrate neuropeptides, both in the FMRFamide family, seem to gate ion channels directly, instead of acting on GPCRs. FMRFamide itself gates a snail Na^+ channel[25] and the nematode peptide KPNFIRF-NH_2 appears to directly gate a Cl^- channel in muscle cell membranes of *Ascaris suum*.[26] The distribution of peptide-gated ion channels in invertebrate species of interest, and their potential as targets for chemotherapy, remains unknown and understudied. Since Cl^- channels gated by small molecule neurotransmitters, particularly glutamate and GABA, are validated targets for anthelmintics and insecticides (see chapter by Martin and Robertson), more work should be devoted to the characterization of these proteins.

Genome sequencing efforts and some functional studies have identified multiple families of peptide GPCRs in invertebrate species, as detailed elsewhere in this volume and reviewed recently.[27,28] There are evident connections with peptide GPCR families in mammals, especially the A and B classes, which include receptors for neuropeptides and peptide hormones. Invertebrates also express multiple G protein subunits. In the context of this book, it is worthwhile to discuss the potential of invertebrate GPCRs as targets for chemicals by providing the perspective of what is known about drugs that target neuropeptide receptors in vertebrates.

Historical Perspectives

Nonpeptide ligand discovery programs arose from the convergence of three research streams. The first and most basic is the science of drug discovery based on screening collections of chemicals for biological activity; this conceptual process developed from the work of Paul Ehrlich, who screened a collection of industrial dyes for activity against trypanosomes in mice in what may have been the first such exercise (see ref. 29 for review). Prior to this conceptual leap, drugs were derived almost exclusively by purification from natural products that had demonstrated medicinal activity.

Ehrlich's work led to the realization that antibacterials such as sulfa drugs and subsequently antibiotics like penicillins, could be found by screening dyestuffs and microbial fermentation extracts, respectively, for inhibitory activity against bacteria in culture. This paradigm propelled the concept of screening of collections of chemicals (or fermentation extracts) into broad use in the pharmaceutical industry, especially postWorld War II (see ref. 30). These efforts were initially carried out at the level of the tissue or organism (whether animal or microbial), were relatively low-throughput (thousands of assays per year) and were not generally used for the discovery of nonpeptide ligands for peptide receptors (with the exception of ligands for opiate receptors as analgesics; see below).

The first crucial step for this discussion about the discovery of nonpeptide ligands for peptide receptors was the development and widespread adoption of mechanism-based, high-throughput screening (HTS) technologies in the mid-1980s (see refs. 31,32 for review). These processes used subcellular assay targets (membrane preparations, cell extracts or purified proteins) in formats that enabled the screening of hundreds of thousands of compounds in a matter of a few months. The technology is based on the premise that discovery of a novel structure with selective affinity for a particular drug target (receptor) can justify a medicinal chemistry approach to generate drug candidates from very early leads. The implementation of this technology finally enabled screening approaches to focus on neuropeptide receptors in the absence of the confounding influences presented by host and pharmaceutical factors.

The second crucial stream was the realization that opiates such as morphine (Fig. 2A) act as agonists at receptors for the enkephalin/endorphin neuropeptides in mammals.[33] This conceptual leap was unanticipated, as the peptides had been unknown—indeed unsuspected—and were discovered only because they displaced radiolabeled opiates from binding sites in brain membranes. The essential conclusion was that nonpeptide ligands for peptide receptors had evolved in nature and that medicinal chemistry programs based on them had generated a wide variety of highly valuable drugs for pain relief and other indications. Development of a receptor/peptide-based concept of analgesia provided the basis for understanding the scope of opiate pharmacology in humans.[34,35] By extrapolation, this discovery provided a path to rapidly expanded therapeutic exploitation of the rich and still growing peptide diversity found in mammalian physiological systems.

The third crucial development was the demonstration that the confluence of the first two streams could generate new lead compounds: novel nonpeptide ligands could be discovered by screening chemical collections for the ability to displace radiolabeled peptides from their receptors in membrane preparations (the inverse of the opiate/enkephalin story). The question became: can one discover new "morphines" for other kinds of neuropeptide receptors? The crucial proof of this concept was the discovery of novel nonpeptide small molecules, based initially on the natural product asperlicin, that are ligands for cholecystokinin (CCK) receptors.[36,37]

Screening for Novel Nonpeptide Ligands

As noted above, the concept that antiparasitic drugs can be discovered by screening collections of chemicals for specific biological activities was pioneered by Ehrlich in mice infected with trypanosomes.[29] Similarly, the discovery of new analgesic compounds that act like opiates began well in advance of an understanding of the enkephalins or their receptors and was accomplished almost entirely through testing in animal models. However, new efforts toward the systematic discovery of nonpeptide ligands for non-opiate peptide receptors were almost completely dependent on the development of mechanism-based screening methods as a key strategic advance in the field of drug discovery.[31] In particular, this method was based on the seminal discovery that one could describe binding sites for neuroactive agents by measuring the binding of radiolabeled derivatives to membrane preparations obtained from cells in the target tissue.[38] The method proved that stereospecific binding of such ligands could be quantified and occurred with affinity that matched the physiological potency of the unlabeled drug. It led to the initial pharmacological characterization of receptors based on chemical rather than biological parameters and revolutionized the study of neuroactive drugs. The concept embodied in this technique was subsequently developed to permit the identification and characterization of binding sites (= receptors) in brain membranes for opiates such as morphine,[35] as well as for a plethora of neurotransmitters and neuropeptides.[31] Indeed, the elaboration of this technology provided the first functional, biochemically based description of neuropeptide receptors.

This assay platform was used to screen extracts of mammalian brain for endogenous substances that could displace radiolabeled morphine from its binding sites in brain membranes; the hypothesis driving this work was that a competition binding assay could identify the endogenous ligand(s) for the opiate receptor.[33] Unexpectedly, the endogenous ligands turned out to be a family of related and previously unknown neuropeptides, the enkephalins, dynorphins and endorphins. These are derived from 3 distinct precursor genes, the products of which are proteolytically processed to generate a diverse set of bioactive peptides that act preferentially on one or more of the multiple subtypes of opiate receptors found in the mammalian CNS.[34] Further work on the pharmacology of this system led to at least a basic understanding of how peptide and nonpeptide ligands for the family of opiate receptors exert their biological effects; this extensive literature provides the primary evidence that screening chemical collections for similar kinds of molecules that target non-opiate peptide receptors is worthwhile.

It is also worth noting that a consideration of the structure of nonpeptidic opiates reveals no readily apparent structural basis to link them with the amino acid sequences of the opiate peptides. Despite some progress in designing nonpeptide ligands for peptide receptors,[24] it was considered unlikely that de novo design protocols could readily identify novel nonpeptide ligands for other peptide receptors (finding new morphines, as it were). As an alternative and concurrent with developments in drug discovery focused on mechanism-based assays, HTS approaches were pursued. Whereas most prior drug discovery strategies involved the exposure of tissues or whole animals to chemical collections in a low-throughput effort (a few thousand compounds per year) to find bioactive molecules, the new wave catching the pharmaceutical industry in the 1980s was to employ assays that screened for compounds which acted upon specific target sites, in a mechanism-based, high-throughput (hundreds of thousands of compounds per year) paradigm. The convergence of the development of mechanism-based, high-throughput screening HTS strategies and the ability to detect binding of radiolabeled peptides to their receptors offered an ideal platform with which to search for novel drugs of potentially enormous therapeutic utility. This technology subsequently expanded with remarkable speed and became a central focus for the discovery of ligands that competed with many kinds of neurotransmitters for their receptors.[31] However, the application of this method for the discovery of nonpeptide ligands for peptide receptors is the topic of interest here.

Discovery of Novel Nonpeptide Ligands through Screening

An early example of a screen for nonpeptide ligands for a peptide receptor illustrates the process and its evolution. It has been known for some time that a number of mammalian neuropeptides play roles in regulating satiety (appetite) and, therefore, weight. As obesity-related diseases became more prevalent in the Western world, an increasing area of emphasis in the pharmaceutical industry was the discovery of new drugs that could safely reduce weight by reducing appetite (increasing feelings of satiety; see ref. 39 for review). A prominent example is CCK, which is present in peripheral tissues (both neuronal and endocrine) and the mammalian CNS. This peptide plays a number of physiological roles in mammals, but the initial therapeutic target was the action of CCK in mediating satiety. As noted, the first example of mechanism-based discovery of novel nonpeptide ligands was the effort of scientists at Merck & Co. who screened fermentation extracts to discover nonpeptide compounds that are ligands for CCK receptors. The goal was the identification of nonpeptide agonists which mimic the satiety-inducing effects of CCK, but with much better pharmaceutical profiles than could be achieved with the peptide itself. As a screening assay, this project used a preparation of bovine tissue with a relatively high abundance of CCK receptors; the assay measured the ability of test substances to displace [^{3}H]CCK from its membrane binding sites.[36]

The initial efforts identified a fungal secondary metabolite, asperlicin (Fig. 2B), which would not have been discovered in any other kind of assay available at the time; it was not active in animal assays, was poorly bioavailable and was not very potent.[36] A considerable amount of subsequent medicinal chemistry work on asperlicin generated simpler compounds based on the benzodiazepine pharmacophore that were much more potent and pharmaceutically tractable (Fig. 2B).[37] Unfortunately, the first compounds discovered were antagonists, not agonists, and thus not useful for mimicking satiety; they were also more selective for peripheral rather than CNS CCK receptors. Although considerable effort has been expended on the discovery of nonpeptide CCK receptor ligands since this initial report,[40,41] it now appears unlikely that agonists will be efficacious for the control of obesity. Antagonist indications, including panic disorders and pancreatic cancer, have not been successfully advanced.[41] The failure to date to exploit peptidergic signaling for obesity therapy may reflect the considerable redundancy in physiological systems that control appetite and metabolism; nonetheless, the therapeutic goal is so enormous that efforts continue.[4]

The discovery of novel CCK receptor ligands, while not yet leading to the development of marketed drugs, generated a considerable amount of activity as a new paradigm; the hope that the vast therapeutic potential represented in neuropeptide physiology could be harvested was a compelling motivation. The initial wave of discovery efforts relied on ligand-displacement or receptor binding assays, which relied upon radioactivity-based assays, an undesirable technology poorly suited for operation in HTS suites shared by many groups. This platform was circumvented by the development of functional, cell-based assays that could instantly distinguish between agonists and antagonists and could detect ligands that acted at sites other than the ligand binding domain.

Functional assays focus on the measurement of downstream effects that occur as a consequence of peptide binding to GPCRs in a cellular context.[43-45] The endpoints of these assays include the accumulation of second messengers (cAMP, Ca^{2+}, inositol phosphates) or the results of expression of reporter genes. GPCR expression systems have even been developed in the yeast, *Saccharomyces cerevisiae*;[44,45] they have been adapted for HTS platforms which measure the ligand-induced expression of reporter genes such as β-galactosidase or HIS3, an enzyme required for histidine biosynthesis; this latter format permits yeast growth to be measured as an index of ligand activation of the heterologous GPCR.[46] Regardless of the particular format, cell-based assays are amenable for operation in extremely high-throughput modes (millions of compounds per month) in ultra-micro formats. Their development has made ligand discovery for peptide receptors almost routine. Early discrimination of compound binding kinetics among receptor subtypes can be quickly attained in these formats, simply by repeating the initial assays in cells transfected with clones expressing the range of potential targets. This kind of analysis can prioritize lead compounds based on intrinsic potency and receptor subtype selectivity, important early determinants of clinical potential. However, finding an interesting hit in such a screen is only the first and in many ways the easiest, aspect of the process of drug discovery. Many concerns complicate the path.

Complicating Issues: Subtype- and Species-Specificity

Drug discovery HTS efforts, whether using competitive inhibition of ligand binding or through functional cell-based assays, led to the identification of many compounds that do not interact with the peptide binding site on the receptor. Such compounds can alter the ability of peptide ligands to bind to or activate the receptor by action at a distance, since GPCRs possess many distinct binding sites for ligands that alter receptor function without occupying the peptide binding site. This phenomenon has several consequences, most notably that design approaches cannot be expected to generate the same diversity of compounds (in terms of site-of-action) as a broad, random HTS might find. This consideration becomes important in the quest for pharmacological discrimination among receptor subtypes. Thus, one might expect that the amino acid residues which comprise the peptide binding site would be highly conserved across receptor sub-types, at least in forming the 3-dimensional shape/charge landscape that captures the ligand with high affinity and selectivity, since the subtypes all recognize the same peptide ligand. Structural conservation of the ligand binding site would make it difficult to identify nonpeptide ligands that occupy the site selectively among related subtypes. It must be reiterated that strong selectivity among peptide receptor subtypes is a generally essential feature of new drugs targeted to these systems.

In contrast, amino acids outside the peptide binding region might be expected to be more variable among receptor subtypes, which could lead to a greater ability to achieve subtype selectivity. This variability may also account for the repeatedly demonstrated species-specific nature of the drug-receptor interaction in neuropeptidergic systems.[47] Since the implications of this concept for discovery of broad spectrum anthelmintics or insecticides that act at peptide receptors are significant, this topic warrants additional consideration.

Early observations from nonpeptide ligand discovery programs revealed that species differences in receptor sequence led to species-specific pharmacology in antagonist binding; a consequence of this phenomenon is that animal studies required to support drug development are difficult to interpret for a human-selective analog. A second consequence is that discovery efforts must be directed at the target of interest, the human homolog of the neuropeptide receptor. These concepts are illustrated by the history of the discovery of nonpeptide ligands for the neurokinin-1 (substance P) receptor. Peptides in the neurokinin family play profound and diverse roles in pain sensation, vasodilation and immune responses, among other physiological processes. A potent nonpeptide antagonist of substance P was discovered by screening a chemical library for substances that displaced radiolabeled peptide from the NK1 receptor subtype in bovine caudate membranes;[48] as was typically the case in such exercises, the lead compound was an antagonist (see below). Subsequent experiments based on the construction of chimeras between the NK1 and NK3 receptors (to which the antagonist did not bind) revealed that the binding sites of the peptide agonist and nonpeptide antagonist were structurally distinct.[49] Further work identified that the histidine residue at position 197 was essential for binding of nonpeptide ligands, but was not involved in binding of the peptide agonist.[50]

The antagonist was potently active in some animal models and showed high affinity for human NK1 receptors, but poor affinity for rat and mouse NK1 receptors (and poor potency in those animals); the rodent and human receptors differ in 22/407 amino acids. Two of these are responsible for the species differences in pharmacology.[51,52] Interestingly, a different nonpeptide antagonist, discovered independently in a similar screening assay using rat brain membranes,[53] showed the opposite preference (higher affinity for rodent than for human NK1 receptors). The generality of these observations was supported by work on nonpeptide antagonists for the CCK-B (gastrin) receptor.[54] The canine and human CCK-B receptors share 90% amino acid identity, but differ significantly in affinity for a series of nonpeptide antagonists. The basis for this difference was traced to the amino acid at residue 319, even within the aliphatic family (valine, leucine, or isoleucine); this area was unimportant for agonist binding.

These examples of pharmacological specificity between the same receptor in different mammalian species are encouraging for the prospects of discovering nonpeptide ligands which can distinguish between even closely related members of peptide receptor families in a single species (humans). As noted previously, the general existence of multiple receptor subtypes for neuropeptides makes robust pharmacological discrimination among them a necessary prerequisite for the advancement of any lead chemical series. Medicinal chemistry must be able to differentiate among closely related receptor subtypes. Fortunately, the extent of amino acid identity across mammalian species for a particular neuropeptide receptor subunit is typically much higher (~90%) than the identity of receptor subtypes for a given neuropeptide within a species. For example, the three main types of opiate receptors (μ, δ and κ) share about 60% amino acid identity at the species level[55] and this parameter ranges between 30-50% for members of the neuropeptide Y receptor subtype family.[56] These analyses suggest that discovery of nonpeptide ligands with high receptor subtype selectivity should not be an insurmountable obstacle.

In this context, it is worth noting that a considerable degree of receptor subtype specificity was obtained for nonpeptide opiate receptor ligands purely through bioassay-guided medicinal chemistry, demonstrating the potential for new drug discovery in the neuropeptide arena. The remarkable array of opiate-like pharmacophores is a testament to the power of this approach (see Fig. 2 for examples). None of the most prominent members of this therapeutic class was discovered through use of an assay that involved a peptide receptor in a membrane preparation or whole-cell format. Within it are ligands that selectively target different receptor subtypes as well as other compounds, even in the same structural class, which have little subtype selectivity. The implications of these observations for the discovery of nonpeptide ligands useful for the control of invertebrate infestations and infections are discussed below.

Complicating Issues: Pharmacology of Agonists and Antagonists

It is essential to recognize that the class of nonpeptide ligands that is most abundantly populated, the opiates, was derived and expanded prior to the identification of opiate receptors and opioid peptides. The analgesic properties of opium, a complex mixture derived from poppy exudates, led to the eventual chemical identification of the active principles, especially morphine. Efforts to make better morphines involved screening in animal or tissue models that sensitively and accurately revealed opiate actions; a quite diverse set of analgesic templates was obtained from this fairly straightforward synthesize-and-assay process (see Fig. 2). Serendipitously, this included the identification of antagonists such as naloxone (Fig. 2) which potently blocked the pharmacological effects of morphine and related compounds. While drugs like naloxone have proven to be invaluable for the acute treatment of toxicity resulting from narcotic overdose, they are otherwise remarkably devoid of pharmacological effects. In the case of opiates, the fact that the lead compounds were derived from a natural source, selected in evolution and by breeding for bioactivity, meant that antagonists were only discovered by medicinal chemistry. Indeed, since chemistry efforts were driven by bioassays in which antagonists were generally inactive, a very pronounced preponderance of agonists was added to the pharmacopeia in the opiate category.

In striking contrast, it has proven much easier to discover peptide receptor antagonists than agonists through HTS and agonists are typically obtained only through modification of discovered antagonists.[57] Why this should be so is not entirely clear, other than the simple hypothesis that it is easier to disrupt the function of a receptor than to activate it. In general, although there are prominent exceptions, neuropeptide agonists may have greater clinical potential than antagonists (witness the example of morphine and naloxone). This may reflect the concept that neuropeptides are more modulators than primary drivers in disease states associated with mammalian neuropeptidergic transmission. Implications for this imbalance in discovery for invertebrate applications are discussed below.

Design

The concept of designing nonpeptide ligands to replicate the specific 3D charge-shape dimensions of the peptide-receptor contact point has had a long history. The conformational flexibility induced by peptide bonds has made it difficult to model peptide structures much larger than a tetramer and even then the universe of possible low energy conformations is dauntingly high. Many approaches have been made to identify the essential pharmacophore in a neuropeptide.[58-60] These include truncated versions, as short peptides that remain active, even if poorly potent, may help to identify the receptor-binding core; alanine-scan series; and D-amino acid scan series. In the last two strategies, residues of the native peptide are sequentially replaced one at a time with either alanine, which maintains the spacing of the peptide backbone but removes possibly important functional groups, or the corresponding D-amino acid, which maintains overall spacing and charge totality but places potentially essential functionality in the incorrect stereo position. If critical residues for receptor can be identified by these steps, rigidity can be introduced with peptide bond mimics or amino acid derivatives that limit conformational space. These 'peptoids' can then be used for modeling to, at least in theory, generate nonpeptide templates for further medicinal chemistry efforts.

As noted above, following the discovery of the opiate receptor family, it was possible to characterize these ligands in terms of affinity for different receptor subtypes. It was also possible to 'map' the enkephalin structure onto the surface of nonpeptide opiates such as morphine. However, it has not proven possible in a robust sense to predict the structures of nonpeptide ligands that mimic the actions of even the pentapeptide enkephalins in a modeling exercise; one could not readily predict that even a structurally simple drug like meperidine (Fig. 2) would be an opiate agonist based on an understanding of the solution conformation of enkephalin. Finally, as noted above, design exercises based on neuropeptides focus only on one site in the receptor; there are many other ways to affect receptor function that are not addressed by this technology.[58,61] However, as noted below, nonpeptide ligands for invertebrate neuropeptide receptors may be better targeted to the peptide binding site to achieve spectrum; design strategies may thus be more important in that arena than for the discovery of drugs that target mammalian neuropeptide receptors.

Current Status

The literature on the field of nonpeptide ligands for peptide receptors for human applications has been reviewed recently[5,62-65] and will not be covered in detail in this chapter. Suffice it to say that a great deal work has been done in pharmaceutical companies and academic laboratories on a wide variety of human neuropeptide systems and that nonpeptide drugs for at least 10 of them have reached the market (Table 1). Breakthrough products in the scope of opiates have not yet appeared, except perhaps for the angiotensin II antagonists used in the treatment of hypertension.[20] Nonetheless, the underexploited therapeutic potential of neuropeptidergic signaling in disease states ensures a continuing effort in drug discovery for neuropeptide receptors.

Applications in Invertebrate Systems: General Considerations

The complexity of functions of neuropeptidergic systems in humans is embodied in the fact that neuropeptides, like other neurotransmitters or neuromodulators, play multiple physiological roles and do so by acting at families of related receptor subtypes with different tissue distribution patterns and downstream signaling pathways. Deriving pharmacological specificity in the face of this complexity is a stern challenge, which is complicated by the fact that therapeutic intervention in humans is designed to restore homeostasis to systems that have become dysfunctional. These factors are not especially germane to considerations of neuropeptide-based therapeutics for invertebrates.

Table 1. Peptide receptors targeted for nonpeptide ligand clinical use or discovery

Ligands in Market	Ligands in R&D
Opiate peptides	Cholecystokinin/gastrin
Angiotensin	Urotensin
Endothelin	Gonadotropin-releasing hormone
Somatostatin[1]	Neurotensin
Vasopressin[1]	Melanocortin
Oxytocin[1]	Neuropeptide Y
Substance P (Neurokinin)	Orexin/hypocretin
Bradykinin	Calcitonin gene related peptide
Motilin[2]	Corticotropin releasing factor
Glucagon-like peptide[1]	Galanin
	Ghrelin
	Gonadotropin releasing hormone
	Melanin concentrating hormone

[1]Including or limited to modified peptides.
[2]Erythromycin, an antibiotic, is a motilin agonist.[71]

The simplest and most obvious difference is that the goal of treatment in invertebrates is to disrupt the normal function rather than to repair it; it is inevitably simpler to wreck a system than to fix one. The second pertains to the need for spectrum in the realm of antiparasitics and pesticides: the value of a compound that affects the function of receptors across a broad phylogenetic swath is much higher than that of a compound which affects very few target invertebrates. In this case, exquisite receptor subtype specificity in a lead compound would be a disadvantage for further development as it would be unlikely to recognize the homologous receptor in distantly related members of the target phylum (Nematoda, Platyhelminthes or Arthropoda).

Screening Targets: GPCRs

Basic research on invertebrate neuropeptide biology has not enjoyed the same kind of financial support by either governments or industry as has work on counterparts in mammalian systems. Consequently, progress in illuminating the molecular pharmacology of neuropeptide-receptor interactions in these organisms has been recent. The sequencing of invertebrate genomes, particularly those of *Caenorhabditis elegans* and *Drosophila melanogaster*, revealed multiple families of neuropeptide precursor genes and multiple GPCRs that were candidate neuropeptide receptors (see refs. 27,28,66). Bioinformatic analyses were incapable of matching peptides to receptors by computational methods. The simplest 'wet lab' approach to matching a collection of orphan receptors with suspected ligands in these model organisms would be to perform bioassays with the peptides in tissues from animals with knock-outs in GPCR-encoding genes compared to wild-type individuals. However, the lack of facile physiological assays in these genetically tractable organisms meant that matching ligands to orphan GPCRs has not been straightforward.

Instead, matching has been accomplished primarily through a process called reverse pharmacology (Fig. 3). In this approach, full-length cDNAs encoding the open reading frame (ORF) of candidate neuropeptide GPCRs are individually expressed in a heterologous system (mammalian cells or yeast). The recombinant cells are then exposed to neuropeptides of interest and activation of the receptor measured by one of several functional assays. As noted in subsequent chapters, this

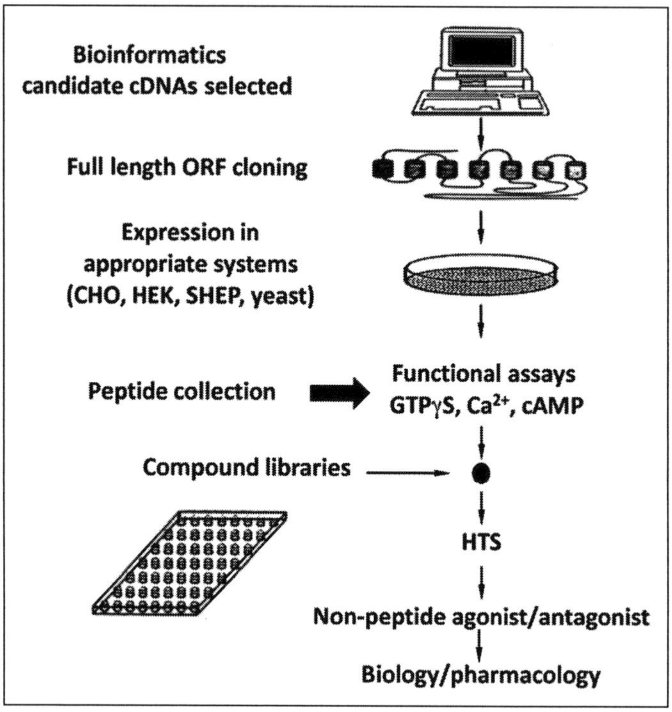

Figure 3. Reverse pharmacology to match peptide ligands with orphan GPCRs.

approach has proven useful for some, but not all, suspected peptide GPCRs from *C. elegans* and *D. melanogaster*; it has been less widely applied to the de-orphanization of GPCRs from target parasites and pests. Of benefit is that the heterologous expression system is generally quite easily adapted for HTS performance, to screen chemical libraries or fermentation extracts for compounds that mimic or block the actions of the peptide ligand on the receptor.

As proposed above, amino acid sequence diversity among related receptors for a specific neuropeptide is likely to be greater outside than inside the ligand binding domain. As an example, consider the family of invertebrate peptides related to FMRFamide, which are broadly distributed among the relevant phyla[7,8,11,66] and have been proposed as suitable targets for chemical intervention.[12-14] Work in several physiological systems has shown that the COOH-terminal RFamide moiety is essential for biological activity and that cross-phylum activity can be observed in FMRFamide-like peptides (see ref. 12 for review). These observations suggest that ligands which bind to the RFamide recognition pocket in invertebrate GPCRs might have the greatest potential for truly broad spectrum activity. Candidates with this binding proclivity could be found be HTS or perhaps by design strategies; whether sufficiently deleterious consequences could be achieved by antagonists which occupy this site is unknown (see below). A few RFamide-containing peptides are found in mammals, so host:parasite selectivity would have to be evaluated empirically for compounds identified in a discovery exercise focused on this site.

It remains an open question whether agonists or antagonists are preferred for activity against invertebrates. Loss-of-function mutations in or RNAi interference suppression of genes encoding neuropeptide precursors and many candidate peptidergic GPCRs have generally modest, though not necessarily undetectable, effects on the behavior of *C. elegans* (see ref. 66 for review); the issue has been less studied in other invertebrates. Conversely, application of many neuropeptides to tissues of worms and flies causes profound physiological effects.[6,7,11,12] These observations suggest that nonpeptide agonists at neuropeptide receptors might be expected to be more useful. The situation is compounded by the evident difficulty in discovering nonpeptide agonists (as opposed to antagonists) through HTS. The use of functional assays in which agonists can be specifically identified from the start should allow the process to focus only on the most promising hits at the onset. Clearly, however, much remains to be learned about the molecular physiology of neuropeptide receptors in situ in invertebrates in order to optimize the search for effective and safe drugs that act on them.

Screening Targets: Peptide Processing

Inhibitors of proteases that process neuropeptide precursors may have unique value for invertebrates, because specificity within invertebrate peptide families (or across them) is desirable rather than being a disadvantage. In the model organisms *C. elegans* and *D. melanogaster*, a limited set of proteases seem to be involved in this biochemistry.[66-68] Phenotypes of knock-out or knock-down (RNA interference) animals are evident in *C. elegans*, particularly for a proprotein convertase encoded in the *egl-3/kpc-2* gene (see ref. 66). In *C. elegans*, loss-of-function mutations in a carboxypeptidase E encoded by the *egl-21* gene show profound defects in motor function (see ref. 66). This enzyme removes basic residues from the cleavage sites that generate neuropeptides from precursor proteins; based on the phenotypic consequences, inhibitors of EGL-21 would be desirable candidate anthelmintics. As carboxypeptidases are present in mammals, selectivity is also an issue for this target (see below).

Another intriguing area for chemical intervention is in peptide amidation, mediated by removal of a COOH-terminal glycine residue by a bifunctional enzyme which leaves the amine in place as an amide. Peptide amidating systems are known in *C. elegans*[66] and *D. melanogaster*[69] and an unusual set of two enzymes has been shown to perform this reaction in an important trematode parasite, *Schistosoma mansoni*.[70] Interference with amidation could be a druggable target with diverse negative consequences for invertebrates as long as the inhibitor is inactive on the homologous human enzyme.

Despite this promise, it is undeniable that much less is known about invertebrate than mammalian neuropeptide processing enzymes. Nevertheless, significantly greater opportunities for drug discovery may be found in them. The disadvantage for mammals in this regard is that very high specificity is demanded when targeting a particular neuropeptidergic system; general interference with these enzymes would be predicted to cause a large number of unwanted side effects. In contrast, inhibition of peptide processing pathways in invertebrates that affect multiple peptidergic systems would be a considerable advantage, as the generally deleterious consequences would be of therapeutic value in this context. The critical factors to consider are spectrum and safety. The binding site for an enzyme inhibitor can be represented by a 3-dimensional landscape of charge and space; when considered with respect to spectrum and selectivity, this landscape includes features which may be understood as *essential* (required for inhibitor binding), *exclusionary* (incompatible with inhibitor binding) and *neutral* (unimportant for binding). Typically, these features are specific to a particular inhibitor (usually discovered in an HTS) and its structural class and may be identified based on crystallographic analysis of the inhibitor-target complex and subsequent modeling of host and additional parasite sequences. Computational methods can be used to predict whether the inhibitor can be expected to demonstrate the requisite spectrum (by analyzing target species, once the relevant cDNA sequence is obtained) and host:parasite selectivity based on the conservation of essential and exclusionary residues.

What to Screen

The need for spectrum in drugs to be used for invertebrate indications demands the initial discovery of compounds that recognize ligand binding features in receptors from organisms that are separated by deep evolutionary distance; a suitably promiscuous ligand is needed instead of a highly specific one. This presents a somewhat unusual task for the medicinal chemist. The likelihood of finding a good starting molecule through HTS operations depends a great deal on the composition of the library used in screening. A significant proportion of anthelmintics and pesticides are derived from natural products. To a large extent, this can be understood from the perspective that secondary metabolites generated by plants and microbes have been preselected by evolutionary pressure for bio-activity against a broad range of invertebrate pests and predators. Unfortunately, natural products are no longer favored for use in HTS in most large pharmaceutical and agrochemical companies; this presents the rare opportunity for bringing together new targets and new chemistry for discovery of new drugs for the control of pests and parasites.

Conclusion

Neuropeptidergic systems have provided a wealth of drugs that are considered essential in human medicine. Most prominent in this regard are the opiates, originally derived from the poppy and mimicked by a stunning array of nonpeptide ligands for central nervous system peptide receptors. These drugs are most commonly used as analgesics; their pain relief efficacy is unequaled. It is even more remarkable to note that the vast majority of useful drugs in this category were discovered and developed in the absence of any information on their receptors. Although basic research has since led to the identification of many other neuropeptides that play crucial roles in mammalian physiology, breakthroughs in their therapeutic exploitation have been hard to deliver. Systems developed for screening for nonpeptide ligands for peptide receptors have been greatly advanced in these efforts, which continue. Neuropeptidergic systems also play crucial roles in insect and helminth physiology and represent exciting targets for the discovery of new compounds with selective activity against invertebrates. In many significant ways, the promise of such agents is greater for invertebrates than vertebrates. Fortunately, much has been learned in the pursuit of nonpeptide ligands for mammalian peptide receptors that can inform and guide similar efforts directed at invertebrates.

Acknowledgements

Mr. Yovany Moreno kindly produced Figures 1 and 2. Work in the author's laboratory is supported by grants from NSERC, the FQRNT Centre for Host-Parasite Interactions, the Canada Research Chairs and the Bill and Melinda Gates Foundation.

References

1. Scharrer B. The neuropeptide saga. Amer Zool 1990; 30:887-95.
2. Hökfelt T, Broberge C, Xu ZQD et al. Neuropeptides—an overview. Neuropharmacol 2000; 39:1337-56.
3. Strand FL. Neuropeptides: Regulators of Physiological Processes. MIT Press, Cambridge, MA 1999.
4. Hökfelt T, Bartfai T, Bloom F. Neuropeptides: opportunities for drug discovery. Lancet Neurol 2003; 2:463-72.
5. Brain SD, Cox HM. Neuropeptides and their receptors: innovative science providing novel therapeutic targets. Brit J Pharmacol 2006; 147:S202-11.
6. Nässel DR. Neuropeptides in the nervous system of Drosophila and other insects: multiple roles as neuromodulators and neurohormones. Prog Neurobiol 2002; 68:1-84.
7. McVeigh P, Kimber MJ, Novozhilova E et al. Neuropeptide signaling systems in flatworms. Parasitol 2005; 131(Suppl):S41-55.
8. McVeigh P, Leech S, Mair GR et al. Analysis of FMRFamide-like peptide (FLP) diversity in phylum Nematoda. Int J Parasitol 2005; 35:1043-60.
9. McVeigh P, Alexander-Bowman S, Veal E et al. Neuropeptide-like protein diversity in phylum Nematoda. Int J Parasitol 2008; 38:1493-503.
10. Masler EP, Kelly TJ, Menn JJ. Insect neuropeptides: discovery and application in insect management. Arch Insect Biochem Physiol 1993; 22:87-111.
11. Gäde G, Goldsworthy GJ. Insect peptide hormones: a selective review of their physiology and potential application for pest control. Pest Man Sci 2003; 59:1063-75.
12. Maule AG, Mousley A, Marks NJ et al. Neuropeptide signaling systems—potential targets for parasite and pest control. Curr Top Med Chem 2002; 2:733-58.
13. Mousley A, Maule AG, Halton DW et al. Inter-phyla studies on neuropeptides: the potential for broad-spectrum anthelmintic and/or endectocide discovery. Parasitol 2005; 131(Suppl):S143-67.
14. Greenwood K, Williams T, Geary T. Nematode neuropeptide receptors and their development as anthelmintic screens. Parasitol 2005; 131:S169-77.

15. Åkerlund M. Oxytocin antagonists in the treatment of preterm labour. Fet Matern Med Rev 2002; 13:31-41.
16. Ashworth DM, Bit AR, Baxter AJ et al. Nonpeptide oxytocin agonists. Drugs Future 2006; 31:345-53.
17. Lottmann H, Baydala L, Eggert P et al. Long-term desmopressin response in primary nocturnal enuresis: open-label, multinational study. Int J Clin Pract 2009; 63:35-45.
18. Decaux G, Soupart A, Vassart G. Nonpeptide arginine-vasopressin antagonists: the vaptans. Lancet 2008; 371:1624-32.
19. Matchar DB, McCrory DC, Orlando LA et al. Systematic review: comparative effectiveness of angiotensin-converting enzyme inhibitors and angiotensin II receptor blockers for treating essential hypertension. Ann Intern Med 2008; 148:16-29.
20. Ram CV. Angiotensin receptor blockers: current status and future prospects. Am J Med 2008; 121:656-63.
21. Zerilli T, Pyon EY. Sitagliptin phosphate: a DPP-4 inhibitor for the treatment of type 2 diabetes mellitus. Clin Ther 2007; 29:2614-34.
22. Knop FK, Vilsbøll T, Holst JJ. Incretin-based therapy of type 2 diabetes mellitus. Curr Prot Pept Sci 2009; 10:46-55.
23. Hansen KB, Knop FK, Holst JJ et al. Treatment of type 2 diabetes with glucagon-like peptide-1 receptor agonists. Int J Clin Pract 2009; 63:1154-60.
24. Jones RM, Boatman PD, Semple G et al. Clinically validated peptides as templates for de novo peptidomimetic drug design as G-protein-coupled receptors. Curr Opin Pharmacol 2003; 3:530-543.
25. Lingueglia E, Deval E, Lazdunski M. FMRFamide-gated sodium channel and ASIC channels: a new class of ionotropic receptors for FMRFamide and related peptides. Peptides 2006; 27:1138-52.
26. Purcell J, Robertson AP, Thompson DP et al. PF4, a FMRFamide-related peptide, gates low-conductance Cl^- channels in Ascaris suum. Eur J Pharmacol 2002; 456:11-17.
27. Meeusen T, Mertens I, De Loof A et al. G protein-coupled receptors in invertebrates: a state of the art. Int Rev Cytol 2003; 230:189-261.
28. Claeys I, Poels J, Simonet G et al. Insect neuropeptide and peptide hormone receptors: current knowledge and future directions. Vit Horm 2005; 73:217-86.
29. Hawking F. History of chemotherapy. In: Schnitzer RJ, Hawking F, eds. Experimental Chemotherapy. New York: Academic Press, 1963:1-24.
30. Drews J. Drug discovery: a historical perspective. Science 2000; 287:1960-63.
31. Williams M. Receptor binding in the drug discovery process. Med Res Rev 1991; 11:147-184.
32. Kenny BA, Bushfield M, Parry-Smith DJ et al. The application of high-throughput screening to novel lead discovery. Prog Drug Res 1988; 51:245-269.
33. Hughes J, Smith T, Kosterlitz H et al. Isolation of two related pentapeptides from the brain with potent opiate agonist activity. Nature 1975; 258:577-9.
34. Janecka A, Fichna J, Janecki T. Opioid receptors and their ligands. Curr Top Med Chem 2004; 4:1-17.
35. Snyder SH, Pasternak GW. Historical review: opioid receptors. Trends Pharmacol Sci 2003; 24:198-205.
36. Chang RSL, Lotti VJ, Monaghan RL et al. A potent nonpeptide cholecystokinin antagonist selective for peripheral tissues isolated from Aspergillus alliaceus. Science 1985; 230:177-9.
37. Evans BE, Rittle KE, Bock MG et al. Methods for drug discovery: development of potent, selective, orally effective cholecystokinin antagonists. J Med Chem 1988; 31:2235-46.
38. Goldstein A, Lowney LI, Pal BK. Stereospecific and nonspecific interactions of the morphine congener levorphanol in subcellular fractions of mouse brain. Proc Natl Acad Sci USA 1971; 68:1742-47.
39. Cooke D, Bloom S. The obesity pipeline: current strategies in the development of anti-obesity drugs. Nature Revs Drug Disc 2006; 5:919-31.
40. Herranz R. Cholecystokinin antagonists: pharmacological and therapeutic potential. Med Res Rev 2003; 23:559-605.
41. Berna MJ, Tapia JA, Sancho V et al. Progress in developing cholecystokinin (CCK)/gastrin receptor ligands that have therapeutic potential. Curr Opin Pharmacol 2007; 7:583-92.
42. Silverman L, Campbell R, Broach JR. New assay technologies for high-throughput screening. Curr Opin Chem Biol 1998; 2:397-403.
43. Pausch MH. G protein-coupled receptors in Saccharomyces cerevisiae: high-throughput screening assays for drug discovery. Trends Biotech 1997; 15:487-94.
44. Dowell SJ, Brown AJ. Yeast assays for G protein-coupled receptors. Receptors Channels 2002; 8:343-52.
45. Thomsen W, Frazer J, Unett D. Functional assays for screening GPCR targets. Curr Opin Biotech 2005; 16:655-65.

46. Wang ZX, Broach JR, Peiper SC. Functional expression of CXCR4 in Saccharomyces cerevisiae in the development of powerful tools for the pharmacological characterization of CXCR4. (2006) Meth Mol Biol 2006; 332:115-27.
47. Berthold M, Bartfai T. Modes of peptide binding in G protein-coupled receptors. Neurochem Res 1997; 22:1023-31.
48. Snider RM, Constantine JW, Lowe JA III et al. A potent nonpeptide antagonist of the Substance P (NK_1) receptor. Science 1991; 251:435-7.
49. Gether U, Johansen TE, Snider RM et al. Different binding epitopes on the NK_1 receptor for substance P and a nonpeptide antagonist. Nature 1993; 362:345-8.
50. Fong TM, Cascieri MA, Yu H et al. Amino-aromatic interaction between histidine 197 of the neurokinin-1 receptor and CP 96345. Nature 1993; 362:350-3.
51. Fong TM, Yu H, Strader CD. Molecular basis for the species-selectivity of the Neurokinin-1 receptor antagonists CP-96,345 and RP67580. J Biol Chem 1992; 267:25668-71.
52. Sachais BS, Snider RM, Lowe JA III et al. Molecular basis for the species selectivity of the Substance P antagonist CP-96,345. J Biol Chem 1993; 268:2319-23.
53. Garret C, Carruette A, Fardin V et al. Pharmacological properties of a potent and selective nonpeptide substance P antagonist. Proc Natl Acad Sci USA 1991; 88:10208-12.
54. Beinborn M, Lee YM, McBride EW et al. A single amino acid of the cholecystokinin-B/gastrin receptor determines specificity for nonpeptide antagonists. Nature 1993; 362:348-50.
55. Satoh M, Minami M. Molecular pharmacology of the opioid receptors. Pharmacol Ther 1995; 68:343-64.
56. Michel MC, Beck-Sickinger A, Cox H et al. XVI. International union of pharmacology recommendations for the nomenclature of neuropeptide Y, peptide YY and pancreatic polypeptide receptors. Pharmacol Rev 1998; 50:143-50.
57. Sugg EE. Nonpeptide agonists for peptide receptors: lessons from ligands. Ann Rep Med Chem 1997; 32:277-83.
58. Marshall GR. Peptide interactions with G-protein coupled receptors. Biopolymers 2001; 60:246-77.
59. Giannis A, Kolter T. Peptidomimetics for receptor ligands—discovery, development and medical perspectives. Agnew Chem Int Ed Engl 1993; 32:1244-67.
60. Hruby VY, Agnes RS. Conformation-activity relationships of opioid peptides with selective activities at opiate receptors. Biopolymers 1999; 51:391-410.
61. Rashid AJ, O'Dowd BF, George SN. Minireview: Diversity and complexity of signaling through peptidergic G protein-coupled receptors. 2004; Endocrinol 145:2645-52.
62. Freidinger RM. Nonpeptidic ligands for peptide and protein receptors. Curr Opin Chem Biol 1999; 3:395-406.
63. Gurrath M. Peptide-binding G protein-coupled receptors: new opportunities for drug design. Curr Med Chem 2001; 8:1605-48.
64. Estriarte M, Rich D. Peptidomimetics for drug design. In: Abraham DJ, ed. Burger's Medicinal Chemistry and Drug Discovery. Hoboken:Wiley-Interscience 2003:1(6):633-85.
65. Blakeney JS, Reid RC, Le GT et al. Nonpeptidic ligands for peptide-activated G protein-coupled receptors. Chem Rev 2007; 107:2960-3041.
66. Li C, Kim K. Neuropeptides ed. The C. elegans Research Community. WormBook, doi/10.1895/wormbook.1.7.1, http://www.wormbook.org 2008.
67. Husson SJ, Mertens I, Janssen T et al. Neuropeptidergic signaling in the nematode Caenorhabditis elegans. Prog Neurobiol 2007; 82:33-55.
68. Isaac RE, Bland ND, Shirras AD. Neuropeptidases and the metabolic inactivation of insect neuropeptides. Gen Comp Endocrinol 2009; 162:8-17.
69. Han M, Park D, Vanderzalm PJ et al. Drosophila uses two distinct neuropeptide amidating enzymes, dPAL1 and dPAL2. J Neurochem 2004; 90:129-41.
70. Mair GR, Niciu MJ, Stewart MT et al. A functionally atypical amidating enzyme from the human parasite Schistosoma mansoni. FASEB J 2004; 18:114-21.
71. Peeters T, Matthijs G, Depoortere I et al. Erythromycin is a motilin receptor agonist. Am J Physiol 1989; 257:G470-4.

Chapter 3

Interaction of Mimetic Analogs of Insect Kinin Neuropeptides with Arthropod Receptors

Ronald J. Nachman* and Patricia V. Pietrantonio

Abstract

Insect kinin neuropeptides share a common C-terminal pentapeptide sequence Phe^1-Xaa_1^2-Xaa_2^3-Trp^4-Gly^5-NH_2 (Xaa_1^2 = His, Asn, Phe, Ser or Tyr; Xaa_2^3 = Pro, Ser or Ala) and have been isolated from a number of insects, including species of Dictyoptera, Orthoptera and Lepidoptera. They have been associated with the regulation of such diverse processes as hindgut contraction, diuresis and the release of digestive enzymes. In this chapter, the chemical, conformational and stereochemical aspects of the activity of the insect kinins with expressed receptors and/or biological assays are reviewed. With this information, biostable analogs are designed that protect peptidase-susceptible sites in the insect kinin sequence and demonstrate significant retention of activity on both receptor and biological assays. The identification of the most critical residue of the insect kinins for receptor interaction is used to select a scaffold for a recombinant library that leads to identification of a nonpeptide mimetic analog. C-terminal aldehyde insect kinin analogs modify the activity of the insect kinins leading to inhibition of weight gain and mortality in corn earworm larvae and selective inhibition of diuresis in the housefly. Strategies for the modification of insect neuropeptide structures for the enhancement of the topical and oral bioavailability of insect neuropeptides and the promotion of time-release from the cuticle and/or foregut are reviewed. Promising mimetic analog leads for the development of selective agents capable of disrupting insect kinin regulated processes are identified that may provide interesting tools for arthropod endocrinologists and new pest insect management strategies in the future.

Introduction

Insect neuropeptides of the insect kinin class share a common C-terminal pentapeptide sequence Phe^1-Xaa_1^2-Xaa_2^3-Trp^4-Gly^5-NH_2 (Xaa_1^2 = His, Asn, Phe, Ser or Tyr; Xaa_2^3 = Pro, Ser or Ala). They have been isolated from a number of insects, including species of Dictyoptera, Orthoptera and Lepidoptera. The first members of this insect neuropeptide family were isolated on the basis of their ability to stimulate contractions of the isolated cockroach hindgut,[1-3] but they are also potent diuretic peptides that stimulate the secretion of primary urine by Malpighian tubules, organs involved in the regulation of salt and water balance.[3-5] In addition, the insect kinins have been implicated in the regulation of digestive enzyme release.[6-8] More recently, insect kinins and/or analogs, have been reported to inhibit weight gain by larvae of the tobacco budworm (*Heliothis virescens*) and corn earworm (*Helicoverpa zea*),[3,9] both serious agricultural pests.

*Corresponding Author: Ronald J. Nachman—Areawide Pest Management Research Unit, Southern Plains Agricultural Research Center, U.S. Department of Agriculture, 2881 F/B Road, College Station, Texas, USA. Email: nachman@tamu.edu

Neuropeptide Systems as Targets for Parasite and Pest Control, edited by Timothy G. Geary and Aaron G. Maule.

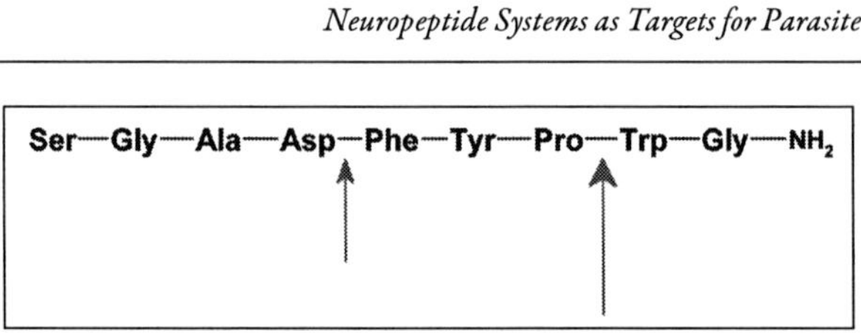

Figure 1. Primary (large arrow) and secondary (small arrow) hydrolysis cleavage sites of tissue-bound peptidases for the natural cricket insect kinin SGADFYPWG-NH_2 (AK-I). From Zubrzak et al (2007).[49]

Unfortunately, insect kinin peptides are unsuitable as research tools for insect neuroendocrinologists and/or pest control agents due to susceptibility to both exo- and endopeptidases in the hemolymph and gut of the insect. These insect neuropeptides are also unsuitably designed for efficient bioavailability, whether delivered via a topical and/or oral route. Members of the insect kinin family are hydrolyzed and therefore inactivated, by tissue-bound peptidases of insects. Two susceptible hydrolysis sites in insect kinins[3] have been reported. The primary site is between the Pro^3 and the Trp^4 residues, with a secondary site N-terminal to the Phe^1 residue in natural extended insect kinin sequences (Fig. 1). Experiments demonstrate that angiotensin converting enzyme (ACE) from the housefly can cleave at the primary hydrolysis site, whereas neprilysin (NEP) can cleave insect kinins at both the primary and secondary hydrolysis sites.[3,10-13] In this chapter, we review what is known about chemical, conformational and stereochemical aspects of the interaction of insect kinins with their receptors and how this knowledge can be harnessed to design and develop biostable mimetic analogs that retain an ability to bind and potentially activate, insect kinin receptors. Strategies for the modification of insect neuropeptides to enhance bioavailability characteristics and the potential to control pest insect populations with mimetic insect kinin analogs that disrupt water and ion balance and/or digestive processes are also discussed.

Functional Analysis of Arthropod Receptors Selective for Insect Kinins

G protein-coupled receptors (GPCR) constitute the largest super family of receptors.[14] GPCR are integral plasma membrane proteins which are characterized by seven hydrophobic transmembrane spanning α-helices, three intracellular loops, three extracellular loops, an amino terminal outside the cell and a carboxy terminal inside the cell.[15,16] On recognizing extracellular primary peptide messengers, GPCR mediate intracellular communication across the cell membrane by coupling to intracellular heterotrimeric G-proteins that activate secondary-messenger pathways. The activated G-protein modulates various intracellular processes that regulate cell function.[17] The ability to selectively regulate GPCR and their signaling pathways may lead to the discovery of novel insecticidal targets.[18]

Holmes et al (2000)[19] used polymerase chain reaction (PCR) and molecular techniques to obtain a cDNA of a novel leucokinin-like peptide receptor, the first known neuropeptide receptor from the Southern cattle tick, *Boophilus microplus* (Acari: Ixodidae). For this, degenerate primers were designed based on the sequence similarity of transmembrane (TM) regions III and VI lymnokinin and vertebrate neurokinin receptors.[19,20] Until the cloning of tick leucokinin-like peptide receptor, this receptor subfamily comprised only one receptor, the lymnokinin receptor, cloned from a mollusc, the pond snail *Lymnaea stagnalis.*[20] The amino acid sequence of the *B. microplus* receptor showed most similarity to the CG10626 *Drosophila melanogaster* gene product and to the lymnokinin receptor from *L. stagnalis,* which has been characterized as a leucokinin-like peptide receptor.[19] The C-terminal end of TM II (residues 103-110) and the first extracellular loop (residues 111-123) are highly conserved among three leucokinin-like peptide receptors the *B. microplus, L. stagnalis* and *D. melanogaster* but not in other neuropeptide receptor subfamilies.[19] Similarly, the first eight residues of the C terminus (337-344) are highly conserved among the leucokinin-like peptide receptors and less conserved among the other neuropeptide receptors.[19] The presence of the mRNA in all the life stages of the tick and diverse tissues reflects the importance of this receptor for various functions and suggests that it can be used as a target for the development of novel and specific acaricides.[19] The receptor from the Southern cattle tick, (*B. microplus*) was stably and functionally expressed in CHO-K1 cells and responded to physiological concentrations of insect kinin peptides as determined by single cell intracellular calcium measurements using confocal fluorescence microscopy.[21] To date no endogenous tick kinin has been identified. Based on the similarity to the tick receptor sequence, Holmes et al predicted the CG10626 *Drosophila* gene product encoded the first known insect leucokinin receptor,[19] which was later confirmed by Radford et al (2002).[22] In the genome of *Drosophila melanogaster*, only one drosokinin receptor is present and only one peptide, drosokinin, is the ligand for this receptor.[22] In the mosquito *Aedes aegypti*, three kinins have been isolated and the single cDNA encoding the three peptides has been cloned.[23,24] The *A. aegypti* kinin receptor is the first cloned from mosquitoes.[25] In *A. aegypti* females the function of insect kinins had been previously investigated in the Malpighian tubule, where they increase fluid secretion and depolarize the transepithelial voltage,[23,24] therefore, cDNA was synthesized from these tissues to isolate the receptor cDNA. The strategy consisted in designing degenerate primers corresponding to transmembrane regions VI and VII of the invertebrate receptors known, the drosokinin, lymnokinin and *Boophilus microplus* myokinin receptors to obtain a specific 280 bp PCR product, followed by 5′ and 3′ RACE to obtain the complete open reading frame sequence. The *Aedes* cDNA sequence was used to improve the prediction of the mosquito *Anopheles gambiae* kinin receptor sequence (ex XP309852; EAA05450 discontinued record) which lacked the first transmembrane region and residues at the C-terminus.[25] The *A. gambiae* receptor symbol is now GPRLKK; the protein Ensembl peptide ID is AGAP010851-PA; and the transcript (ID AGAP010851-RA, see *Anopheles gambiae* genome in www.vectorbase.org) sequence is identical to that predicted by Pietrantonio et al (2005).[25] The *A. aegypti* cDNA encodes a 584 amino acid residue protein of predicted molecular mass of 65.2 kDa.

The mosquito kinin receptor cDNA was expressed in CHO-K1 cells for functional analysis by intracellular calcium measurements. In addition to analysis by confocal fluorescence cytometry mentioned above, the receptor-expressing cells were analyzed by calcium bioluminescence in a plate assay using the calcium reporter aequorin.[25] These analyses showed the mosquito kinin receptor is indeed a multiligand receptor triggering intracellular calcium release in response to the three Aedes kinins and kinin analogs, similarly to the *Boophilus* kinin receptor. The two methodologies showed that the decreasing rank order of potency on the *Aedes* receptor is Aedes kinin 3 followed by Aedes kinin 2 and Aedes kinin 1. These three neuropeptides are significantly different in their EC_{50}. In *Drosophila,* in addition to its localization and enrichment in Malpighian tubules,[26] the receptor is present in the nervous system, hindgut and in gonads of males and females as shown by western blot and immunohistochemistry.[22] In females of *Aedes aegypti,* the receptor transcript was confirmed by PCR in the hindgut and head and in the ovaries by in situ hybridization (Pietrantonio and Jagge, unpublished results). It is clear that in these dipterans, the kinin receptor has additional functions in addition to the initially discovered role in chloride transport in the Malpighian tubule.[28,29] The kinin receptor in the central nervous system has more recently been implicated in pre-ecdysis behavior in *Drosophila melanogaster* and *Manduca sexta.*[30,31] Analysis of the first coleopteran genome of the beetle *Tribolium castaneum* revealed this species is devoid of the kinin receptor.[32] Thus, it appears different orders of insects have evolved diverse (and perhaps equivalent) signaling pathways at least for specific molecular events in diuresis and metamorphosis, functions in which insect kinins are involved in the Diptera.

Chemical, Conformational and Stereochemical Aspects of Receptor Interaction

Myotropic and diuretic assays of tissues in vitro, as well as assays using stably expressed insect kinin receptors, show that the full biological activity of the insect kinins resides in the C-terminal pentapeptide, which is the active core.[5,33,34] Exceptions are the housefly Malpighian tubule fluid secretion assay[35] and the plate assay for an expressed insect kinin receptor from the mosquito *Aedes aegypti,* where the C-terminal pentapeptide core is less potent by several orders of magnitude. Diuretic, myotropic and/or receptor-interaction activity in these assays is completely lost when the C-terminal amide of the insect kinins is replaced with a negatively charged acid moiety.[36,37] Within the core pentapeptide, the aromatic residues Phe[1] and Trp[4] are the most important for activity whereas a wide range of variability is generally tolerated at position 2, from acidic to basic residues and from hydrophilic to hydrophobic.[5,33] The expressed insect kinin receptor from the mosquito *Aedes aegypti* represents a particular exception, as it clearly prefers an aromatic residue in position 2, which is consistent with the presence of aromatic residues in position 2 of all three of the native *Aedes* kinins.[37]

Despite the steric bulk in the backbone of the Aib-containing insect kinin analog FF(Aib)WGa, it nevertheless elicits a very strong calcium bioluminescence response in both tick and mosquito receptors.[37] This is in agreement with the potent activities of Aib containing analogs observed in a cricket Malphigian tubule fluid secretion assay, an in vivo housefly diuretic assay and a cockroach hindgut myotropic assay.[3,10] In the mosquito receptor, it is statistically equipotent with the positive-control, agonist analog FFFSWGa; whereas in the tick receptor it is an order of magnitude more potent than this same agonist. Therefore, it is the most potent peptide analog yet reported for the tick receptor. The steric bulk of the Aib residue also restricts the number of conformations available to the backbone of this analog and provides some insight into the conformation adopted by the insect kinins at the two receptors. A previous solution conformation study using both NMR spectroscopic data and molecular dynamics calculations concludes that the analog adopts only two major turn conformations. These consist of a turn over residues Phe[1] through Trp[4], comprising 60% of the population and another over residues Phe[2] through Gly[5], comprising the remaining 40%.[11,38,39] Similarly, NMR spectroscopic data and molecular dynamics calculations on an active head-to-tail, cyclic analog (*cyclo*(AFFPWG)) reveal the presence of two major turn types within the active core region of the insect kinins, analogous with those observed in the Aib-containing analog. The more rigid of the two conformations featured a *cis*Pro in the third position of a Type-VI β-turn over core residues 1-4, or Phe-Phe-**Pro**-Trp. ROESY spectra supported a well-defined Cβ-*exo*/Cγ-*endo* pucker for the *cis*Pro ring that was observed in unrestrained molecular dynamics for this cyclic analog. The other less rigid turn system involved a *trans*Pro and encompassed residues 2-5, or Phe-**Pro**-Trp-Gly. From unrestrained molecular dynamics calculations, the most favorable *cis*Pro structure had an intramolecular energy about 7 kcal/mole lower than the most favorable *trans*Pro structure, consistent with NMR data that indicated that the *cis*Pro 1-4 turn structure was the predominant conformation in solution by a 60:40 ratio.[11,39,40] This is in agreement with systematic studies on linear peptides with Pro[3] in which the flanking aromatic residues promote the formation of Type VI β-turns in aqueous solution. Such turns are further enhanced when small, hydrophyllic residues (i.e., Asp, Ser, Thr, **Gly** or Asn) follow the aromatic-Pro-aromatic motif,[77] as occurs in the cyclic insect kinin analog. The molecular modeling studies further indicate that interactions between the aromatic sidechains in positions 1 and 4 help to stabilize the turn over residues 1-4 containing the *cis*Pro configuration, which might otherwise be expected to be less energetically favorable than *trans*Pro.[11,39]

In an effort to provide definitive evidence that the most populous *cis*Pro Type VI β-turn over residues Phe[1] through Trp[4] represented the 'active conformation' for receptor interaction, insect kinin analogs incorporating restricted conformation components that preferentially mimic a *cis* peptide bond and a Type VI β-turn were synthesized and evaluated. NMR studies with insect kinin analogs incorporating either the tetrazole or 4-aminopyroglutamate (APy) (Fig. 2), moieties

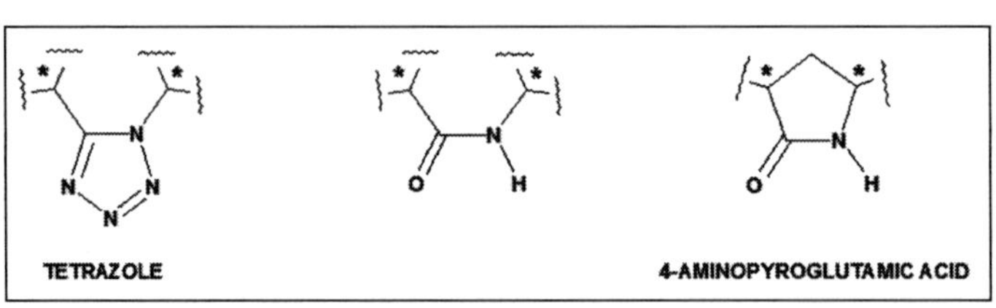

Figure 2. A comparison of the structures of the tetrazole (ψ(CN_4), left) and 4-aminopyroglutamic acid (APy; right) motifs, mimics of the *cis*-peptide bond (middle) and a Type VI β-turn. From Nachman et al (2004).[41]

Table 1. ***Estimated potencies (EC_{50}) and percentage of maximal bioluminescence response of different analogs in reference to FFSWGa tested on tick (BmLK3) receptor expressing cell lines.[44] EC_{50} are an estimate of the concentration required to induce a half-maximal response. The tick receptor response is compared with activity of the APy analogs[41,42] and tetrazole analogs[40,42] in a cricket diuretic assay.***

	Tick Receptor, (BmLK3 Cell Line)		**Cricket Diuretic Assay**	
Analogs	**EC_{50} (μM)**	***Maximal Response (%)**	**Stimulation of Malpighian Tubule Fluid Secretion EC_{50} (10^{-8} M)**	**Maximal Response (%)**
Amino Pyroglutamate analogs				
1518, Ac-RF(APy)WGa (R,S)	1.6	65	0.7	93
RN2, Ac-RF(APy)WGa (S,S)	11.1	48	14	93
RN3, Ac-RF(APy)WGa (S,R)	N.D.	0	12	96
RN4, Ac-RF(APy)WGa (R,R)	N.D.	45	7	83
Tetrazole analogs				
FF_{ψ}**(CN_4)**(dA)WGa (L,D)	N.D.	0	(43)$^{\Psi}$	
FF_{ψ}**(CN_4)**AWGa (L,L)	N.D.	32	34	100
$F(dF)_{\psi}$**(CN_4)**AWGa (D,L)	N.D.	47	2.2$^{\Phi}$	94$^{\Phi}$
$F(dF)_{\psi}$**(CN_4)**(dA)WGa (D,D)	N.D.	0	58	100
Positive control				
FFSWGa	0.57	100	-	-

*Maximal response is the maximal bioluminescence response of each analog expressed as a percentage of the maximal response of the active core agonist FFSWGa (positive control) at 10 μM. N.D.: The analog was tested but was either not very active or was not active at lower molarities; thus an EC_{50} could not be determined. $^{\Psi}$The analog demonstrates partial antagonism of the native achetakinins, limiting the response to 50% of the maximum. $^{\Phi}$Nachman et al, unpublished data.

that mimic one turn over the other, indicate a predominant population of a β-turn involving the Phe^1 to Trp^4 region.[40,41] These restricted conformation analogs in the natural L,L (or S,S) configuration demonstrate significant retention of both diuretic activity in the cricket Malpighian tubule fluid secretion assay and interact with the expressed insect kinin receptor from the tick *Boophilus microplus* (Table 1).

A more detailed investigation of the stereochemical requirements for the interaction of the restricted-conformation analogs provided evidence for optimal turn-region stereochemistry in both the cricket Malpighian tubule fluid secretion assay and the expressed insect kinin receptor from the tick *Boophilus microplus*. For this investigation, all four stereochemical variants of tetrazole (L,L; L,D; D,L; and D,D) and APy (*2S,4S*; *2R,4S*; *2S,4R*; and *2R,4R*) moieties were incorporated into the C-terminal region of the insect kinin sequence (see Table 1). In the in vitro cricket diuretic assay, the most active agonists were the tetrazole analog FFψ(**CN_4**)AWGa (D,L) (Nachman and Coast, unpublished data) and APy analog Ac-RF(APy)WGa (*R,S*) with EC_{50}'s of 22 and 7 nM, respectively.[42] The optimal stereochemistry for the two turn mimic moieties in the cricket diuretic assay could therefore be identified as (D,L) and (2R,4S), respectively. The other three stereochemical analog variants (*2S,4S*)-, (*2S,4R*)- and (*2R,4R*)-APy retain agonist activity, but were less potent than the (*2R,4S*) variant by at least an order of magnitude. Conformational studies in aqueous solution indicate that the (*2R,4S*)-APy analog is considerably more flexible than the other three variants, which may explain its greater potency.[42] The (L,L)- and (D,D)-tetrazole stereochemical variants retain agonist activity, but were less potent than the (D,L) variant by at least an order of magnitude; whereas, in contrast, the (L,D)-tetrazole analog demonstrated an ability to antagonize the diuretic response of natural achetakinins.[40] It was suggested that the change in stereochemistry of the α-carbon at the N-terminal end of the tetrazole moiety from L to D appears to inhibit the activation response by interfering with the electrostatic interaction that occurs between the side chains of the Phe[1] and Trp[4] that allows these two critical side chains to present an optimal aromatic surface to the receptor.[41]

As with the cricket diuretic assay, an evaluation of the restricted conformation analogs in the expressed tick receptor demonstrated that the APy moiety was a superior mimic of the active conformation of the insect kinins over the tetrazole. Three of the four APy analogs demonstrated significant agonist activity. The (*2R,4S*)- and (*2S,4S*)-APy analogs demonstrated EC_{50} values of 1.6 μM in the bioluminescence plate assay (Maximal response: 65%) and 11 μM (Maximal response: 48%), respectively.[43] Analog (*2R,4R*)-APy was not active enough to determine an EC_{50} value and reached a maximal response of 45% only at concentrations of 10 μM and higher; whereas the (*2S,4R*)-APy analog proved essentially inactive. Only two of the tetrazole analogs demonstrated any appreciable activity; (D,L) (Maximal response: 47%) and (L,L) (Maximal response: 32%) and neither were active enough to allow the calculation of an EC_{50} value. Unlike the cricket diuretic assay, both the (D,D)- and (L,D)-tetrazole analogs were essentially inactive on the tick receptor; and no antagonist activity was observed for the (L,D) analog. Nonetheless, the optimal stereochemistry for activation of the tick receptor proved to be the same as for agonist activity in the cricket diuretic assay—(*2R,4S*) for the APy moiety and (D,L) for the tetrazole moiety.[43]

It should be noted that the NOVOstar bioluminescence method used to evaluate the response of the insect kinin analogs on expressed receptors reported in these studies is less sensitive as compared with a less practical confocal fluorescence cytometry method previously employed.[21] The bioluminescence plate assay is between 50 to 70-fold less sensitive. This difference should be taken into account when estimating the potency that these analogs would likely demonstrate in in vitro or in vivo physiological bioassays. For example, an analog that is active at 1 μM in the bioluminescence assay would be active at 20 nM or lower in bioassays.[44]

The critical nature of the sidechains of Phe[1] and Trp[4] and the *cis*Pro, Type VI turn conformation to the activity of the insect kinins was also confirmed by evaluation of a small series of pseudotetrapeptide analogs that featured only these minimal constructs. These 'minimalist' analogs, based on an amino piperidinone carboxylate scaffold, retained very weak, but statistically significant diuretic activity in the in vitro cricket Malpighian tubule secretion bioassay.[45]

In summary, structure/conformation-activity data suggest that a receptor interaction model for the insect kinins can be proposed in which the C-terminal pentapeptide region adopts a Type VI turn over residues Phe[1] to Trp[4] and that the aromatic side chains of Phe[1] and Trp[4] are oriented towards the same region and interact with the receptor (Fig. 3). Conversely, the side chain of residue 2 lies on the opposite face pointing away from the receptor surface, which explains why this position is more tolerant to changes[46] (Fig. 3).

Interaction of Biostable Insect Kinin Analogs with Receptors and Activity in In Vitro and In Vivo Bioassays

Members of the insect kinin family are hydrolyzed and therefore inactivated, by tissue-bound peptidases of insects. Two susceptible hydrolysis sites in insect kinins[3] have been reported. The primary site is between the Pro[3] and the Trp[4] residues, with a secondary site N-terminal to the Phe[1] residue in natural extended insect kinin sequences (Fig. 1). Experiments demonstrate that angiotensin converting enzyme (ACE) from the housefly can cleave at the primary hydrolysis site, whereas neprilysin (NEP) can cleave insect kinins at both the primary and secondary hydrolysis sites.[3,10-13]

Incorporation of β-amino acids can enhance both resistance to peptidase attack and biological activity[47,48] and this strategy has not been previously applied to insect neuropeptides. Recent work has described the synthesis of a number of analogs of the insect kinin C-terminal pentapeptide core

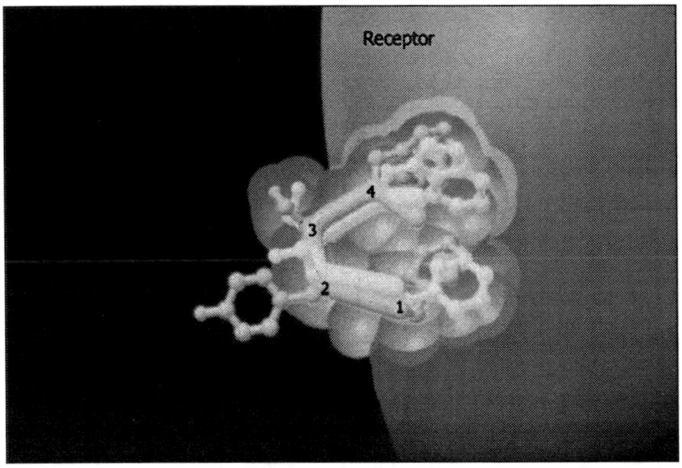

Figure 3. The currently accepted receptor-interaction model of an insect kinin C-terminal pentapeptide core region in a 1-4 turn, the active conformation. The side chain aromatic groups of Phe[1] and Trp[4] of the core region are both important for an agonist response. The indole group of Trp[4] is the most critical for receptor binding. From Nachman et al (2002).[40]

in which the critical residues Phe[1], Pro[3] and Trp[4] and or adjacent residues, are replaced with β^3-amino acid and/or their β^2-amino acid counterparts. β^3-amino acids feature an additional methylene group incorporated between the alpha carbon and the carboxyl group of the natural α-amino acid structure, whereas in β^2-amino acids the additional methylene group is placed between the alpha carbon and the amino group.[47,48]

Several noteworthy β^3-amino acid insect kinin analogs are listed in Table 2. The examples in Table 2 represent the replacement of only noncritical α-amino acids with β-amino acids. In the single-replacement analog 1460 (Ac-RFF(β^3**Pro**)WGa), a β^3-amino acid is placed adjacent to the primary peptidase-susceptible peptide bond (between Pro[3]-Trp[4]) of the insect kinins.[3] In the two double-replacement analogs, β-amino acids are placed in proximity to both primary and secondary peptidase-susceptible sites (Fig. 3). It was anticipated that these β-amino acids would afford some measure of resistance to hydrolysis by the peptidases that degrade and therefore inactivate, the natural peptides. Indeed, analogs 1460, 1577 (Ac-R(β^3**Phe**)FF(β^3**Pro**)WGa) and 1578 (Ac-RF(β^3**Phe**)-(β^3**Pro**)WGa) exhibit significantly enhanced resistance to peptidases that attack at the susceptible sites[49] (Table 2). Double-replacement analog 1578 demonstrates the greatest

Table 2. β-amino acid-containing insect kinin analogs—enhancement of peptidase resistance[49]

		% Hydrolysis	
Peptide Analog		**Human NEP**	***Drosophila* ACE**
LK I	DPAFNSWGa	99.8	81.6
1460	Ac-RFF(β^3**Pro**)WGa	15.4	10.7
1577	Ac-R(β^3**Phe**)FF(β^3**Pro**)WGa	0	3.5
1578	Ac-RF(β^3**Phe**)-(β^3**Pro**)WGa	2	16.5

Table 3. ***Estimated potencies (EC_{50}) and percentage of maximal bioluminescence response of different analogs in reference to FFFSWGa and the minimum active core sequence FFSWGa, tested on mosquito (E10) and tick (BmLK3) receptor expressing cell lines.[44,49] EC_{50} are an estimate of the concentration required to induce a half-maximal response.***

	Mosquito Receptor (E10 Cell Line)		Tick Receptor (BmLK3 Cell Line)		Cricket Malpighian Tubule Fluid Secretion	
Analogs	**EC_{50} (10^{-8}M)**	***Maximal Response (%)**	**EC_{50} (10^{-8}M)**	***Maximal Response (%)**	**EC_{50} (10^{-10}M)**	***Maximal Response (%)**
Single Beta Analogs						
1460 Ac-RFF(β^3**Pro**)WGa	36.7	50	28.7	72	0.3	100
1656 Ac-RFFP(β^2**Trp**)Ga	N.D.	0	68.5	57	400	97
Double Beta Analogs						
1577 Ac-R(β^3**Phe**)FF (β^3**Pro**)WGa	65.3	65	49.5	64	1	100
1578 Ac-RF(β^3**Phe**-β^3**Pro**)WGa	N.D.	7	68.7	62	10	100
Positive Controls						
FFFSWGa	68.5	100	27.0	100	-	-
FFSWGa Insect kinin active core	N.D.	29	59.0	100	-	-
AK-III DQGFNSWGa	-	-	-	-	3	100

*Maximal response is the maximal bioluminescence response of each analog expressed as a percentage of the maximal response of the known agonist FFFSWGa (positive control) at 10 μM. for tick and mosquito receptors; or to achetakinin-III (AK-III) in the cricket diuretic assay. N.D.: The analog was tested but was either not sufficiently active to determine an EC_{50} value.

resistance to the peptidases ACE and NEP among those listed in Table 2. Under conditions in which the natural insect kinin LK-I is degraded 100% by NEP and 82% by ACE, analog 1578 shows no degradation by NEP and only 4% by ACE. These three analogs are also blocked at the N-terminus with an acetyl (Ac) group, which confers resistance to hydrolytic degradation by an additional class of peptidases, the aminopeptidases.[50]

Despite the incorporation of an additional methylene group (-CH_2-) within the backbone of the C-terminal pentapeptide core region, the single replacement of amino acids with their β-amino acid counterparts led to significant retention of activity in expressed kinin receptors from the tick *Boophilus microplus* and mosquito *Aedes aegypti*, very evolutionarily distant arthropod species. The single-replacement analog 1460, that involves modification of the Pro^3, was the most potent of the β-amino acid analogs in both mosquito and tick receptors[44] (Table 3). Analog 1460 proved to be more active than the positive control agonist FFFSWGa in the mosquito and considerably more active than the minimum active core sequence FFSWGa. In the tick, 1460 was equipotent with the positive control agonist FFFSWGa and more active than the C-terminal pentapeptide active core. Analog 1460 therefore is a nonselective, biostable agonist for these two receptor systems. However, it is clear that the mosquito receptor is considerably more sensitive to modifications at Trp^4 than the tick receptor. In particular, the mosquito receptor is intolerant to the replacement of Trp^4 with β^2Trp, as 1656 (Ac-RFFP(β^2**Trp**)Ga) is inactive[44] (Table 3). This complete loss of activity by the β^2Trp analog 1656 in the mosquito receptor cell line may result from an increase in the distance between the α-carbons of the critical aromatic residues at positions 2 and 4 over what is found in the natural peptide. The tick receptor remains relatively tolerant to the introduction of a methylene group between the α-carbon and the amino group of Trp^4, as the potency of 1656 was not statistically different from the C-terminal pentapeptide core analog FFSWGa and the analog retained 57% of the maximal response.[44] By virtue of this difference in receptor-interaction requirements, analog 1656 is a selective agonist for the tick receptor. Although not as yet tested, analog 1656 would be expected to exhibit significantly enhanced peptidase resistance as well.

Double replacement analog 1577 features modification of noncritical regions; the noncritical Pro_3 residue and the residue just outside of the critical C-terminal pentapeptide core region. Accordingly, this analog retains significant activity in both the mosquito and tick receptors, demonstrating potency that exceeds that of the minimum active core and is not significantly different from that of the positive control agonist FFFSWGa.[44] It is a nonselective, biostable agonist of both arthropod receptors.

Double replacement analog 1578 features modification of noncritical residues Phe^2 and Pro^3 that effectively changes the distance between the sidechains of the critical residues Phe^1 and Trp^4. Like analog 1656, this modification is tolerated to a much greater extent by the tick receptor than the mosquito receptor. Consequently, the analog 1578 demonstrates retention of activity in the tick receptor assay, with a potency that is not significantly different from that of the active core analog FFSWGa; but remains essentially inactive in the mosquito receptor system[44] (Table 3). Thus, biostable agonist analog 1578 demonstrates selectivity between the two expressed arthropod receptor cell lines.

The β-amino acid insect kinin analogs have not as yet been evaluated in in vitro mosquito or tick diuretic assays. However, data for these analogs has been reported in an in vitro insect kinin Malpighian tubule fluid secretion assay from the cricket *Acheta domesticus*[49] and are listed in Table 3 as well. As with the two expressed arthropod receptors for the insect kinins, analog 1460 was the most potent β-amino acid analog in the cricket fluid secretion bioassay, demonstrating a potent EC_{50} of 0.3×10^{-10}M and a 100% maximal response. This analog exceeds by an order of magnitude the activity of at least one of the native achetakinins, AK-III ($EC_{50} = 3 \times 10^{-10}$M) and essentially matches the activity of the most potent of achetakinins. Although inactive in the mosquito insect kinin receptor, the analog containing β^2Trp (1656) retains activity ($EC_{50} = 40 \times 10^{-10}$M; 97% maximal response) in the in vitro cricket diuretic assay, although considerably reduced from the native achetakinins.[49] This analog demonstrates a degree of selectivity for the expressed tick receptor over receptors of the mosquito and cricket. As with the two expressed arthropod insect kinin receptors, the double-replacement analog 1577 retained potent activity in the cricket diuretic assay, with an EC_{50} value of 1×10^{-10}M and a100% maximal response[49] (Table 3). While virtually inactive in the expressed mosquito receptor, the double-replacement analog 1578 retained activity in the cricket fluid secretion assay with an EC_{50} of 10×10^{-10}M and a 100% maximal response. The assay results obtained with the β-amino acid insect kinin analogs suggest that the receptor associated with the Malpighian tubules of the cricket is more similar to the expressed insect kinin receptor from the tick than that of the mosquito.

Another set of insect kinin analogs that demonstrate enhanced resistance to peptidases feature a replacement of the third position of the C-terminal pentapeptide core (Pro or Ser) with α-aminoisobutyric acid (Aib), a sterically hindered α,α-disubstituted amino acid which effectively protects the primary tissue-bound peptidase hydrolysis site.[3,10] Incorporation of a second Aib residue adjacent to the secondary peptidase hydrolysis site, as with analog **(Aib)**FS**(Aib)**WGa (analog 781), further enhances the biostability.[3] For instance, over a one hour period a natural helicokinin is completely degraded by either peptidases bound to Malpighian tubule tissue from the moth *Helicoverpa zea* or a pure preparation of the peptidase NEP. Conversely, the double-Aib analog **(Aib)**FS**(Aib)**WGa remains to the extent of 100% and 90%, respectively, over a three to four hour period.[3,10] In the cricket Malpighian tubule assay, analog 781 exceeds the activity of the native achetakinins by an order of magnitude. Analogous Aib-containing insect kinin analogs also demonstrate activity that either approaches or exceeds that of the positive control FFFSWGa in expressed receptors for mosquito *Aedes aegypti* and the tick *Boophilus microplus*, respectively.[78]

Due to the different structure-activity requirements of the insect kinin receptor found with the Malpighian tubule fluid secretion assay in the housefly (*Musca domestica*), the in vitro activity of analog 781 is over four orders of magnitude less than that of the native muscakinin. Nonetheless, the in vivo diuretic activity of 781 is *equipotent* with that of the native muscakinin. Evidence indicates that analog 781 demonstrates a longer hemolymph residence time in the housefly than the peptidase-susceptible muscakinin and it is this extended presence that likely explains the remarkable in vivo activity observed for this analog.[3] Therefore, biostable characteristics can enhance the in vivo activity of insect kinin analogs.

While in vivo activity studies of biostable insect kinin analogs have not as yet been completed for the cricket, tick or mosquito, some in vivo results have been obtained in larvae of the corn earworm moth, *Helicoverpa zea*.[3] Injection of helicokinins into developing larvae of the related moth *Heliothis virescens* has been observed to inhibit weight gain.[9] A helicokinin-II analog (VRFSSWGa) and a biostable Aib helicokinin analog **pQ**RFS(**Aib**)WGa (Hek-Aib) were injected daily (0.5 nmoles) into 5-day old *H. zea* larvae for 5 or 6 days until pupation occurred. The helicokinin analog demonstrated a developmental trend that reached a peak on day 5 posttreatment, with a statistically insignificant 20% weight reduction as compared with controls. In contrast, the biostable, helicokinin Aib analog elicited a stronger, statistically significant effect spanning days 4-7, reaching a peak at day 5 of about a 50% reduction in mean larval weight as compared with controls. The time of pupation was delayed by a factor of 25%.[3]

Seinsche et al (2000)[9] demonstrated that the weight gain inhibition they observed in *H. virescens* is accompanied by an increase in the excretion of water in the feces, consistent with the diuretic activity previously observed in crickets,[27] flies,[10,51] as well as the lepidopteran *H. virescens*.[9] However, the higher excretion of fluid alone could not sufficiently explain the observed loss of weight in insect kinin-treated tobacco budworm larvae. The authors further speculated that the insect kinins could have induced a starvation signal in the *Heliothis* larvae, resulting in mobilization of energy stores and a decreased efficiency in exploiting digested nutrients.[9] For instance,[52] demonstrated that the levels of insect kinin-like immunoreactivity in the haemolymph increased 10-fold in crickets starved for 48 h without access to water. They speculated that the insect kinins may play a role in energy mobilization during starvation. Support for this idea comes from work by[53] who found that nanomolar doses of insect kinins led to increases in lipid concentration similar to that caused by the adipokinetic hormone and also inhibited protein synthesis. To this should be added the likely contribution made by the reported ability of insect kinins to inhibit release of protease and amylase digestive enzymes from the lepidopteran midgut,[7,8] which would have prevented the diet from being efficiently digested. Together with the increased excretion of fluid and induction of a starvation response, an inhibition of digestive enzyme release may have led to the weight losses observed in both *H. virescens*[9] and *H. zea*[10] treated with insect kinins and/or analogs.

C-Terminal Aldehyde Analogs of Insect Kinins

Aldehydes can form reversible imine bonds with amino groups. Peptide analogs containing reversible binding moieties at the C-terminus, such as an aldehyde, have been reported to inhibit various classes of proteolytic enzymes.[54-56] It has been further postulated that a C-terminal aldehyde moiety could form a covalent, reversible Schiff base (imide linkage) with the amino group of a Lys residue[57] in an insect kinin receptor pocket, thereby modifying the ligand-receptor interaction characteristics of the resulting insect kinin analog. Enhanced and/or modified receptor-interaction characteristics could lead to an enhancement and/or modification of in vivo activity. Evaluations of two C-terminal aldehyde kinin analogs, R-LK-CHO (Fmoc-RFFPWG-H) and V-LK-CHO (Boc-VFFPWG-H), in developmental and diuretic assays have been reported.[34,58] Both aldehyde analogs demonstrated in vitro stimulation of fluid secretion in isolated cricket Malpighian tubules in the physiological concentration range and full efficacy, thereby providing evidence that they could interact with an insect kinin receptor site.

R-LK-CHO (EC_{50} = 250 nM) was approximately 10 fold less active than V-LK-CHO (EC_{50} = 30 nM) in this cricket diuretic assay, but both were several orders of magnitude less active than their parent hexapeptide insect kinin analogs.[34] Regardless of the presence of an aldehyde or an amide group at the C-terminus, the V-containing analogs were 10-fold more potent than the R-containing analogs in this in vitro diuretic assay system.

H. Zea *Larval Weight Gain Inhibition Bioassay*

Injection of R-LK-CHO into 5-day old *H. zea* larvae induced statistically significant reductions in weight gain in comparison with control animals on days 2 and 4-6 at the 5 nm dose, but not at 500 pm. Day 6 larvae experienced a significant reduction in weight gain at the 5 nm dose, with treated animals observed to be about 65% of the weight of controls. No significant difference in mortality was observed between treated and control groups. Thus, the incorporation of the aldehyde group in the C-terminus did appear to enhance the activity of the insect kinins in this in vivo weight gain inhibition assay.[34] A previous study had determined that the observed inhibition of weight gain in *H. zea* larvae on days 3-6 post treatment with the *normal* insect kinin VRFSSWGamide was not statistically significant. Significant weight gain was not observed in this previous study until a peptidase-resistant, Aib-containing insect kinin analog (pQRFS(Aib) WG-amide) was synthesized and evaluated in the in vivo larval assay. Corn earworm larvae treated with this fortified analog were observed to be about 50-60% of the weight of control animals,[3] with no significant increase in mortality.

The other aldehyde analog, V-LK-CHO, demonstrated a more pronounced effect than R-LK-CHO in the *H. zea* larval weight gain inhibition assay. V-LK-CHO induced significant reductions in weight gain on days 2 and 4 through 6, after initiation of the treatment at *both* the 500 pm and 5 nm dose. At day 6, treated larvae were observed to be 65% and 40% that of the weight of the control animals at doses of 500 pm and 5 nm, respectively (Fig. 4). In order to reduce weight gain to 65% of that of control animals, treatments of 500 pm for V-LK-CHO and 5 nm of R-LK-CHO were required, a 10-fold difference in potency (Fig. 4). Notably, in those animals treated with V-LK-CHO a significant increase in mortality was observed at both doses (45%{500 pm}and 67%{5 nm})[34] (Fig. 4).

The significant increase in mortality observed in larvae treated with V-LK-CHO is not likely a result of some general toxic effect of the aldehyde moiety itself, as increased mortality was not observed in R-LK-CHO, which also features a C-terminal aldehyde.[34] However, the enhanced reductions in larval weight gain induced by the insect kinin C-terminal aldehyde analogs are consistent with a modified interaction of these analogs with the insect kinin receptor over that occurring with the normal C-terminal amide peptides.

In Vitro and In Vivo Housefly Diuretic Bioassays

As mentioned previously, insect kinin analogs in which the C-terminal amide was replaced by an aldehyde moiety retained an ability to stimulate fluid secretion by cricket Malpighian tubules. Although a reliable in vivo assay for diuresis in crickets did not exist, such an assay was available for houseflies.[51,59] Thus, the insect kinin aldehydes V-LK-CHO, R-LK-CHO and analogs were evaluated for activity on housefly Malpighian tubules (Table 4).

Neither of the two aldehyde analogs V-LK-CHO and R-LK-CHO stimulated fluid secretion, although notably tubules exposed to R-LK-CHO did not respond when subsequently challenged with a supramaximal concentration (10 nM) of native muscakinin (Musdo-K). In contrast, the same concentration of Musdo-K elicited a marked diuretic response in tubules that had first been exposed to V-LK-CHO. The inhibitory effect of R-LK-CHO on the diuretic activity of 10 nM Musdo-K was dose-dependent with an IC_{50} of 12 μM (Fig. 5), which compares favorably with the EC_{50} of an N-terminal truncated Musdo-K analog of similar length.[35,58]

Other aldehyde analogs were also tested on housefly tubules for diuretic activity and their ability to inhibit stimulation of fluid secretion by 10 nM Musdo-K. Of these, Fmoc-RAHPWG-H, which closely resembles R-LK-CHO, had diuretic activity with an EC_{50} of 2.3 μM. On the other hand, an Aib containing aldehyde analog, Fmoc-RFF(**Aib**)WG-H, had no effect on fluid secretion

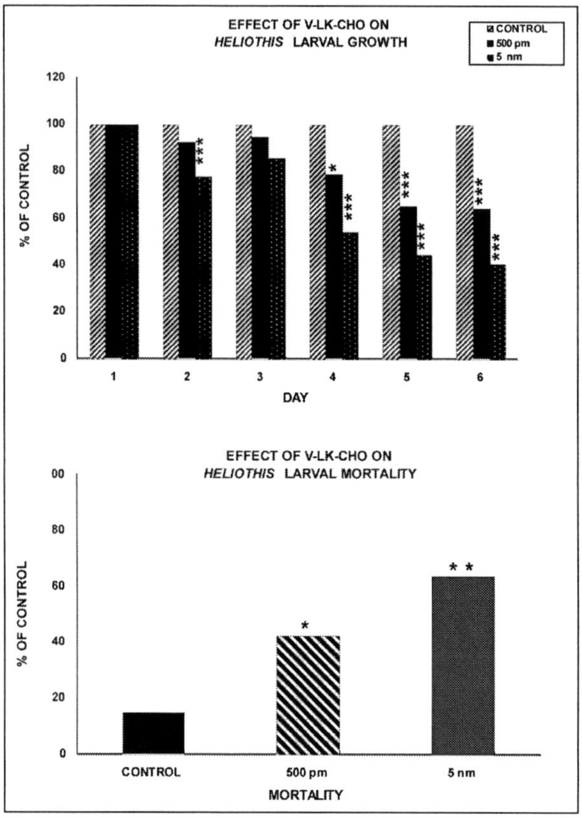

Figure 4. The graph at the top illustrates the effects of daily injections of 500 pmoles of the insect kinin C-terminal aldehyde analog V-LK-CHO (Boc-VFFPWG-H) on inhibition of weight gain in 5-day old larvae of the corn earworm *H. zea* expressed as a percentage of saline injected control animals. Each test group contained at least 10 animals and the experiment was repeated five times (***$P < 0.001$; **$P < 0.01$; *$P < 0.05$; indicates the difference from the control group is statistically significant). The graph at bottom illustrates the percent mortality observed in these *H. zea* larvae injected with either 500 pm or 5 nm of V-LK-CHO for 5 days compared with saline injected controls. From Nachman et al (2003).[34]

Table 4. *In vitro diuretic activity of C-terminal insect kinin analogs in the housefly Musca domestica*[34]

Insect Kinin Analog	Stimulation of Malpighian Tubule Fluid Secretion– EC_{50} (10^{-6}M) (% Maximal Response)
Boc-Val-Phe-Phe-Pro-Trp-Gly-H (V-LK-CHO)	Inactive
Fmoc-Arg-Phe-Phe-Pro-Trp-Gly-H (R-LK-CHO)	(12*) (Inhibitory)
Fmoc-Arg-**Ala-His**-Pro-Trp-Gly-H	2.3 (100)
Fmoc-Arg-Phe-Phe-**Aib**-Trp-Gly-H	Inactive

*IC_{50} for inhibitory activity

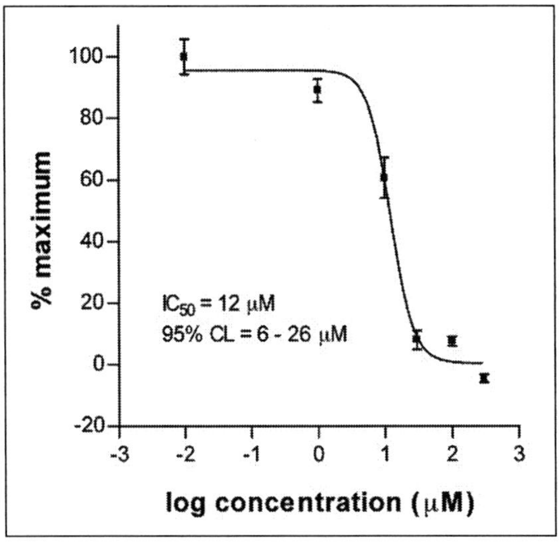

Figure 5. Dose-response curve for R-LK-CHO showing the percentage inhibition of the diuretic activity of 10 nM Musdo-K. Data points are the means of 5-7 determinations and vertical lines ±1 S.E.M. From Nachman et al (2007).[58]

and, in common with V-LK-CHO, it did not inhibit the diuretic activity of 10 nM Musdo-K.[58] The inhibitory activity of R-LK-CHO is thus highly specific and not a generalized effect of the aldehyde moiety.

The ability of R-LK-CHO to prevent stimulation of fluid secretion by Musdo-K was of considerable interest, because the analog might be useful as a probe to investigate the role of the native kinin in the control of diuresis in vivo. By using a high sensitivity flow through humidity analyzer the episodic excretion of urine from intact houseflies was recorded.[59] R-LK-CHO inhibits the in vivo activity of Musdo-K as evidenced by a marked reduction in the amount of urine voided over the initial 45min post-injection period in flies injected with 50 pmol each of the analog and the kinin compared with the kinin alone. R-LK-CHO was also shown to reduce the volume of urine voided during a diuresis initiated by the injection of 3 μL saline alone, although a larger dose (250 pmol) of the aldehyde was needed for a pronounced effect, probably because of the expanded hemolymph volume[58] (Fig. 6). The diuretic response to hypervolemia is partly attributable to the release of Musdo-K from neurohaemal sites into the circulation and the inhibitory effect is consistent with a selective effect of R-LK-CHO at the kinin receptor.

Injection of 1 μL distilled water is a more effective stimulant of diuresis than is the injection of 3 μL of saline, with total urine loss over 3h being more than double that in saline injected flies. This difference in part reflects an autonomous response of the Malpighian tubules, which secrete at higher rates when the osmotic concentration of the bathing fluid is reduced, as would occur after injecting flies with distilled water.[51] R-LK-CHO would not necessarily be expected to have any effect on the autonomous response of Malpighian tubules to haemolymph dilution and yet it reduced the total amount of urine voided over 3h from flies injected with 1 μL of distilled water by almost 50% (Fig. 7).

The markedly reduced urine output from flies injected with 1 μL distilled water containing 50 pmol R-LK-CHO suggests this analog has a toxic effect on Malpighian tubules and in support of this we have show that it blocks stimulation of fluid secretion by thapsigargin, a SERCA inhibitor and by ionomycin, a calcium ionophore.[58] Kinin neuropeptides use Ca^{2+} as a second messenger to open a paracellular or transcellular chloride conductance pathway. This is mimicked by thapsigargin and ionomycin, which increase the level of intracellular calcium by promoting Ca^{2+} release for intracellular stores and the influx of Ca^{2+} from the bathing fluid, respectively. The ability of

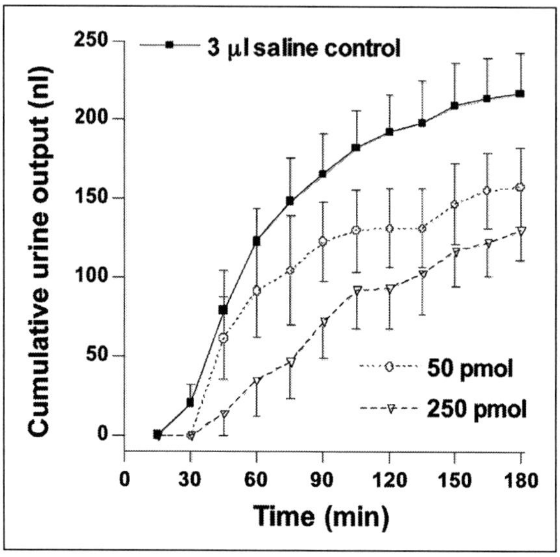

Figure 6. R-LK-CHO attenuates the in vivo diuretic response to hypervolemia induced by the injection of 3 µL of saline. The volume of urine excreted by individual flies was measured over 15 min periods for 3 h in insects injected with saline alone (solid squares, solid line) or saline containing either 50 pmol (open circles, dotted line) or 250 pmol (open triangle, dashed line) of the aldehyde analog. Data points show the means ± 1 S.E.M of the cumulative urine output in 10 (saline alone) and 6 (+ R-LK-CHO) flies. From Nachman et al (2007).[58]

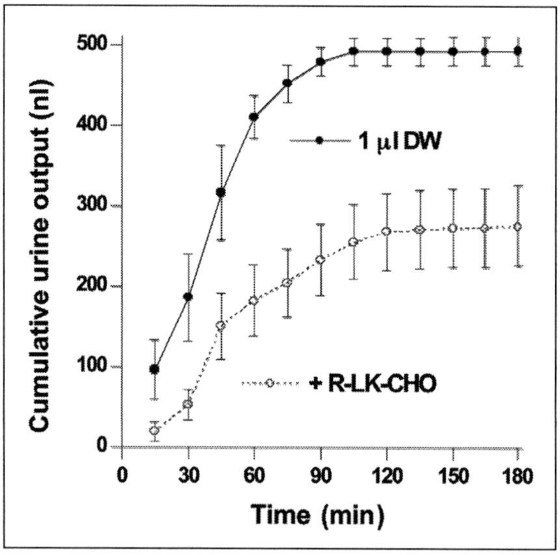

Figure 7. R-LK-CHO significantly reduces the amount of urine excreted by flies injected with 1 µL of distilled water. The volume of urine excreted by individual flies was measured over 15 min periods for 3 h in insects injected with distilled water alone (solid squares, solid line) or with distilled water containing 50 pmol (open circles, dotted line) of the aldehyde analog. Data points show the means ± 1 S.E.M of the cumulative urine output in 7 (distilled water) and 9 (+ R-LK-CHO) flies. From Nachman et al (2007).[58]

R-LK-CHO to block the activity of these pharmacological probes shows it cannot be an antagonist of the kinin receptor, but must act downstream of the second messenger pathway.

At present, the cellular action(s) of R-LK-CHO on Malpighian tubules that leads to prevention of stimulation of fluid secretion by Musdo-K, thapsigargin and ionomycin is unknown. It is clearly not a generalized toxic effect, because the same analog has diuretic activity in the cricket Malpighian tubule assay. Moreover, closely related aldehyde analogs tested on housefly tubules either have no activity (V-LK-CHO and Fmoc-RFF(**Aib**)WG-H) or stimulate fluid secretion (Fmoc-RAHPWG-H). R-LK-CHO may act as a 'magic bullet' and bind with the kinin receptor on housefly tubules, become internalized and thereby gain access to intracellular processes that couple a rise in intracellular calcium levels to the opening of the chloride conductance pathway.[58]

Compounds in the hemolymph, including pesticide toxins, are actively transported into the lumen of the Malpighian tubules and their rate of elimination is dependent on the rate of fluid secretion.[60,61] At high rates of excretion, the toxins do not reach the high concentrations that would allow them to diffuse back into the hemolymph down a concentration gradient. An agent capable of selective depression of fluid secretion would be expected to allow pesticides to achieve higher concentrations in the hemolymph; and, in turn, likely reduce the amount of toxin required to kill an insect. Whatever the mode of action, the selective activity of R-LK-CHO on housefly tubules represents an important milestone and lead in the long-term goal of the development of environmentally-friendly insect management agents based on the insect kinins.

Nonpeptide Mimetic Agonists/Antagonists of Expressed Insect Kinin Receptors

Perhaps the ultimate goal in the search for biostable, bioavailable analogs would be the design and/or discovery of nonpeptide mimetic agonists or antagonists of the insect kinins. The availability of expressed insect kinin receptors can accelerate the discovery process through the evaluation of nonpeptide libraries. A recent biorational approach has based the selection of a nonpeptide library on the presence, within its structure, of the side chain moiety of the most critical residue of the peptide (Nachman, unpublished). As discussed in an earlier section, the most critical residues for the interaction of the insect kinins with expressed receptors from the tick *Boophilus microplus* and mosquito *Aedes aegypti* have been determined to be the Phe[1] and Trp[4] within the C-terminal pentapeptide core region.[11,37,39] Data obtained from an in vitro Malpighian tubule fluid secretion assay indicate that a C-terminal pentapeptide insect kinin analog in which the Phe is replaced with an Ala demonstrates an antagonist response against native achetakinins, whereas the analog in which Trp is replaced with Ala is devoid of activity.[62] Furthermore, a C-terminal aldehyde analog in which Ala replaces Phe retains weak activity in an in vitro cricket diuretic assay.[58] This would suggest that Trp, which contains an indole side chain moiety, represents the most critical amino acid for the binding of insect kinins with the receptor. Consequently, a 400 member nonpeptide library based on the imidopyrido**indole** ('Ipi') scaffold[63] was constructed (Fig. 8) and evaluated in expressed insect kinin receptors from the tick *Boophilus microplus* and mosquito *Aedes aegypti*. One of the Ipi

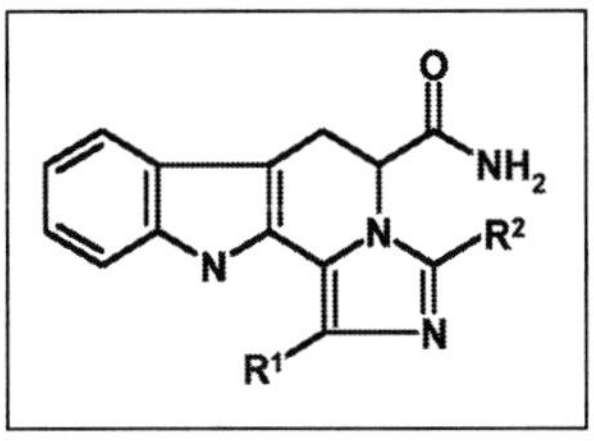

Figure 8. Structure of a rationally-designed, recombinant nonpeptide library based on an imidopyridoindole ('Ipi') scaffold. The scaffold contains an indole group (highlighted with a circle), which appears in the side chain of Trp, the most critical residue within the insect kinin active core region for receptor interaction.

recombinant library analogs demonstrated significant activity in both of the expressed arthropod insect kinin receptor assays (Nachman and Pietrantonio, unpublished data) and preliminary studies indicate it demonstrates statistically significant activity in in vitro Malpighian tubule secretion assays of the mosquito *Aedes aegypti* and cricket *Acheta domesticus* at μM concentrations (Nachman and Coast, unpublished data). The completion of full dose-response curves for the in vitro diuretic assays and an evaluation of in vivo biological activity await future experiments. Future evaluations of this and other rationally designed, indole-containing libraries may identify other nonpeptide mimetic agonists as well as antagonists. The data further underscores the preeminent importance of the indole moiety of Trp^4 to the interaction of the insect kinins with their receptors.

Prospects for Enhanced Topical and/or Oral Bioavailability

Insect neuropeptides in general are not suitably designed to efficiently penetrate either the outer cuticle or the digestive tract of insects. Nonetheless, studies have shown that at least one class of insect neuropeptides can be modified to enhance bioavailability characteristics.

Topical Activity

Topical experiments have not been conducted with the insect kinins, but have been conducted with a couple of other insect neuropeptide families. Topical application of members of the adipokinetic hormone (AKH) family in mixed aqueous/organic solvent to the cuticle of the cricket *Gryllus bimaculatus* did lead to a significant AKH-like increase in hemolymph lipids.[64] The AKH family as a group of peptides are particularly hydrophobic, a factor that may aid their penetration through the hydrophobic cuticular waxes. On the other hand, experiments involving topical application of aqueous solutions of members of the pyrokinin/PBAN family did not produce significant pheromone production in the tobacco budworm moth *Heliothis virescens*.[65,66] Structural modification to produce pyrokinin analogs that feature amphiphilic properties greatly enhances their ability to both penetrate the hydrophobic cuticle and also to maintain the aqueous solubility required to reach their target receptor once they encounter the hemolymph.[65,66] The pyrokinin/PBAN family shares a common C-terminal pentapeptide FXPRLa and regulates a number of physiological processes, including regulation of sex pheromone biosynthesis in females of Lepidoptera.[67-69] The development of a series of pseudopeptide analogs of this neuropeptide family began with the addition of various hydrophobic groups to the N-terminus of the C-terminal pentapeptide active core, which in conjunction with the polar/charged Arg side chain, confer an amphiphilic property. Hydrophobic groups appended to the N-terminus included fatty acids of various chain lengths, cholic acid, carboranylpropionic acid and aromatic acids.[46,65,70-73] Many of these amphiphilic analogs showed greater in vivo potency in a pheromonotropic assay than the native 33-membered pyrokinin PBAN when delivered via injection in female *H. virescens* moths. In studies involving topical application, neither PBAN nor its C-terminal pentapeptide active core elicited pheromone production when applied at 1-2 nmoles/female. By contrast, amphiphilic analogs induced significant pheromone production 15 min after topical application of aqueous solutions to the lateral abdominal surface of the moths with ED_{50} values ranging from 60 to 500 pmoles per female and ED_{max} values of 60-2000 pmoles per female.[71] It is worth emphasizing that this result was achieved without the assistance of organic solvent mixtures. When applied to dissected pieces of *H. virescens* cuticle, 24-hour recoveries of a series of amphiphilic pyrokinin analogs ranged from 5-70%. In addition, prolonged pheromone production exceeding 20 hours following a single topical application of an amphiphilic pyrokinin analog to *H. virescens* moths was observed. In addition, the nature of the hydrophobic moiety was observed to influence the duration of the slow release of a given amphiphilic pyrokinin analog. The results demonstrated that the insect cuticle could serve as a reservoir for the time-release of a physiologically active, amphiphilic analog of an insect neuropeptide.[71] The development of topically-active, amphiphilic analogs of the insect kinins would be a logical extension of these studies.

One amphiphilic pyrokinin analog, **2Abf-Suc**-FTPRLa, featured an appended brominated fluorine aromatic ring as the hydrophobic moiety and demonstrated highly unusual in vivo activity following delivery via injection.[73] Unlike other amphiphilic analogs, a single injection of 500 pmoles of this brominated fluorine (2Abf) pyrokinin analog into female *H. virescens* moths induced a highly unnatural response; continuous production of high levels of pheromone for as long as 20 hours.[73] While such a result might be expected from the time-release of an amphiphilic analog following topical application, the observed prolonged pheromone production following injection suggested that the 2Abf analog might have a strong affinity for and/or interaction with, the pheromone receptor. Indeed, recent studies on an expressed pyrokinin/PBAN receptor from *H. virescens*, the Abf analog proved to be more active than the native 33-membered PBAN neuropeptide and considerably more active than the parent C-terminal pentapeptide fragment (Nachman and Adams, unpublished data). However, the analog had an interesting side effect. That is, it induced mortality in 100% of the treated moths. The LC_{50} value for this potent toxic side effect was found to be 0.7 pmoles and 100% mortality could be achieved with a 5 pmole dose. Related analogs such as **7Abf-Suc**-AARAAa and another pyrokinin analog that featured only the fluorine ring, both of which retained similar amphilphilic and solubility properties, did not demonstrate any toxicity.[73] Therefore, the toxicity was not a result of the presence of the 2Abf moiety. Furthermore, the toxic effect was highly specific to the presence of the pyrokinin sequence. Although the mechanism of the insecticidal activity of the 2Abf analog in *H. virescens* is not known, it is hypothesized that the specific nature of the toxicity results from an interaction of receptor sites for the pyrokinin/PBAN class of insect neuropeptides.[73] This class of neuropeptides has a wide range of known biological activities in insects in addition to pheromonotropic activity, including hindgut and oviduct myotropic activity, pupariation, induction of egg diapause, diapause break and melanotropic activity.

Oral Activity

No oral activity data has been reported for the insect kinin class of neuropeptides. Generally, oral activity for unmodified insect neuropeptides is poor to nonexistent. Small quantities of members of the pyrokinin/PBAN[74] and the proctolin classes of neuropeptides[75] have been reported to survive exposure to the digestive enzymes and pH of the digestive tract and penetrate through to the hemolymph to reach their target receptors. In addition, small quantities (<3%) of A-type allatostatins have been shown to be transported across dissected foregut tissue of the moth *Manduca sexta*.[76] An early attempt to feed the pyrokinin PBAN to adult females of the moth *H. zea* reported very low and inconsistent levels of pheromone production that were not progressively dose-dependent.[74] In other experiments, no statistically significant pheromone production was observed in starved adult females of the related moth species *H. virescens* 1-2 hours after ingestion of a sugar solution of 50 pmoles/μL of PBAN or the C-terminal pentapeptide core FTPRLa. However, biostable amphiphilic, pyrokinin analogs **Hex**-FT(**Hyp**)RLa (901) and **Hex**-FT(**Oic**)RLa (904) demonstrated an ability to penetrate the dissected portions of the insect digestive tract as well as significant oral activity.[3] The components hydroxyproline (Hyp) and octahydroindole-2-carboxylate (Oic) were incorporated as sterically-hindered replacements for Pro to enhance resistance to tissue-bound peptidases that would be encountered in hemolymph. Indeed, these analogs proved to be completely resistant to degradation by peptidase bound to Heliothine Malpighian tubule tissue over a 120 min period, whereas a natural pyrokinin was completely degraded in 30 min. Direct penetration of the two analogs through dissected cockroach foregut and midgut were investigated. The digestive system of the cockroach was chosen because the guts of adult moths are not of sufficient size or stability to allow for practical delivery of peptide analog solutions. Indeed, Figure 9 shows that out of a total of 2.5 nmoles placed within the lumen of a sealed foregut, 800 nmoles (over 30%) of Oic analog 904 penetrated the tissue preparation. It is interesting to note that Oic analog 904 demonstrates time-release properties, as equal amounts were recovered over the 0-4 hour period as over the 4-24 hour period. The majority of Hyp analog 901 penetrated in the first 0-4 hour period.[46] The lumen of the insect foregut features a cuticular component, which could explain why the time-release effect is similar to that observed for the outer cuticle for these amphiphilic analogs. It also suggests that the foregut can serve as a reservoir for the time-release delivery of neuropeptide analogs in insects, thereby bypassing the hostile, peptidase-rich environment of the midgut.

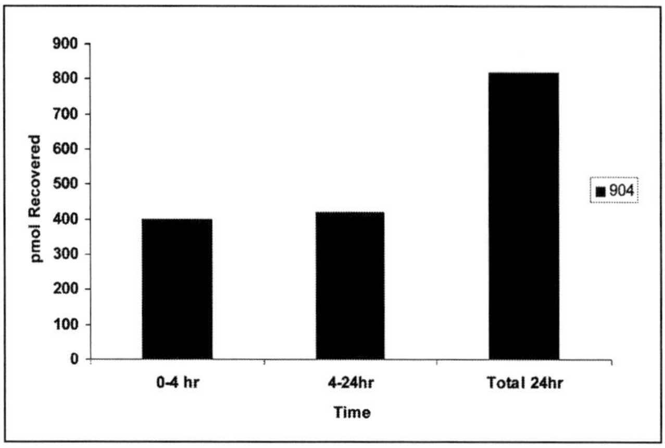

Figure 9. Amount of Oic-pyrokinin analog 904 penetrating isolated, ligated cockroach foregut preparations over indicated time periods from an initial 2.5 nmoles. From Nachman et al (2002b).[46]

These in vitro penetration studies of analogs 901 and 904 were followed by in vivo oral pheromonotropic activity trials in adult female *H. virescens*.[46] Pheromone production was monitored following ingestion of 30 μL of a sugar solution containing 50 pmoles/μL of either 901 or 904 at 1.5, 3, 4 and 6 hours post feeding (Fig. 10). A statistically significant increase in pheromone titer was observed at 1.5 hours postfeed with 901 with a 17% maximal response. Oral administration of the analog 904 induced statistically significant levels of pheromone at 1.5, 3 and 4 hours postfeed, but not at 6 hours. Optimal pheromone production was achieved at 3 hours, with a highly significant ~60% maximal response.[46] The shift in the pheromone spike from 1.5 hours for 901 to 3 hours post-feed for 904 is consistent with the greater time-release effect observed for the direct penetration of the more hydrophobic 904 in both ligated fore-and midgut preparations.

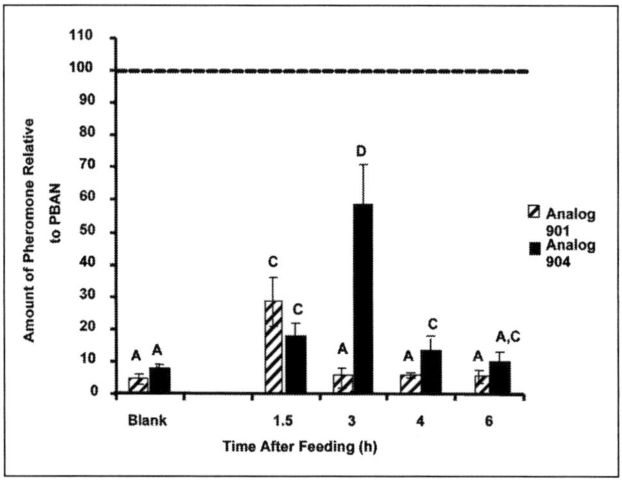

Figure 10. Amount of pheromone, relative to maximal levels produced by injected PBAN, produced by Hyp-pyrokinin analog 901 and Oic-pyrokinin analog 904, at 1.5, 3, 4 and 6 hours following oral administration. The dotted line at 100% denotes maximal production of pheromone by injected PBAN (positive control). From Nachman et al (2002b).[46]

The development of orally-active, biostable amphiphilic analogs of the insect kinins would be a logical extension of these studies.

Conclusion

Evaluation of truncated analogs, an Ala-replacement series and restricted conformation analogs incorporating turn-mimic components on recombinant arthropod expressed receptors for the insect kinins and biological assays in arthropods provided a structure/conformation-activity profile for this important class of neuropeptides. The insect kinins require a C-terminal pentapeptide as the minimal core sequence for activity, although full activity in expressed mosquito receptors and in housefly Malpighian tubule assays requires at least a C-terminal hexapeptide fragment. Replacement of residues Phe^1 and Trp^4 with Ala led to peptide analogs devoid of agonist activity and are, therefore, the most critical for an agonist response. However, while the analog in which Ala replaces Phe^1 is completely inactive, the analog in which Ala replaces Trp^4 demonstrates antagonism in in vitro cricket diuretic assays. Furthermore, a novel insect kinin analog featuring a C-terminal aldehyde group in which Phe^1 is replaced with an Ala retains weak activity in in vitro cricket diuretic assays. These results suggest that the Trp^4 is the most important for receptor binding. The evidence further suggests that the insect kinins adopt a *cis*Pro Type VI turn over residues 1-4, allowing the side chains of Phe^1 and Trp^4 to form an optimal aromatic surface during successful receptor interaction. A rationally-designed nonpeptide recombinant library based on the 'indole' side chain functional group of $Trp,^4$ the most critical of amino acids in the insect kinin core region, was synthesized and evaluated in expressed insect kinin receptors from the tick *Boophilus microplus* and mosquito *Aedes aegypti*. An insect kinin mimetic analog from this imidazopyridoindole library demonstrated a significant interaction with both expressed receptors as well as retention of fluid secretion activity in Malpighian tubules of the mosquito and cricket. Further evaluation of this and other rationally designed nonpeptide libraries can potentially yield other biostable agonists and antagonists of the insect kinins. Primary and secondary tissue-bound peptidase hydrolysis sites have been identified in the C-terminal region and analogs containing either β-amino acids or α-aminoisobutyric acid (Aib) adjacent to these sites retain significant bioactivity and demonstrate greatly enhanced biostability. Two biostable analogs show a selective agonist response in the expressed tick receptor assay over the expressed mosquito receptor.

Another class of insect kinin analogs containing a C-terminal aldehyde demonstrates unusual properties. One C-terminal aldehyde analog enhances the weight-gain inhibition activity of the insect kinins in *H. zea* larvae to such an extent that it results in increased mortality. Another 'magic bullet' C-terminal aldehyde analog selectively targets housefly Malpighian tubules, the major organ of diuresis in insects and leads to marked inhibition of urine release. An agent capable of selective depression of fluid secretion would be expected to allow pesticides to achieve higher concentrations in the hemolymph; and, in turn, likely reduce the amount of toxin required to kill an insect.

While neuropeptides are not generally designed for penetration of the outside cuticle or the gut wall in large quantities, enhancement of bioavailability has been demonstrated in at least one other class of insect neuropeptides. Amphiphilic analogs of the pyrokinin/PBAN family of insect neuropeptides have shown an ability to efficiently penetrate in vitro preparations of insect cuticle and foregut, as well as demonstrate potent activity in in vivo pheromonotopic bioassays when administered via topical or oral routes. Amphiphilic analogs contain both polar and apolar components that confer surfactant-like characteristics. Similar modification of the insect kinins to impart amphiphilic character would present a likely path to the development of mimetic analogs with enhanced bioavailability.

In conclusion, the studies presented here have led to the identification of interesting tools for arthropod endocrinologists and promising mimetic analog leads in the development of selective, environmentally friendly arthropod pest control agents capable of disrupting insect kinin regulated processes.

Acknowledgements

We wish to acknowledge financial assistance from a grant from the USDA/DOD DWFP Initiative (#0500-32000-001-01R) (RJN), the North Atlantic Treaty Organization (NATO) Collaborative Research Grant (#LST.CLG.979226) (RJN), a NRI/USDA grant number 2003-35302-13678 (PVP) and a Texas Advanced Technology/Research Grant (#000517-0103-2001) (RJN).

References

1. Holman GM, Cook BJ, Nachman RJ. Isolation, primary structure and synthesis of Leucokinins VII and VIII: The final members of this new family of chephalomyotropic peptides isolated from head extracts of Leucophaea maderae. Comp Biochem Physiol 1987; 88C(1):31-4.
2. Coast GM. The regulation of primary urine production in insects. In: Coast GM, Webster SG, eds. Recent Advances in Arthropod Endocrinology. Cambridge: Cambridge University Press, 1998:189-209.
3. Nachman RJ, Strey A, Isaac E et al. Enhanced in vivo activity of peptidase-resistant analogs of the insect kinin neuropeptide family. Peptides 2002a; 23:735-45.
4. Coast GM, Holman GM, Nachman RJ. The diuretic activity of a series of cephalomyotropic neuropeptides, the achetakinins, on isolated malpighian tubules of the house cricket, Acheta domesticus. J Insect Physiol 1990; 36(7):481-8.
5. Nachman RJ, Holman GM. Myotropic insect neuropeptide families from the cockroach Leucophaea maderae: Structure-activity relationships. In: Menn JJ, Masler EP, eds. Insect Neuropeptides: Chemistry, Biology and Action. Washington, DC: American Chemical Society, 1991:194-214.
6. Sajjaya P, Deepa Chandran S, Sreekumar S et al. In vitro regulation of gut pH by neuropeptide analogues in the larvae of red palm weevil, Rhynchophorus ferrugineus. J Adv Zool 2001; 22(1):26-30.
7. Harshini S, Nachman RJ, Sreekumar S. Inhibition of digestive enzyme release by neuropeptides in larvae of Opisina arenosella (Lepidoptera: Cryptophasidae). Comp Biochem Physiol 2002; B132:353-8.
8. Harshini S, Manchu V, Sunitha VB et al. In vitro release of amylase by culekinins in two insects: Opisinia arenosella (Lepidoptera) and Rhynchophorus ferrugineus (Coleoptera). Trends Life Sci 2003; 17:61-4.
9. Seinsche A, Dyker H, Losel P et al. Effect of helicokinins and ACE inhibitors on water balance and development of Heliothis virescens larvae. J Insect Physiol 2000; 46:1423-31.
10. Nachman RJ, Isaac RE, Coast GM et al. Aib-containing analogues of the insect kinin neuropeptide family demonstrate resistance to an insect angiotensin-converting enzyme and potent diuretic activity. Peptides 1997; 18:53-7.
11. Nachman RJ, Roberts VA, Holman GM et al. Concensus chemistry and conformation of an insect neuropeptide family analogous to the tachykinins. In: Epple A, Scanes CG, Stetson MH, eds. Progress in Comparative Endocrinology. New York: Wiley-Liss, Inc, 1990:342:60-6.
12. Cornell MJ, Williams TA, Lamango NS et al. Cloning and expression of an evolutionary conserved single-domain angiotensin converting enzyme from Drosophila melanogaster. J Biol Chem 1995; 270:13613-9.
13. Lamango NS, Sajid M, Isaac RE. The endopeptidase activity and the activation by Cl^- of angiotensin-converting enzyme is evolutionarily conserved: purification and properties of an antiotensin-converting enzyme from the housefly, Musca domestica Nazarius. Biochem J 1996; 314:639-46.
14. King FD, Wilson S. Recent advances in 7-transmembrane receptor research. Curent Opin Drug discovery Development 1999; 2:83-95.
15. Von Zastrow M. Mechanisms regulating membrane trafficking of G protein-coupled receptors in the endocytic pathway. Life Sci 2003; 74:217-24.
16. Marchese A, Chen C, Kim YM et al. The ins and outs of G protein-coupled receptor trafficking. Trends Biochem Sci 2003; 8:369-76.
17. Bondensgaard K, Ankersen M, Thogersen H et al. Recognition of privileged structures by G-protein coupled receptors. J Med Chem 2004; 47:888-99.
18. Vanden Broeck J. Insect G protein-coupled receptors and signal transduction. Arch Insect Biochem Physiol 2001; 48:1-12.
19. Holmes SP, He H, Chen AC et al. Cloning and transcriptional expression of a leucokinin-like peptide receptor from the southern cattle tick, Boophilus microplus (Acari: Ixodidae). Insect Mol Biol 2000; 9:457-65.
20. Cox KJA, Tensen CP, Van der Schors RC et al. Cloning, characterization and expression of a G-protein-coupled receptor from Lymnaea stagnalis and identification of a leucokinin-like peptide, PSFHSWSamide, as its endogenous ligand. J Neurosci 1997; 17:1197-205.
21. Holmes SP, Barhoumi R, Nachman RJ et al. Functional analysis of a G protein-coupled receptor from the southern cattle tick Boophilus microplus (Acari: Ixodidae) identifies it as the first arthropod myokinin receptor. Insect Mol Biol 2003; 12:27-38.
22. Radford JC, Davies SA, Dow JAT. Systematic G-protein-coupled receptor analysis in Drosophila melanogaster identifies a leucokinin receptor with novel roles. J Biol Chem 2002; 277:38810-7.

23. Hayes TK, Pannabecker TL, Hincley DJ et al. Leucokinins, a new family of ion transport stimulators and inhibitors in insect Malpighian tubules. Life Sci 1989; 44(18):1259-66.
24. Veenstra JA, Pattillo JM, Petzel DH. A single cDNA encodes all three Aedes leucokinins, which stimulate both fluid secretion by the Malpighian tubules and hindgut contractions. J Biol Chem 1997; 272(16):10402-7.
25. Pietrantonio PV, Jagge C, Taneja-Bageshwar S et al. The mosquito Aedes aegypti (L.) leucokinin receptor is a multiligand receptor for the three Aedes kinins. Insect Mol Biol 2005; 14(1):55-67.
26. Wang J. Kean L. Yang J et al. Function-informed transcriptome analysis of Drosophila renal tubule. Genome Biol 2004; 5(9):R69. Epub 2004.
27. Coast GM. Fluid secretion by single isolated Malpighian tubules of the house cricket, Acheta domesticus and their response to diuretic hormone. Physiol Entomol 1988; 13:381-91.
28. O'Donnell MJ, Rheault MR, Davies SA et al. Hormonally controlled chloride movements across Drosophila tubules is via ion channels in stellate cells. Am J Physiol 1998; 274 (Regulatory Integrative Comp Physiol 43): R1039-49.
29. Wang S, Rubenfeld AB, Hayes TK et al. Leucokinin increases paracellular permeability in insect Malpighian tubules. J Exp Biol 1996; 199:2537-42.
30. Kim Y-J, Žitňan D, Galizia CG et al. A command chemical triggers an innate behavior by sequential activation of multiple peptidergic ensembles. Curr Biol 2006a; 16:1395-407.
31. Kim Y-J, Žitňan D, Cho K-H et al. Central peptidergic ensembles associated with organization of an innate behavior. Proc Natl Acad Sci USA 2006b; 103(38):14211-6.
32. Li B, Predel R, Neupert S et al. Genomics, transcriptomics and peptidomics of peptide and protein hormones in the red flour beetle Tribolium castaneum. Genome Res 2008; 18(1):113-22.
33. Nachman RJ, Roberts VA, Holman GM et al. Leads for insect neuropeptide mimetic development. Arch. Insect Biochem Physiol 1993; 22:181-97.
34. Nachman RJ, Coast GM, Douat C et al. A C-terminal aldehyde insect kinin analog enhances inhibition of weight gain and induces significant mortality in Helicoverpa zea larvae. Peptides 2003; 24:1615-21.
35. Coast GM, Zabrocki J, Nachman RJ. Diuretic and myotropic activities of N-terminal truncated analogs of Musca domestica kinin neuropeptide. Peptides 2002; 23:701-8.
36. Nachman RJ, Coast GM, Holman GM et al. Diuretic activity of C-terminal group analogues of the insect kinins in Acheta domesticus. Peptides 1995; 16:809-13.
37. Taneja-Bageshwar S, Strey A, Zubrzak P et al. Comparative structure-activity analysis of insect kinin core analogs on recombinant kinin receptors from Southern cattle tick Boophilus microplus (Acari: Ixodidae) and mosquito Aedes aegypti (Diptera: Culicidae). Arch Insect Biochem Physiol 2006; 62(3):128-40.
38. Moyna G, Williams HJ, Nachman RJ et al. Conformation in solution and dynamics of a structurally constrained linear insect kinin pentapeptide analogue. Biopolymers 1999; 49:403-13.
39. Roberts VA, Nachman RJ, Coast GM et al. Consensus chemistry and beta-turn conformation of the active core of the insect kinin neuropeptide family. Chem Biol 1997; 4:105-17.
40. Nachman RJ, Zabrocki J, Olczak J et al. cis-Peptide bond mimetic tetrazole analogs of the insect kinins identify the active conformation. Peptides 2002c; 23:709-16.
41. Nachman RJ, Kaczmarek K, Williams HJ et al. An active insect kinin analog with 4-aminopyroglutamate, a novel cis-peptide bond, type VI β-turn motif. Bioploymers 2004; 75:412-19.
42. Kaczmarek K, Williams HJ, Coast GM et al. Comparison of insect kinin analogs with cis-peptide bond motif 4-aminopyroglutamate identifies optimal stereochemistry for diuretic activity. Biopolymers (Peptide Sci) 2007; 88:1-7.
43. Taneja-Bageshwar S, Strey A, Kaczmarek K et al. Comparison of insect kinin analogs with cis-peptide bond, type VI β-turn motifs identifies optimal stereochemistry for interaction with a recombinant insect kinin receptor from the Southern cattle tick Boophilus microplus. Peptides 2008; 29(2):295-301.
44. Taneja-Bageshwar S, Strey A, Zubrzak P et al. Identification of selective and nonselective, biostable β-amino acid agonists of recombinant insect kinin receptors from the Southern cattle tick Boophilus microplus and mosquito Aedes aegypti. Peptides 2008; 29(2):302-9.
45. Kamoune L, De Borggraeve WM, Verbist BMP et al. Structure based design of simplified analogues of insect kinins. Tetrahedron 2005; 61:9555-62.
46. Nachman RJ, Teal PEA, Strey A. Enhanced oral availability/pheromonotropic activity of peptidase-resistant topical amphiphilic analogs of pyrokinin/PBAN insect neuropeptides. Peptides 2002b; 23:2035-43.
47. Cheng RP, Gellman SH, DeGrado WF. β-Peptides: From structure to function. Chem Rev 2001; 101:3219-32.
48. Juaristi E, Soloshonok VA, eds. Second Edition of Enantioselective Synthesis of beta-Amino Acids. New York: Wiley, 2005.
49. Zubrzak P, Williams H, Coast GM et al. Beta-amino acid analogs of an insect neuropeptide feature potent bioactivity and resistance to peptidase hydrolysis. Biopolymers (Peptide Science) 2007; 88(1):76-2.
50. Gregory H, Hardy PM, Jones PM et al. The antral hormone gastrin. Nature 1964; 204:931-3.

51. Coast GM. Diuresis in the housefly (Musca domestica) and its control by neuropeptides. Peptides 2001; 22:153-60.
52. Chung JS, Goldsworthy GJ, Coast GM. Haemolymph and tissue titers of achetakinins in the house cricket Acheta domesticus: effect of starvation and dehydration. J Exp Biol 1994; 193:307-19.
53. Goldsworthy GJ, Coast GM, Wheeler CH et al. The structural and functional activity of neuropeptides. In: Crampton JM, Eggelton P, eds. Royal Entomological Society Symposium on Insect Molecular Science. London: Academic Press, 1992:205-25.
54. Chapman KT. Synthesis of a potent, reversible inhibitor of interleukin-1b converting enzyme. Bioorg Med Chem Lett 1992; 2:613-8.
55. Fehrentz JA, Heitz A, Castro B et al. Aldehydic peptides inhibiting rennin. FEBS Lett 1984; 167:273-6.
56. Sarubbi E, Seneei PF, Angelestro MR et al. Peptide aldehydes as inhibitors of HIV protease. FEBS Lett 1993; 319:253-6.
57. Lehninger AL. Biochemistry: the molecular basis for cell structure and function. New York: Worth Publishers, Inc, 1970:80.
58. Nachman RJ, Fehrentz JA, Martinez J et al. A C-terminal aldehyde analog of the insect kinins inhibits diuresis in the housefly. Peptides 2007; 28:146-52.
59. Coast GM. Continuous recording of excretory water loss from Musca domestica using a flow-through humidity meter: hormonal control of diuresis. 2004; 50:455-68.
60. Maddrell SHP. The functional design of the insect excretory system. J Exp Biol 1981; 90:1-15.
61. O'Donnell MJ, Maddrell SHP. Paracellular and transcellular routes for water and solute movements across insect epithelia. J Exp Biol 1983; 106:231-53.
62. Nachman RJ, Holman GM, Haddon WF. Leads for insect neuropeptide mimetic development. Arch Insect Biochem Physiol 1994; 22:181-97.
63. Reixach N, Crooks E, Ostresh JM et al. Inhibition of β-amyloid-induced neurotoxicity by imidazopyridoindoles derived from a synthetic combinatorial library. J Struct Biol 2000;130:247-58.
64. Lorenz MW, Zemek R, Kodrik D et al. Lipid mobilization and locomotor atimulation in Gryllus bimaculatus by topically applied adipokinetic hormone. Physiol Entom 2004; 29:146-51.
65. Nachman RJ, Teal PEA, Radel P et al. Potent pheromonotropic/myotropic activity of a carboranyl pseudotetrapeptide analog of the insect pyrokinin/PBAN neuropeptide family administered via injection or topical application. Peptides 1996; 17(5):747-52.
66. Abernathy RL, Teal PEA, Meredith JA et al. Induction of moth sex pheromone production by topical application of an amphiphilic pseudopeptide mimic of pheromonotropic neuropeptides. Proc Nat Acad Sci USA 1996; 93:12621-5.
67. Holman GM, Wright MS, Nachman RJ. Insect neuropeptides: Coming of age. In: Grimwade AM, ed. ISI Atlas of Science, Plants and Animals, Philadelphia: Institute for Scientific Information, Inc., 1988:212, 129-36
68. Raina AK, Jaffe H, Kempe TG et al. Identification of a neuropeptide hormone that regulates sex pheromone production in female moths. Science 1989; 244:796-8.
69. Predel R, Nachman RJ. The FXPRLamide (Pyrokinin/PBAN) peptide family. In: Kastin A, ed. Handbook of Biologically Active Peptides. Elsevier, 2006:207-13.
70. Nachman RJ, Teal PEA, Ujvary I. Comparative topical pheromonotropic activity of insect pyrokinin/PBAN amphiphilic analogs incorporating different fatty and/or cholic acid components. Peptides 2001; 22:279-85.
71. Nachman RJ, Teal PEA. Amphiphilic mimics of pyrokinin/PBAN neuropeptides that induce prolonged pheromonotropic activity following topical application to a moth. In: Konopinska D, Goldsworthy G, Nachman RJ et al, eds. Insects: Chemical, Physiological and Environmental Aspects 1997. Wroclaw: University of Wroclaw Press, 1998:145-54, 293.
72. Teal PEA, Nachman RJ. Prolonged pheromonotropic activity of pseudopeptide mimics of insect pyrokinin neuropeptides after topical application or injection into a moth. Regul Pept 1997; 72:161-7.
73. Teal PEA, Nachman RJ. A brominated-fluorene insect neuropeptide analog exhibits pyrokinin/PBAN-specific toxicity for adult females of the tobacco budworm moth. Peptides 2002; 23:801-6.
74. Raina AK, Rafaeli A, Kingan TG. Pheromonotropic activity of orally administered PBAN and its analogs in Helicoverpa zea. J Insect Physiol 1995; 40:393-7.
75. Bavoso A, Falabella P, Goacometti R et al. Intestinal absorption of proctolin in Helicoverpa armigera (Lepidoptera noctuidae) larvae. Redia 1995; 78:173-85.
76. Audsley N, Weaver RJ. In vitro transport of an allatostatin across the foregut of Manduca sexta larvae and metabolism by the gut and hemolymph. Peptides 2007; 28:136-45.
77. Yao J, Feher VA, Espefo BE et al. Stabilization of a type VI turn in a family of linear peptides in water solution. J Mol Biol 1994; 243:736-53.
78. Taneja-Bageshwar S, Strey A, Isaac RE et al. Biostable agonists that match or exceed activity of native insect kinins on recombinant arthropod GPCRs. Gen Comp Endocrin 2009; 162:122-128.

CHAPTER 4

Neuropeptides in Helminths:
Occurrence and Distribution

Nikki J. Marks and Aaron G. Maule*

Abstract

Nematode neuropeptide systems comprise an exceptionally complex array of ~250 peptidic signaling molecules that operate within a structurally simple nervous system of ~300 neurons. A relatively complete picture of the neuropeptide complement is available for *Caenorhabditis elegans*, with 30 *flp*, 38 *ins* and 43 *nlp* genes having been documented; accumulating evidence indicates similar complexity in parasitic nematodes from clades I, III, IV and V. In contrast, the picture for parasitic platyhelminths is less clear, with the limited peptide sequence data available providing concrete evidence for only FMRFamide-like peptide (FLP) and neuropeptide F (NPF) signaling systems, each of which only comprises one or two peptides. With the completion of the *Schmidtea medditeranea* and *Schistosoma mansoni* genome projects and expressed sequence tag datasets for other flatworm parasites becoming available, the time is ripe for a detailed reanalysis of neuropeptide signaling in flatworms. Although the actual neuropeptides provide limited obvious value as targets for chemotherapeutic-based control strategies, they do highlight the signaling systems present in these helminths and provide tools for the discovery of more amenable targets such as neuropeptide receptors or neuropeptide processing enzymes. Also, they offer opportunities to evaluate the potential of their associated signaling pathways as targets through RNA interference (RNAi)-based, target validation strategies. Currently, within both helminth phyla, the *flp* signaling systems appear to merit further investigation as they are intrinsically linked with motor function, a proven target for successful anti-parasitics; it is clear that some nematode NLPs also play a role in motor function and could have similar appeal. At this time, it is unclear if flatworm NPF and nematode INS peptides operate in pathways that have utility for parasite control. Clearly, RNAi-based validation could be a starting point for scoring potential target pathways within neuropeptide signaling for parasiticide discovery programs. Also, recent successes in the application of *in planta*-based RNAi control strategies for plant parasitic nematodes reveal a strategy whereby neuropeptide encoding genes could become targets for parasite control. The possibility of developing these approaches for the control of animal and human parasites is intriguing, but will require significant advances in the delivery of RNAi-triggers.

Introduction

Twenty years ago almost nothing was known about the occurrence and distribution of peptide signaling molecules in helminth nervous systems. This was despite the publication of a physical map of the nervous system of the free-living nematode, *Caenorhabditis elegans* by White and coworkers,[1] the first ultrastructural reconstruction of an entire metazoan nervous system and one of the foundation stones to the subsequent exploitation of *C. elegans* as a model

*Corresponding Author: Aaron G. Maule—Parasitology, School of Biological Sciences, Queen's University Belfast, Medical Biology Centre, 97 Lisburn Road, Belfast BT9 7BL, UK. Email: a.maule@qub.ac.uk

Neuropeptide Systems as Targets for Parasite and Pest Control, edited by Timothy G. Geary and Aaron G. Maule.

organism for laboratory-based research. Although several authors had reported the occurrence of neurosecretory vesicles in parasitic nematodes in the 1960s,[2,3] it was not until indirect immunocytochemical methods were applied to helminths in the mid-late 1980s, that data started to accumulate on the occurrence and distribution of neuropeptides in nematode and platyhelminth nervous systems.[4-13] These and subsequent studies provided information that revolutionized our view of the chemical complexity of helminth nervous systems and provided strong evidence that neuropeptides played a major role in neuronal signaling processes in worms.

Although invaluable in demonstrating the widespread occurrence of neuropeptides in worms and indeed generating data on the complexity and organization of nervous systems in numerous helminth parasites, these studies did not provide information on the primary structures of these peptides. This breakthrough followed the sterling efforts of Tony Stretton and colleagues who managed to extract, purify and sequence small quantities of neuropeptides from large-scale tissue extracts of the gastrointestinal parasite of pigs, *Ascaris suum*.[14] Subsequent studies on this and other nematodes generated a constant flow of new peptide sequences that served to expand our knowledge of their peptide complements.[15-29] One noticeable feature at this time was that almost all the primary sequence data were for multiple members of one neuropeptide family, the FMRFamide-related peptides (FaRPs) or FMRFamide-like peptides (FLPs). During the same period, the structures of a relatively small number of flatworm neuropeptides were determined.[30-35] Again, FLPs featured amongst the flatworm neuropeptide sequence data, but a second family, designated neuropeptide F (NPF) was also much in evidence.

In the 1990s, the molecular biological and genomics revolutions were making a significant impact on nematode neuropeptide discovery. The first reported nematode *flp* gene was significant in that it encoded multiple copies of distinct FLPs and was alternatively spliced to facilitate variation in the resultant peptide products.[36] However, even then, the extent of neuropeptide diversity in nematodes was not realized. In the late 1990s, several groups reported the occurrence of numerous neuropeptide genes, including insulin-like peptide encoding *ins* and multiple *flp* genes in *C. elegans*.[37-40] The completion of the *C. elegans* genome (at time of writing, still the only complete [to the last base] metazoan genome sequence) provided the opportunity for more comprehensive neuropeptide gene discovery efforts and allowed the identification of a wide range of non-*ins* and non-*flp* neuropeptide genes that were grouped together and designated *nlp* (neuropeptide-like protein) genes.[41,42] Taken together, these findings resulted in a broad reevaluation of the importance and complexity of neuropeptides to neuronal signaling processes in nematodes and presented *C. elegans* neurobiologists with a bewilderingly diverse catalog of peptide signaling molecules for which there was very little known about the associated biology. Further, this knowledge base provided a tool for the interrogation of expressed sequence tag (EST) datasets which have been accumulating for numerous, important pathogenic nematodes since the turn of the century. Recently, this has facilitated the discovery of *flp* genes in a wide range of parasitic nematodes revealing significant conservation in the *flp* complements across phylum Nematoda.[43,44]

In direct contrast to the situation in nematodes, progress in flatworm peptide discovery stalled through the late 1990s and early years of this century, largely because the laborious biochemical methods used to generate individual peptide sequences were now out of vogue and could not compete with the rapidly evolving genomics approaches. Although our knowledge of flatworm neuropeptides is still trailing far behind that for nematodes, significant EST deposits for a relatively small number of flatworms and genome sequence data for the blood fluke *S. mansoni* and the turbellarian *S. mediterranea*, at least threaten to rectify this situation and provide much needed data on neuropeptide complements in platyhelminths.

For those working in helminth parasite control, interest in neuropeptide signaling systems stems from several observations. Firstly, unlike the situation in mammals, where neuropeptides are most commonly involved in the modulation of synaptic function, invertebrate neuropeptides act as both modulators and fully fledged neurotransmitter molecules. This latter function has much appeal as a target system for chemotherapies aimed at disrupting normal parasite behaviors. Secondly, there are stark differences in the neuropeptides found in invertebrates and vertebrates such that target proteins ensconced within the associated signaling pathways are likely to be distinct from host proteins. Thirdly and most importantly, many of the leading anthelmintics (especially those used to treat nematode parasites) disrupt neuromuscular function, primarily by acting at receptors within classical transmitter signaling pathways.[45,46] Currently, none of these drugs act on the neuropeptide signaling processes which impact neuromuscular function, exposing the potential of these systems as target sources. Several recent reviews provide comprehensive coverage of neuropeptides in flatworms and nematodes.[45,47-50] This chapter aims to provide an overview of current knowledge on neuropeptides and their expression in helminths and how this might relate to targets for parasite control.

Nematode Neuropeptides

All nematodes have around 300 neurons (the *C. elegans* hermaphrodite has 302 neurons), making their nervous system the most structurally simple amongst the triploblastic metazoans. Nematodes appear to compensate for these structural limitations by employing a huge number of intercellular signaling molecules, of which the largest structural class is the neuropeptides. In nematodes, these neuropeptides currently fall within three distinct groupings, the insulin-like peptide family encoded on *ins* genes, the FMRFamide-like peptide family encoded on *flp* genes and a group encompassing all other neuropeptide families designated the neuropeptide-like proteins and encoded on *nlp* genes. Presently, these groupings encompass 102 distinct genes that encode in excess of 250 different neuropeptides—a quite staggering complement for relatively simple animals with ~300 nerve cells.

Nematode FMRFamide-Like Peptides (FLPs)

Nematodes show exceptional diversity with respect to the complement of FLPs they express. The number of different FLPs identified (so far) in *C. elegans* stands at >70 and these are encoded on a battery of 30 distinct genes (designated *flp-1* to *flp-28*, *flp-32* and *flp-33*); note that *flp-32* has only been predicted such that its transcript awaits identification.[43,44,50,51] Bioinformatic and PCR/DNA sequencing studies have provided the bulk of the available data on *C. elegans* FLPs,[36,40,43,51,52] and until recently, only a small number of studies contributed peptide sequence information.[18,23-28] More recently, the powerful peptidomic approaches adopted by the Schoofs laboratory have provided peptide structural data that support many of the earlier predictions on peptide sequences and which have uncovered some additional peptides, e.g., a *flp-33* encoded peptide.[50,53]

C. elegans has been the most important tool in our understanding of the nematode FLP complement, although studies on other species have contributed to this knowledge base. Multiple FLPs have been structurally characterized from the large gastrointestinal parasite of swine and Clade III nematode, *A. suum*.[14,16,17,29,54] Small numbers of FLPs have been structurally isolated from *Panagrellus redivivus*,[15,19-21] the free-living, Clade III nematode (*C. elegans* belongs to Clade V) and the sheep parasite from Clade V, *Haemonchus contortus*.[22,27] A screen of the growing expressed sequence tag (EST) datasets for parasitic nematodes revealed that there was much similarity in the *flp* genes across phylum Nematoda.[43] Indeed, all but one (*flp-20*) of the *C. elegans flp* genes were identified in the EST dataset of at least one parasitic nematode. All in all, this effort uncovered a total of 33 *flp* genes encoding over 90 distinct FLPs in 33 different nematode species from Clades I, III, IV and V; there was no evidence of significant inter-Clade or inter-lifestyle variation in FLP complement. The range of FLP signature sequences that have been identified in nematodes are outlined in Table 1. One feature that stands out is the large number of species in which most *flp* genes have been identified. Only *flp-29*, *-30* and *31* do not appear to be expressed in *C. elegans*. Indeed, *flp-30* and *31* have only been identified in plant parasitic nematodes of Clade IV and could play a role specific to plant-nematode interactions. *flp-29* has been identified in parasites of Clades III and V and yet appears to be absent from *C. elegans* (also Clade V) such that it could play a role in parasite-specific activities.

Most FLPs appear to be associated with intercellular communication in nematodes and they have been shown to have wide-ranging functions. In particular, many are associated with the modulation of motor activities or sensory modalities such that the appeal of proteins associated with these signaling pathways as drug targets is strong. The specific functions of nematode FLPs are discussed in detail in Chapter 5 and will not be discussed further here.

***Table 1.* Nematode FMRFamide-like peptide (FLPs) signature sequences and their occurrence in nematode genera**

Gene Name	Amino Acid Sequence/ C-Terminal Signature	Length of N-Terminal Extension	Genera in Which Transcript/Peptide/Peptide with Corresponding Mass Has Been Identified
flp-1	x PNFLRF.NH_2 *PNFMRY.NH_2*	0-15 0-1	*Ancylostoma, Ascaris, Caenorhabditis, Globodera, Heterodera, Meloidogyne, Necator, Onchocerca, Parastrongyloides, Strongyloides, Wuchereria*
flp-2	2x EPIRF.NH_2	3	*Ancylostoma, Caenorhabditis, Necator, Ostertagia*
flp-3	x $SPX_oGTMRF.NH_2$	0-15	*Ascaris, Caenorhabditis, Heterodera, Meloidogyne*
flp-4	2x $PX_iFIRF.NH_2$	2-6	*Ancylostoma, Ascaris, Caenorhabditis, Heterodera*
flp-5	3x KFIRF.NH_2	2-7	*Ancylostoma, Caenorhabditis, Globodera, Heterodera, Meloidogyne, Necator, Pratylenchus*
flp-6	x KSAYMRF.NH_2	0	*Ancylostoma, Ascaris, Brugia, Caenorhabditis, Globodera, Heterodera, Meloidogyne, Necator, Ostertagia, Onchocerca, Strongyloides, Teladorsagia*
flp-7	x $PX_oX_iRSXX_oX_oRF.NH_2$	1	*Ancylostoma, Caenorhabditis, Globodera, Heterodera, Meloidogyne, Ostertagia, Strongyloides*
flp-8	x $KNEFX_oRF.NH_2$	0	*Ancylostoma, Ascaris, Caenorhabditis, Necator, Onchocerca, Xiphinema*
flp-9	2x KPSFVRF.NH_2	0	*Ancylostoma, Ascaris, Caenorhabditis, Necator, Ostertagia*
flp-10	$YX_oRF.NH_2$	4	*Ancylostoma, Caenorhabditis, Xiphinema*
flp-11	x RNXLVRF.NH_2 NGAPQPFVRF.NH_2	2-7 0	*Ancylostoma, Ascaris, Caenorhabditis, Globodera, Haemonchus, Heterodera, Meloidogyne, Necator, Ostertagia, Pratylenchus, Radopholus, Strongyloides, Teladorsagia, Wuchereria*

continued on next page

Table 1. Continued

Gene Name	Amino Acid Sequence/ C-Terminal Signature	Length of N-Terminal Extension	Genera in Which Transcript/Peptide/Peptide with Corresponding Mass Has Been Identified
flp-12	x (R/K)NKFEFIRF.NH_2	1	*Ancylostoma, Ascaris, Caenorhabditis, Globodera, Heterodera, Meloidogyne, Necator, Onchocerca*
flp-13	x PLIRF.NH_2 PX_oX_oRF.NH_2	4-5 4-5	*Ancylostoma, Ascaris, Caenorhabditis, Globodera, Haemonchus, Heterodera, Meloidogyne, Necator, Onchocerca, Ostertagia, Pratylenchus, Pristionchus, Strongyloides*
flp-14	x KHEYLRF.NH_2 *KHEFVRF.NH_2*	0 0	*Ancylostoma, Ascaris, Brugia, Caenorhabditis, Globodera, Meloidogyne, Necator, Onchocerca, Parastrongyloides, Pratylenchus, Radopholus, Strongyloides, Teladorsagia, Trichinella*
flp-15	GPQGPLRF.NH_2 GP(S/T)GPLRF.NH_2	1-2	*Ancylostoma, Caenorhabditis, Necator, Nippostrongylus, Ostertagia, Teladorsagia*
flp-16	x AQTFVRF.NH_2 *GQTFVRF.NH_2*	0 0	*Ancylostoma, Ascaris, Caenorhabditis, Globodera, Haemonchus, Heterodera, Meloidogyne, Necator, Onchocerca, Ostertagia, Parastrongyloides, Pratylenchus, Radopholus*
flp-17	KSAFVRF.NH_2 KSQYIRF.NH_2	0 0	*Ancylostoma, Caenorhabditis, Haemonchus, Necator, Ostertagia, Strongyloides, Xiphinema*
flp-18	x PGVLRF.NH_2 *PGX_oX_oRF.NH_2*	2-9 2-9	*Ancylostoma, Ascaris, Caenorhabditis, Globodera, Haemonchus, Meloidogyne, Necator, Onchocerca, Ostertagia, Pristionchus, Strongyloides, Trichinella*
flp-19	x WXX_iQX_oRF.NH_2	0-3	*Caenorhabditis, Dirofilaria, Heterodera, Meloidogyne, Necator, Pratylenchus, Strongyloides, Teladorsagia*
flp-20	X_oX_oMRF.NH_2	0	*Ancylostoma, Ascaris, Caenorhabditis, Haemonchus, Ostertagia, Parastrongyloides, Pratylenchus*

continued on next page

Table 1. Continued

Gene Name	Amino Acid Sequence/ C-Terminal Signature	Length of N-Terminal Extension	Genera in Which Transcript/Peptide/Peptide with Corresponding Mass Has Been Identified
flp-21	$GPRPLRF.NH_2$	2-3	*Ancylostoma, Ascaris, Brugia, Caenorhabditis, Haemonchus, Meloidogyne, Necator, Ostertagia, Pratylenchus, Pristionchus, Radopholus, Strongyloides, Teladorsagia*
flp-22	x $KWMRF.NH_2$	4-7	*Ancylostoma, Caenorhabditis, Globodera, Heterodera, Ostertagia, Parastrongyloides, Pratylenchus, Pristionchus, Radopholus, Strongyloides, Teladorsagia*
flp-23	$QDFLRF.NH_2$	4	*Caenorhabditis, Teladorsagia*
flp-24	$VPSA(G/A)DMMX_oRF.NH_2$	0	*Ancylostoma, Ascaris, Brugia, Caenorhabditis, Necator, Ostertagia, Onchocerca*
flp-25	2x $YD(F/Y)(V/I)RF.NH_2$	1-8	*Caenorhabditis, Globodera, Meloidogyne, Necator, Strongyloides*
flp-26	$EFNADDLTLRFG.NH_2$ $AFX_iPX_iXLX_iLRF.NH_2$	0 4-7	*Ancylostoma, Caenorhabditis, Necator*
flp-27	$G(G/S)RMRF.NH_2$	2	*Ancylostoma, Caenorhabditis, Heterodera, Meloidogyne, Necator, Radopholus*
flp-28	$VLMRF.NH_2$	0	*Caenorhabditis, Pristionchus*
flp-29	$IX_oMRF.NH_2$	0	*Ancylostoma, Ascaris, Haemonchus, Ostertagia*
flp-30	$QMREPLRF.NH_2$	0	*Meloidogyne*
flp-31	$LYRPRGPPRF.NH_2$	0	*Meloidogyne, Pratylenchus*
flp-32	$AMRNSLVRF.NH_2$	0	*Caenorhabditis, Teladorsagia*

*Single letter amino acid annotations are used throughout. Amino acid signature sequences: X_i = hydrophilic residue; X_o = hydrophobic residue; X = variable residue; amino acids in parentheses represent alternatives. Signatures shown are the most commonly encoded peptide(s) on the associated transcripts. Sequences in italics appear on some transcripts. Note that, where possible (i.e., when ESTs from multiple species were available) single amino acid variations in one EST were ignored as the chances of these representing single nucleotide errors in the EST sequences appeared high. Note that the delineation between the *flp-11* and *flp-32* transcripts is unclear from the available EST data for most species; a number of the transcripts tentatively identified as *flp-11s* could actually be *flp-32*. A *flp-33* encoded peptide ($APLEGFEDMSGFLRTIDGIQKPRF.NH_2$) has been identified in *C. elegans* (Husson et al 2007)[50] but it's occurrence in other nematodes is unknown.

Nematode FLP Distribution/Expression

Much evidence relating to the expression of FLPs in nematodes has accumulated from immunocytochemical studies using polyclonal antisera that cannot distinguish between the multiple similar epitopes the FLP family presents. Although most of these studies do not inform on the distribution of individual FLPs in nematodes, they have provided an overview of FLP distribution in a variety of species.[55-61] The take-home message from these efforts has been that FLPs are widely expressed in nematode nervous systems (reports range from 50 to 75% of neurons) and that they occur in all neuronal subtypes (inter-neurons, motor neurons, pharyngeal neurons and sensory neurons). FLP-immunopositive neurons are common in the brain (circumpharyngeal nerve ring) and associated ganglia as well as in the innervation of muscular organs such as the ovijector/vulva and pharynx.

Most is known about *flp* expression in *C. elegans* which has been monitored using upstream promoter regions of *flp* genes fused to green fluorescent protein (GFP) coding regions in transgenic animals following germ-line transformations.[51,52,62] The expression of *flp-1* to *flp-23* has been examined, although no expression was observed for *flp-9*, *flp-14*, *flp-16* or *flp-23*. While the vast majority of *flps* were expressed exclusively in neurons, some were also expressed in non-neuronal cells: *flp-2* and *flp-11* were expressed in head muscles; *flp-5* and *flp-15* were expressed in pharyngeal muscles; *flp-11* and *flp-15* were expressed in socket and/or sheath cells which form the amphidial channels; *flp-10* was expressed in vulval muscles; *flp-11* and *flp-2* were expressed in uterine cells. Multiple neuronal cells expressed more than one *flp* gene, but no two *flp* genes have identical expression patterns providing a mosaic of *flp* expression. The extent of expression varied considerably with between 2 and 44 neurons expressing individual *flps*. In total, the 19 *flp* genes for which robust expression was recorded (faint GFP expression was ignored) occurred in 53% of *C. elegans* neurons; since the expression of 4 *flp* genes was not determined and another 10 *flp* genes remain to be examined, this is likely to be an underestimate of overall *flp* expression in the worm. Expression onset for most *flps* is during embryogenesis with the earliest expression, during gastrulation, being noted for *flp-15*; most *flps* display expression onset at the comma stage, a mid-stage in gastrulation where the embryo appears slightly folded within the egg. It is unclear why some *flps* display such early expression, but they may play roles in aspects of neuronal development. In some neurons, selected *flp* genes are coexpressed with classical neurotransmitters. However, no consistent relationship between the expression of any individual *flp* and any classical transmitter was reported.

A gene expression fingerprint of *C. elegans* embryonic motor neurons revealed the expression of multiple *flp* genes (*flp-2, 4, 5, 8, 9, 12, 13, 14, 15, 16, 18* and *19*) using GFP fused to the UNC-4 transcription factor that encodes a paired-class homeodomain protein that is expressed in thirteen embryonic motor neurons.[63] Subsequent work used microarray-based methods to monitor gene expression in *C. elegans* neurons.[64] Findings revealed that 20 of the 23 *flp* genes examined displayed enriched expression in a larval pan-neural dataset; those not showing enrichment included *flp-14*, *flp-20* and *flp-23*. Furthermore, a subset of five *flp* genes (*flp-2, 4, 5, 12* and *13*) were enriched in the A-class cholinergic neuron subset, with *flp-13* being the most-highly enriched. These efforts offer a powerful approach to understanding the relationships between gene expression and neural function in *C. elegans* and have clearly emphasized the association between *flps* and motor function in nematodes.

In situ hybridization (ISH) has been employed to examine the expression of 5 *flp* genes in the larval (J2 stage) potato cyst nematode (PCN).[65] As with the pattern of expression observed in *C. elegans* using GFP reporter constructs, ISH indicated neuronal expression with variable patterns for each gene examined. Although the absence of a neuronal map for *G. pallida* and the inability to identify the neuronal axons using ISH made cell identification difficult, comparisons with the expression patterns for the homologous *flps* in *C. elegans* revealed positional differences in the cells expressing some genes. Although *Gp-flp-1* expression was confined to the retrovesicular ganglion (RVG; comprises cell bodies of interneurons and motorneurons situated just posterior to the ventral ganglion), *flp-1* expression was reported in the RVG as well as ventral cord interneurons, amphidial neurons and a pharyngeal motorneuron. *Gp-flp-6* staining was identified in phasmid-like (chemosensory) cells as well as in the lumbar ganglion of PCN J2s; diffuse *Gp-flp-6* staining occurred around the circumpharyngeal nerve ring and pharynx (metacorpal bulb). In contrast, *flp-6* expression in *C. elegans* was not reported in phasmids or the lumbar ganglion but was confined to amphidial cells and a pharyngeal interneuron. Similarly, differences were apparent in the expression data for *flp-12* and *Gp-flp-12*. The former occurred in a variety of ring neurons, interneurons and motor neurons associated with the head whereas the latter was identified in the RVG and neurons associated with the preanal area and lumbar ganglion. *Gp-flp-14* expression was detected in head motorneurons and nerve ring interneurons but the expression of *flp-14* has not been determined for *C. elegans*. Clearly, a snapshot of *flp* gene expression in PCN J2s suggests that *flp* expression in this plant parasite differs from that seen in *C. elegans*. While the benefits of work on *C. elegans* are unquestionable, these data highlight the importance of performing research on target parasite species.

Another approach to unraveling FLP signaling networks in nematodes has involved the application of mass spectrometric methods to *A. suum*.[54,66,67] This approach has relied on mass spectrometric identification interfaced with chemical derivatization of individual FLPs from neuronal structures including the circumpharyngeal nerve ring (CNR), ventral ganglion, RVG, dorsal ganglion, lateral line ganglia (LLG), ventro-dorsal commissures and segments of the dorsal and ventral nerve cords. These efforts mapped the expression of ~40 neuropeptides and revealed that there were similarities and differences in the FLP peptide complements of each neural structure. FLPs were the most abundant neuropeptides identified and, not surprisingly, the CNR expressed the most diverse range of FLPs (peptides with signatures common to products from the following *C. elegans* genes: *flp-1, 3, 4, 6, 8, 9, 11, 12, 13, 14, 16, 18* and *21*). All but one (*flp-3*) of the peptides identified in the CNR was also identified in the ventral ganglion; in addition, the ventral ganglion also expressed ILMRFamide (*As-flp-29*). Not surprisingly, the RVG appeared to express a less diverse complement of FLPs than the CNR and ventral ganglion with no signals being detected for *flp-6, 13, 21* or *29*. The dorsal ganglion expressed peptides matching those expressed on *flp-3, 4, 6, 13, 14* and *18* whereas the LLG expressed peptides predicted to be encoded on genes homologous to *flp-4, 6, 8, 11, 12, 13, 14, 16, 18* and *21*. Although the peptide complement of the dorsal cord appeared less complex than that of the ventral cord, both expressed peptides corresponding to those encoded on *flp-1, 4, 8, 11, 12, 13, 14, 18* and *21*; a mass corresponding to a *flp-16* product was only detected in the ventral cord. The ventro-dorsal commissures yielded little in the way of strong peptide signals although peaks matching *flp-9* and *flp-21* products were detected; the authors voiced caution as this tissue is likely to contain nonneural hypodermal tissue. It is evident from this work that many *flp* genes are expressed across the main neural processes in *A. suum*. Immunocytochemical methods which employed a monoclonal antibody have been used to investigate the distribution of the *As-flp-8* gene products (AF1; KNEFIRFamide) in *A. suum*.[29,57] A small subset of neurons in the head (including pharyngeal neurons) and neurons in the dorsal and ventral nerve cords were immunopositive. It is noteworthy that *flp-8* was not reported to be expressed in pharyngeal or nerve cord neurons in *C. elegans*.

The available literature provides data on *flp* expression that were derived using different techniques from only a few nematode species and reveals a rather complex and incomplete picture which indicates both similarities and differences in inter-species *flp* expression patterns. For example, the RVG of *A. suum* expresses peptides corresponding to predicted products of *flp-1, 3, 4, 8, 9, 11, 12, 14, 16* and *18*. In contrast, using GFP-reporter data, *flp-1, 2, 7, 10, 11, 13* and *21* were detected in the *C. elegans* RVG; the only matched expression being for genes *flp-1* and *flp-11*. The small amount of ISH data from *G. pallida* indicated the expression of *Gp-flp-1* and *Gp-flp-12* in the RVG, corresponding to the *A. suum* data. However, *Gp-flp-14* was not detected in the *G. pallida* RVG but was reported in the *A. suum* RVG. This difference could be due to the distinct life stages being compared as larval *C. elegans* have 16 RVG neural cells whereas adults have 20 and the data for *A. suum* was derived from adults whereas that for *G. pallida* was derived from J2 larvae. Clearly, too little is known about the cellular complement and organization of these structures to make unequivocal statements regarding *flp* expression.

A recent study on the *afp-6* gene (= *As-flp-11*) employed mass spectrometry, ICC and ISH to determine the expression of *As-flp-11* in *A. suum*.[54] Both ICC (using affinity purified antisera to both AMRNALVRFamide and NGAPQPFVRFamide) and ISH (using an *As-flp-11* specific riboprobe) localized expression to a single RIS-like cell in the ventral ganglion; in *C. elegans* RIS is a GABAergic interneuron which is known to express glutamate, dopamine and serotonin receptors.[68-70] This contrasts markedly with the expression reported for *flp-11* in *C. elegans* which was widespread and did not include the RIS neuron. The accumulation of data on *flp* gene expression across different nematodes is providing an extremely complex picture where highly diverse and species specific peptide expression patterns are superimposed upon an anatomically simple and structurally rigid nervous system.

Nematode Insulin-Like Peptides (INSs)

INSs play key roles in development and metabolism across the metazoa (for more details, see Chapter 5). Forty genes encoding peptides which belong to this family (*daf-28* and all the known *ins* genes) have been reported in the literature (see Table 2).[37,38,50,71-74] In vertebrates, insulin is composed of two polypeptide chains (A and B chains) which are linked by two disulfide bonds; an additional disulfide bond occurs within the A chain. Proinsulin, formed when the signal peptide is removed from preproinsulin, comprises the A and B chains and an interconnecting C peptide which is removed by endopeptidases during maturation. In contrast to the highly complex situation in *C. elegans*, only 10 insulin-related peptides are known from humans. Also, unlike the structural similarity of the human insulin-related peptides, there is much variation in the organization of *C. elegans* INS peptides. The *C. elegans* peptides possess A and B chains, but most commonly lack the intervening C peptide (although it is present in INS-1 and INS-18). Variations in the arrangement of the disulfide bonds have enabled the delineation of three distinct classes of *C. elegans* INS peptides, the α, β and γ (see Fig. 1). At time of writing, virtually nothing is known about INS peptides in parasitic nematodes. However, even a cursory glance at the EST datasets for parasitic nematodes reveals a bountiful supply of INSs and much scope for work in this area. Although their involvement in developmental processes, aging regulation and the control of dauer formation in *C. elegans* does not immediately strengthen their candidature as targets for parasite control, there is an obvious need to know more about their role in parasites and the potential of their signaling pathways as drug targets.

Nematode INS Distribution/Expression

Data relating to *ins* gene expression have been examined for 15 *ins* genes (*ins-1, 2, 3, 4, 5, 6, 7, 8, 9, 11, 18, 20, 21, 22, 23*) in *C. elegans* using upstream promoter regions fused to GFP coding regions;[73] currently there are no immunocytochemical or *in situ* hybridization data for *ins* genes or their products. Most *ins* genes examined were found to be expressed in a variety of neurons across numerous life stages (embryos, larvae and adult worms), although no expression was identified for *ins-20*. Excluding *ins-20*, expression was detected for all the *ins* genes examined in all larval stages (L1 to L4) and adults except *ins-2*, which was not detected in L4s and adults. All except *ins-9* were detected in the embryos. Although expression of most *ins* genes was particularly prevalent in sensory neurons (all were detected in some sensory neurons), especially the amphidial and labial neurons, thirteen *ins* genes (all except *ins-9*) were also expressed in other (nonsensory) neurons including dorsal and ventral nerve cord neurons, tail neurons and pharyngeal neurons. An elevated expression in *C. elegans* embryonic and larval stage motor neurons was also reported for some *ins* gene mRNAs using microarray profiling techniques.[63,64] For example, a larval pan-neural dataset revealed enhanced expression of *ins-1, 3, 17, 18, 21, 22, 23, 24, 26, 30* and *daf-28*.[64] Also, six *ins* genes are expressed in nonneuronal tissues, including the hypodermis (*ins-4*), intestine (*ins-1, ins-18*), pharynx (*ins-2*), vulva (*ins-2, ins-5, ins-8*) and vulval muscle (*ins-1*).[73] Unfortunately, there are no data on the structure and expression of INSs in parasitic nematodes. However, it would be interesting to compare INS expression between different nematodes to see if the complexities and differences seen with the FLPs are also apparent in this neuropeptide family.

Table 2. **Caenorhabditis elegans *prepro-insulin like precursor sequences***

Gene Name	Prepro Insulin-Like Protein Amino Acid Sequences*
daf-28	MNCKLIAIFAVLVLSVSAHLGAQAAAANFKAEGPLSRAVRVPGVAVRACGRRLVPYVWSVCGDACEPQEGIDIATQCCTYQCTAEYIQTACCPRLLL
ins-1	MYWFRQVYRPSFFFGFLAILLLSSPTPSDASIRLCGSRLTTTLLAVCRNQLCTGLTAFKRSADQSYAPTTRDLFHIHHQQKRGGIATECCEKRCSFAYLKTFCCNQDDN
ins-2	MNAIIFCLLFTTVTATYEVFGKGIEHRNEHLIINQLDIIPVESTPTPNRASRVQKRLCGRRLILFMLATCGECDTDSSEDLSHICCIKQCDVQDIIRVCCPNSFRK
ins-3	MKLSVVLALFIIFQLGAASLMRNWMFDFEKELEHDYDDSEIGFHNIHSLMARSRRGDKVKICGTKVLKMVMVMCGGECSSTNENIATECCEKMCTME DITTKCCPSR
ins-4	MFSFFTYFLLSALLLSASCRQPSMDTSKADRILREIEMETELENQLSRARRVPAGEVRACGRRLLLFVWSTCGEPCTPQEDMDIATVCCTTQCTPSYIKQACCPEK
ins-5	MHSIVALMLIGTILPIAALHQKHQGFILSSSDSTGNQPMDAISRADRHTNYRSCALRLIPHVWSVCGDAANHKTESMSLKNVAPLIAAPITSKKSAAHLTNWIPE
ins-6	MNSVFTIIFVLCALQVAASFRQSFGPSMSEESASMQLLRELQHNMMESAHRPMPRARRVPAPGETRACGRKLISLVMAVCGDLCNPQEGKDIATECCG NQCSDDYIRSACCP
ins-7	MPPIILVFFLVLIPASQQYPFSLESLNDQIINEEVIEYMLENSIRSSRTRRVPDEKKIYRCGRRIHSYVFAVCGKACESNTEVNIASKCCREECTDDFIRKQCCP
ins-8	MSPIILIFFLVFIPFSQQHTSLEESLNDRIISEEVVEMLSEKEIRPSRVRRVPEQKNKLCGKQVLSYVMALCEKACDSNTKVDIATKCCRDACSDEFIRHQCCP
ins-9	MIVTLIVFLVIGLQMAHLSQVSGNNENGFLNPFDLSQWSEEILHRQYHHHHHHHHGNRARRTLETEKIYRCGRKLYTDVLSACNGPCEPGTEQDLSK LCCGNQCTFVEIRKACCADKL
ins-10	MSLHFSTIQKTILLISFLLLVTLAPRTSAAFPFQICVKKMEKMCRIINPEQCAQVNKITEIGALTDCCTGLCSWEEIRISCCSVL
ins-11	MSSYRQTLFILIILIVIILFVNEGQGAPHHDKRHTACVLKIFKALNVMCNHEGDADVLRRTASDCCRESCSLTEMLASCTLTSSEESTRDI
ins-12	MQSNITASLFIALLIFGVISAAPSHEKTHKKCSDKLYLAMKSLCSYRGYSEFLRNSATKCCQDNCEISEMMALCVVAPNFDDDLLH
ins-13	MKLLHIFIIFLLFQSCSNKMCQYSKKKYKICGVRALKHMKVYCTRGMTRDYGKLLVTCCSKGCNAIDIQRICL
ins-14	MLTHLKFLLLVSLFINFAVSSEDIKCDAKFISRITKLCIHGITEDKLVRLLTRCCTSHCSKAHLKMFCTLKPHEEEPHHEI
ins-15	MKLLPLIVVFALLAVISESYSGNDFQPRDNKHHSYRSCGESLSRRVAFLCNGGAIQTEILRALDCCSTGCTDKQIFSWCDFRKLTRKEKQNYSGLIFRNLNRTQIVSFLLCHVE
ins-16	MQSLPILACLLTLSVFAPEIHGRELKRCSVKLFDILSVICGTESDAEILQKVAVKCCQEQCGFEEMCQHANLKIDKI
ins-17	MFSTRGVLLLLSLMAAVAAFGLFSRPAPITRDTIRPPRAKHGSLKLCPPGGASFLDAFNLICPMRRRRRSVSENYNDGGGSLLGRTMNMCCETGCEFTDIFAICNPFG
ins-18	MVHRLFIVLIAIILVAKSTAISLQQADGRMKMCPPGGSTFTMAWSMSCSMRRRKRDVGRYFEKRALIAPSIRQLQTICCQVGCNVEDLLAYCAPI
ins-19	MIFYLTTYLVTMSPLFLILLLLVSTTYPYIIDSSESYEVLMLFGYKRTCGRRLMNRINRVCVKDIDPADIDPKIKLSEHCCIKGCTDGWIKKHICSEEVLNF GFFEN
ins-20	MDKPSYLSSKEAWKMLNELLKEPKHHHHHHRHKGYCGVKAVKKLKQICPDLCSNVDDNLLMEMCSKNLTDDDILQRCCPE
ins-21	MKTYSFFVLFIVFIFFISSSKSHSKKHVRFLCATKAVKHIRKVCPDMCLTGEEVEVNEFCKMGYSDSQIKYICCPE

continued on next page

Table 2. Continued

Gene Name	Prepro Insulin-Like Protein Amino Acid Sequences*
ins-22	MHTTTILICFFIFLVQVSTMDAHTDKYVRTLCGKTAIRNIANLCPPKPEMKGICSTGEYPSITEYCSMGFSDSQIKFMCCDNQ
ins-23	MFVLLIILSIILAQVTDAHSELHVRRVCGTAIIKNIMRLCPGVPACENGEVPSPTEYCSMGYSDSQVKYLCCPTSQ
ins-24	MRSPTLFLLLLLVPLALCHVFSEPADLELKSYQALEKSLKEMGLIRANQGPQKACGRSMMMKVQKLCAGGCTIQNDDLTIKSCSTGYTDAGFISACC PSGFVF
ins-25	MLFKIIILFFLLLQLSEAKPEAQRRCGRYLIRFLGELCNGPCSGVSSVDIATIACATAVPIEDLKNMCCPNL
ins-26	MRALVAILCLMALCHAAMLDELEMQKEVQEFHHMNGMLQEFMNKGLIGNHHHGTKAGLTCGMNIIERVDKLCNGQCTRNYDALVIKSCHRGVS DMEFMVACCPTMKLFIH
ins-27	MKFFRLILLCALVLTTMAFLAPSTAAKRRCGRRLIPYVYSICGGPCENGDIIIEHCFSGTTPTIAEVQKACCPELSEDPTFSS
ins-28	MMRSFFVLLALLAIVTSTASPTCGRALLHRIQSVCGLCTIDAHHELIAIACSRGLGDKEIIEMCCPI
ins-29	MFCKFVFLIFLLISLSVATADFGAQRRCGRHLVNFLEGLCGGPCSEAPTVELASWACSSAVSIQDLEKLCCPSNLA
ins-30	MSSHALVLFLLLFLLPVALGHFLSKPAPDPRITFNRKLAETLKELQDMGLIQAPREPVVAAQGAKKTCGRSLLIKIQQLCHGICTVHADDLHETACMKGLTDSQLINSCCPPIPQTPFVF
ins-31	MKMPLILLLLVAAASAFVHHFDHSMFARPEKTCGGLLIRRVDRICPNLNYTYKIEWELMDNCCEVVCEDQWIKETFCRAPRFNFFGPSFKALERSCGPKLFTRVKTVCGEDINVDNKVKISDHCCTEGGCTDDW- IKENVCKQTRFNFFRQFLDSPQRSCGPQLFKRVNTLCNENINVENNVSVSKSCCESAAGCTDDWIKKNVCTQHKPFVFRPGFY
ins-32	MTSILLILLLVITVTGMFQELSDLQNLHRFLEGLQGSSSLAVKSRSRRELICGRRLSKTVTNLCVEMNPQKEEDIATKCCKNKGCSREYIKSIMCPDE
ins-33	MANTCLILLLLLVIFVTVGFSMPRIFRASENGVNSSDEVSEELSYSPEEAMDLVKQVIKVREQRRHRRHRHHGQKHCGTKIVRKLQMLCPKMCTISDDTLLTEMCSHSLFDDEIQLRCCPKEDE
ins-34	MLHHKTLIIALLLTLFISGIDSLPFRKHNNHRHLKNQKAQQLKEEATEAPTPAPTTTKAPSGSATTTTTVKTTAAPLAQVNPQCLRRLTLLARGVCRQPCQPSDKPKTSAQQLLQLACSARRPTNEIISYCCPEKSG
ins-35	MKQIFLVILAACLLAIILASPTGKHHKMDENAFGINNRHCQRALKVYSFAICGAICQNYEKILMEGCGSTVMLTMQRTKLICCPEPVDSDELFN
ins-36	MNIGKCSIIFLLFCVFGSILSRAIRKRHPEGKLVIRDCKRYLIMYSRTICKEKCEKFDERNDITFSINLQFIFTDLLVEGCHSNQTLSNERTRELCCPNAGSN
ins-37	MAAFLPIALSIAMLTVLTNANPIHPVPNAAFLPYRSCGSHLVHRAFEACSGKKDRSSDVDLWKMCCKDECTDLDIKESLCKYASQGYGVKFEEEAEEIDMVSFAAEGFKKSCGHDIVVKTVNVPTKSLKQCARTR- GWRRRPRRIPDKFNRVQNYACINCVKIK
ins-38	MNLFLLVCIAFAIITVTSFTPDEKSQRSHVFSYKKHCGRRIVSLVQACDRIDHDLSIDCCTQNCSSEFVKKIMCPSKL
ins-39	MNTFFFLAVLLVFCSAEQMTAKKFSTKTSSPIPELQEVFATVAADEFPFHKANTQPLAIYLNISTPQDCIHKIFRMTISFCSQVECQNMEAMQKICNTTTPTIKHVGELCCPEFFEQVKDDFVTLL

*Single letter amino acid annotations are used throughout. Underlined sequences are predicted peptide products (see Pierce et al 2001)[73].

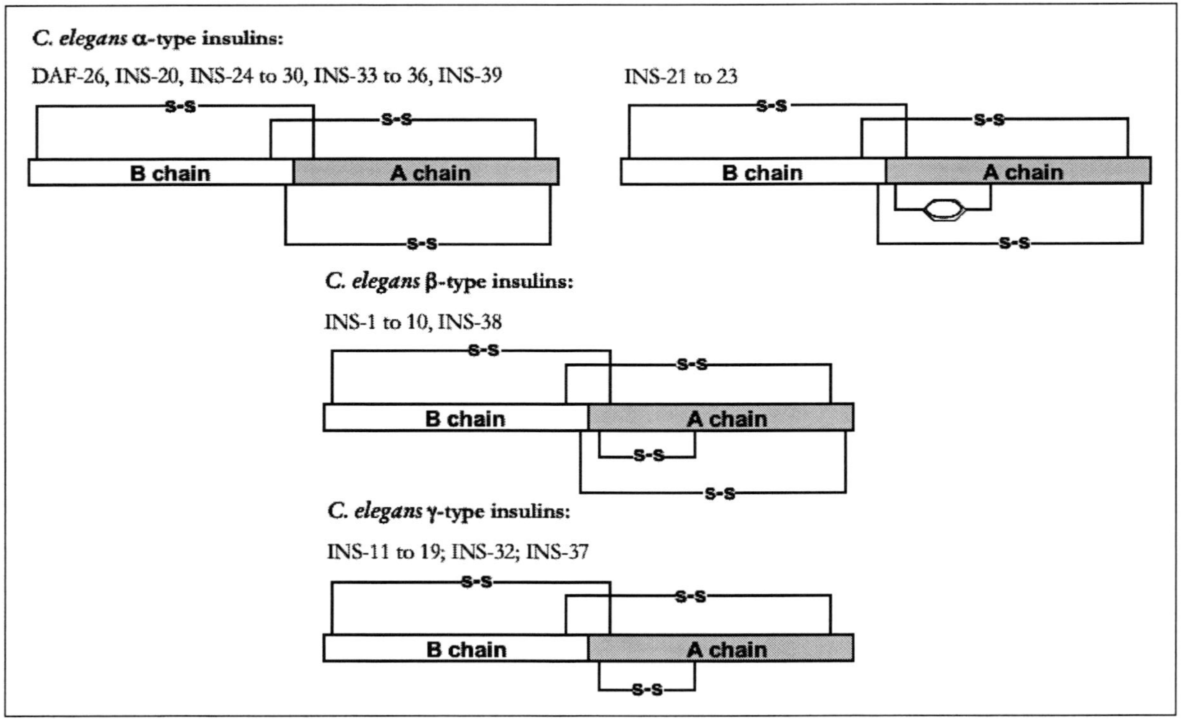

Figure 1. Schematic showing the predicted mature insulin-like peptide (INS) structures from *Caenorhabditis elegans*. The A-chain appears grey and the B-chain appears white. Three groups of INSs have been identified based on the predicted arrangement of the disulfide bonds. *C. elegans* γ-type INSs possess three disulfide bonds which is typical of vertebrate insulins. Both the α- and β-type INSs possess an additional inter-chain disulfide bond. In the α-type INSs the internal A-chain disulfide is either missing or has been replaced by aromatic amino acid interactions. Note that INS-31 has three repeats of each peptide chain and may constitute an additional class. Figure adapted from Duret et al[37] (1998), Pierce et al[73] (2001) and Husson et al[50] (2007).

Nematode Neuropeptide-Like Proteins (NLPs)

In contrast to the FLP and INS family groups, the NLP category has developed as a repository for all the neuropeptides in nematodes which do not qualify as INSs or FLPs. The consequence of this is that the NLPs comprise a wide range of peptide families that display quite disparate structures (and presumably functions). Regardless of this fact, the NLPs have been referred to as a peptide family when they actually represent an amalgamation of multiple peptide families who's only common feature is neuronal expression; note that some NLPs were designated as such and do not display neuronal expression (*nlp-23, 26, 29, 30, 31*) such that they may not be *bona fide* NLPs, unless neuronal expression is subsequently proven. Indeed, this has been further confirmed for peptides encoded by *nlp-27, 29* and *31* as they were identified as infection-inducible anti-microbial peptides, although the modulation of expression by exposure to bacterial or fungal pathogens was only established for *nlp-29* and *31*; this role was proposed for *nlp-27* encoded peptides based on their structural similarities to *nlp-31* encoded peptides.[75]

The NLPs were originally described as comprising 32 genes that were reported to encode neuropeptides belonging to 11 distinct families.[41,42] These included novel neuropeptide families that were not previously identified in other species as well as peptides that displayed structural similarities with neuropeptide families from other invertebrate phyla. The former included: peptides with DRV C-terminal signatures encoded by *nlp-4*; GGxYamides encoded by *nlp-10*; LxDxamides encoded by *nlp-11*; LQFamides encoded by *nlp-12*; MRxamides encoded by *nlp-17*; histidine rich peptides encoded by *nlp-16*; peptides with a nonterminal GxRLPN motif encoded by *nlp-19*; and, a range of glycine-rich, FG containing peptides with variable sequences encoded on *nlp-26*. The latter included: those with a GFxGF motif, described as orcokinin-like peptides (orcokinin is a myotropic neuropeptide from the crayfish, *Orconectes limosus*) and encoded by *nlp-3, 8, 14* and *15*; those with a FRPamide signature, designated myomodulin-like and encoded by *nlp-2, 22* and *23*; those with MSFamide signatures, designated buccalin-like (buccalin is a neuropeptide from the mollusk *Aplysia californica* that modulates acetylcholine-induced myoexcitation) and encoded by *nlp-1, 7* and *13*; those with a MGL/Famide signature, assigned allatostatin-like and encoded by *nlp-5* and *6*; those with a YGGWamide signature were identified as similar to the APGWamide from *A. californica* and were encoded by *nlp-24, 25, 27, 28, 29, 30, 31* and *32*; those with the N-terminal signature GGARAFamide were identified in other nematode ESTs and in predicted products from *nlp-9* and *21*; and, C-terminal FAFA signatures, identified in other nematode species and encoded by *nlp-18* and *20*.

The original NLP identification was based largely on genome searches that relied on pattern finding approaches that exploited the nature of invertebrate neuropeptide genes to encode multiple copies of similar peptides. However, the authors of this approach stated that singly encoded neuropeptides that did not display homology with other identified neuropeptides would not be identified using these methods.[42] It seems likely therefore that other NLPs await discovery and some evidence for this evolved from two dimensional nanoscale liquid chromatography and tandem mass spectrometry approaches.[53] These efforts identified the occurrence of 21 peptides predicted from known *nlp* genes and peptide products from an additional seven (*nlp-35* to *nlp-41*) 'probable' *nlp* genes.[53] In addition, other putative *nlp* genes (*nlp-33, 34* and *42*) have been annotated on WormBase (http://www.wormbase.org/db/gene/gene_class?name=nlp;class=Gene_class) (see Table 3). The genes *nlp-33* and *34* both encode peptides with YGGY sequences similar to those of *nlp-24, 25* and *27* to *32*. The peptides identified by tandem mass spectrometry and encoded by *nlp-35, 36, 37, 39, 40* and *41* all appear to encode novel peptide families. The peptide products of *nlp-38* include three putative peptides with LWamide C-termini, including two GLWamides and one SLWamide. Interestingly, all three of these peptides encompass a signature sequence (WxxxxxxWamide) that is reminiscent of the arthropod B-type allatostatins (also designated myoinhibiting peptides).

Many of the NLPs reported in *C. elegans* have also been predicted in the closely related *C. briggsae*.[53] The free-living bacterivore, *Pristionchus pacificus* was reported to have ESTs encoding FAFA and YGGWamide peptides.[42] Examination of EST data for parasitic nematodes revealed

Table 3. Caenorhabditis elegans *prepro-neuropeptide-like protein precursor sequences*

Gene Name	Prepro Neuropeptide-Like Protein Amino Acid Sequences*
nlp-1	MKATFVLACLLVIAAVSHADLLPKR**MDANAFRMSFG**KRSVSNPAEAKRMDPNAFRMSFGKRSAEQNEQANKEDKATSDKLYDDTKFEEMKR **MDANAFRMSFG**KRSDAHQAADDQVEYVNDDFSLPEQKRMDANAFRMSFGKR**VNLDPNSFRMSFG**KRSTVGYNLDARNYFVGLGRR
nlp-2	MRATLVLFALLCAVYSEAVPLQVYRPDESSAVDVVVLENSPELYDSEDEDEWKQEEEFTEGAMGKRSIALGRSGFRPGKRSMDNFHTVDVSDLIMKRSMAMGRLGLRPGKRSMAYGRQGFRPGKR- SMAYGRQGFRPGKRSMAYGRQGFRPGKRSNDMKEVFPQHVPEIYII
nlp-3	MSKIVACLVLLALSVMCVYSAPYEFRAKRAINPFLDSMGKRAVNPFLDSIGKRSFRPDMITEEKRYFDSLAGQSLGKRSNNRYEMLENYY
nlp-4	MPMDIFSEAQMFRYEDFMDRWTNEKGEFDFKAYVTTDAYMITPYVVIIHRCWESKEVPHRCDRKIVQPIPPLNVTVNSKTPDEKKVYNFGKVDL RFKNPAEVTSKMRSLILFVILLVAFAAARPVSEEVDRVKRDYDPRTEAPRRLPADDDEVDGEDRVKRDYDPRTDAPIRVPVDPEAEGEDRVKRDLSFSPNTEAPQPPPKHHPVPIESAN
nlp-5	MLMKIMVLMGIANIAASFVVSSVRFQSAPMRALIEINRELAKRSVSQLNQYAGFDTLGGMGLGKRSEPDQAGEKRALSTFDSLGGMGLGKRSSSSSRVFVYDKRALQHFSSLDTLGGMGFGRK
nlp-6	MLSFSRLAFVLLVSACVMAMA**APKQMVFGFG**KRSMADISDMPEDVPMKRYKPRSFAMGFGKRAAMRSFNMGFGKRSAEPQIDVDSF
nlp-7	MYIKAALLIVVLFGVASQITSA**LYLKQADFDDPRMFTSSFG**KRSAIESEPQAYPKSYRAIRIQRRSMDDLDDPRLMTMSFGKR**MILPSLADLHRYTMYD**KRGSDIDDPRYFLFSNRLTCRC
nlp-8	MSQKLLPISPLQLLFLQCLLIGFTAA**YPYLIFPASPSSGDSRRLV**KRAFDRFDNSGVFSFGAKRFDRYDDETAYGYGFDNHIFKRSADPYRFMSVPTKKAFDRMDNSDFFGAKRKR**SFDRMGGTEFGLM-** KRSAPESREQLINNLAESIITLRRAREAESSPESQRTIITYDD
nlp-9a	MDRFATRFIALLLVLLQIGSIFA**TPIAEAQGAPEDVDDRRELE**KRGGARAFYGFYNAGNSKRDQAAALPYYLYEKRGGGRAFNHNANLFRFDKRGGGRAFAGSWSPYLERFYDYKRSSYPVYFSDNSYY
nlp-9b	MDRFATRFIALLLVLLQIGSIFA**TPIAEAQGAPEDVDDRRELE**KRGGARAFYGFYNAGNSKRDQAAALPYYLYEKRGGGRAFNHNANLFRFDKRGGGRAFAGSWSPYLERDNSYY
nlp-10	MWYIALLLAVIATSVTAQKADDEPIVFLVRVPIDEMDDDSSLLESYYHPRDILSKRAIPFNGGMYGKRSTMPFSGGMYGKRSGQIFAQRRAAIPFSGGMYGKRSLVPQSYSNNENQIKRGAMPFSGGMYGR
nlp-11	MMSTLALVSLAIFGIAVVCAAPKPATVPVANEEDYLAALYGFEAPGSQFKGAPLQSKRHISPSYDVEIDAGNMRNLLDIGKRSAPMASDYGNQFQMY NRLIDAGKKKR**SPAISPAYQFENAFGLSEALERAG**RR
nlp-12	MLRHHSCALLMLILVFVEVFATQSPTFDRQDRDYRPLQFGKRDGYRPLQFGKRDYRPLQFGKRSSGSSGPVVLEPIWEWQ
nlp-13	MQRSLQIFCIMSAIAMAYSQGSRDDNQSAKRNDFSRDIMSFGKRSGNTADLYDRRIMAFGKRQPSYDRDIMSFGKR**SAPSDFSRDIMSFG**KR **SSSMYDRDIMSFG**KR**SPVDYDRPIMAFG-** KRAEDYERQIMAFGRRK
nlp-14	MLHLIVLLVALSSAVTAGRPRRALDGLDGSGFGFDKR**ALNSLDGAGFGFE**KRALNSLDGQGFGFEKRALDGLDGAGFGFDKR**ALNSLDGAGFGFE**KR ALDGLDGSGFGFDKR**ALNSLDGAGFGFE**KR**ALNSLDGAGFGFE**KRALDGLDGAGFGFDKR**ALNSLDGAGFGFE**KRALDGLDGAGFGFDKR ALNSLDGNGFGFDKRTFKHSSNKLRSVFRNLKGFKQH
nlp-15	MPSSSSSSSFFAAVLLVIVMMSTVESAAVRLRPVGSLFFLNRPHEKR**AFDSLAGSGFDNGFN**KRAFDSLAGSGFGAFNKRAFDSLAGSGFGAFNKRAFDSLAGSGFSGFDKRAFDSLAGQGFTGFEKR- AFDTVSTSGFDDFKL
nlp-16	MSFRFLILLLLASTLLVNSFPIGKDNESDESEVEVDTTTEAENIEVESQEASLDEQLNEVSEDPTREKRGLYSSERTEEEVEISHGMHHREKRHSEHLPHPDPPSHTAKRSTEHHRVKRSEGHPHEKK- ATHSPEGHIVAKDDHHGHEKRSDSHHGHQKRSVDEHHGHQKKNAEDHHEHQKRSEHVEHQAEMHEHQKRSTQEVSGHPEHHLVKRSGGEGGHRHHRSTDQGLDEDEPEDEIQTDENDEVTEEPGSRKRRNTDTPM- PSFPSDHDSEHSNSVAIRVKRVSRAGSSHKVRTLNKNRGNSKAGETTQNDSLTPNSGGVFNS

continued on next page

Table 3. *Continued*

Gene Name	Prepro Neuropeptide-Like Protein Amino Acid Sequences*
nlp-17	MFSKSILFCLLVLVFNVFGANFENDQDVMRPPFQALKR**GSLSNMMRIG**KRQMSRQQEYVQFPNEGVVPCESCNLGTLMRIGRR
nlp-18	MNANVYSIVYFLSFLVLCISAQLHADSGATEVDGIVDKR**SPYRAFAFA**KR**SDEENLDFLE**KRARYGFAKR**SPYRTFAFA**KRASPYGFAFAKRGQFSSFA
nlp-19	MLLRGVCLALLILVTIVQCQNDNDLKEEKRRIGLRLPNFLRFKDPDALMIHKRRIGLRLPNMLKFKDSSNMYHLEKRRMGMRLPNIIFLRNEKKNVLEY
nlp-20	MQVTLLALLLLIVPFFAFAASQYSDDDSELMSNNERFARDLELRKKFAFAFAKRSAGDADVVIEARSGPQAHEGAGMRFAFAKRRAPKEFARFARASFA
nlp-21a	MRNSLFTTLFFGLAALVMVLNA**QYTSELEEDE**KRGGARAMLHKRGGARAFSADVGDDYKRGGARAFYDEKR**GGARAFLTEM**KR**GGARVFQGFEDE** KRGGARAFMMDKRGGGRAFGDMMKRGGARAFVENS- KRDEDWVIRPFEDDRLEKR**GGGRSFPVKPGRLDD**
nlp-21b	MRNSLFTTLFFGLAALVMVLNAQYTSELEEDEKRGGARAMLHKRGGARAFSADVGDDYKRGGARAFYDEKRGGARAFLTEMKRGGARVFQGFEDE KRGGARAFMMDKRGGGRAFGDMMKRGGARAFVENS- KRDEDWVIRPFEDDRLGVF
nlp-22	MRSIIVFIGLTIFALDILLVQTSALGLQGGIDVFRGLGVVDQVDFNQILHRANYLRNTREGRLRYWRLRTLPIMKKSIAIGRAGFRPGKRTTDELTGFPIGV
nlp-23 X	MAAHLVIWMALLGVSAHALELGFYSEHPTSKDDSYESQLLDSNDLDQDDAVDLATPMIVIPNDEDEIYADEEPEPLTMEKRLYISRQGFRPAKRSMAIGRAGMRPGKRAFAAGWNRG- KRSMPFAESYYPLYRNDFSE
nlp-24	MNKLLLLFVLVAVIAMASAQWGGGPYGGYGPRGYGGGYGGGRYGGYGGRGPYGGYGGRGPYGYGGRGPYGGGGLVGALLG
nlp-25	MQSLIALLLLLVCFLHLAQCQWGGGWNNGGGYGNPYGGYGRGGGYGGGYGGGFGAQQAYNVQNAARIGTEVAEGVLVAEEVSEAIGK
nlp-26 X	MNFFVIFSLLLVGLVSAQFGFGGQQSFGGRGGQFGGMQRGGFNGNRGGFGQQSQFGGRGGNQFGGRGGSQFNGRGGNQFGGRGGFGFGKK
nlp-27 x	MISTSSLLILVVLLACFMAASAQWGYGGMPYGGYGGMGGYGMGGYGMGYRRRMWGSPYGGYGGYGGYGGWGK
nlp-28	MISTSSILILVFLLACFMATSAQWGYGGYGRGYGGYGGYGRGMYGGYGRGMYGGYGRGMYGGWGK
nlp-29 X	MISTSSILVLVVLLACFMAASAQWGYGGYGRGYGGYGGYGRGMYGGYGRGMYGGYGRGMYGGYGRGMYGGWGK
nlp-30 X	MISTSSILILVVLLACFMAASAQWGYGGYGRGYGGYGGYGRGYGGYGRGYGGYGRGMWGRPYGGYGWGK
nlp-31 X	MISTSSILVLVVLLACFMAANAQWGYGGYGRGYGGYGGYGRGYGGYGGYGRGYGGYGRGMYGGYGRPYGGYGWGK
nlp-32	MRQFNLLLVFCLIALTALPVFSFPNGLTMDSIDMEPMGAFDENGAADESPRVKRYGGWGGRGGWGRGGGRGYGGRGGGWGGRGGGWGRGGGG RGFYGGGRRGWGK
nlp-33	MISTSLLLVVLLFAILAIVDAQWGYGGPYGGYGGGYGGGPWGYGGGWRRRHWGGYGGGPWGGYGGGPWGGYYGK
nlp-34	MISAKLFFIVLLLIACLLAVAEAQWGYGYRPYGYGGYGGWRRPYYRPWGYRPYGYGWGK
nlp-35	MPRVSSFIVFLTFMVALLAVSHA**AVVSGYDNIYQVLAPRF**RRARMMSFDAEAPQYLQHLLQNLKPRFRRSV

continued on next page

Table 3. Continued

Gene Name	Prepro Neuropeptide-Like Protein Amino acid Sequences*
nlp-36	MSVDLKQQLELADYLGALAVWCIFFGVLFILSVIFNFVCIKKDDDVTALERWGYKKNIDMKLGPHRR**SMVARQIPQTVVADH**
nlp-37	MSSRISVSLLLLAVVATMFFTANVVDATPRSQGNMMRYGNSLPAYAPHVLYRFYNSRQFAPINKR**NNAEVVNHILKNFGALDRLGDVG**K
nlp-38	MQLIHFIVGLAMLISLSLAASDDRVLGWNKAHGLWGKRSVQEASQDKR**TPQNWNKLNSLWG**KRSASSFDDDYTTENGDDDVTMLYKR**SPAQWQ RANGLWG**R
nlp-39	MKLLILFSLFAIFFGVIALDSPIPEFYSTGRSTRVPSHRHRIPRLGKR**EVPNFQADNVPEAGGRV**RRYVGPPLKLVILI
nlp-40	MKLVILLSFVATVAVFA**APSAPAGLEEKL**RALQEQLYSLEKENGVDVKQKEQPAAADTFLGFVPQKRMVAWQPMKRSMINEDSRAPLLHAIEARLAEVLRAGERLGVNPEEVLADLRARNQFQ
nlp-41	MLGLVCKILVVLCLSVLCLTVSA**APGLFELPSRSV**RLIRSDPSAYDGYENSFYRGYGSDNQQFRFNSPQNW
nlp-42	MRVQVVTLLAVLLAVLQFTSAAGNYYSGYPSDRTMKRSALLQPENNPEWNQLGWAWGKRSAGMEIPHRAARALHPVKKNPDWQDLGFAWGRK

*Single letter amino acid annotations are used throughout. Underlined sequences are predicted peptide products (see Nathoo et al 2001)[42]; peptides in boldface were identified by mass spectrometry (Husson et al 2005)[53]; a or b after the gene name designate splice variants; the amino acid blocked in grey (*nlp-40*) indicates that peptides with and without the terminal R residues were identified; *X* indicates nonneuronal expression and therefore some doubt that these are neuropeptides; *x* indicates that although neuronal expression has been reported, this peptide has been predicted to be an anti-microbial peptide (Couillault et al 2004); C-terminal G residues represent known or putative amidation sites. Note that homologs of many of these NLPs were identified in *C. briggsae* (Husson et al 2005)[53] and ESTs encoding related peptides were identified in a range of parasitic nematodes (see Nathoo et al 2001)[42].

that peptides structurally similar to many of the NLPs were also present across the nematode clades:[42] *Ancylostoma caninum* ESTs encoded peptides from the FAFA, GFGX, GGxYamide, MSFamide, FRPamide, GGARAF, MGL/Famide and LxDxamide NLP families; *Brugia malayi* ESTs included FRPamide and YGGWamide peptides; *Globodera pallida/rostochiensis* ESTs encoded FAFA, GFGX and GGARAF peptides; *H. contortus* ESTs encoded MSFamide and MRXamide peptides; *Heterodera glycines* ESTs encoded FAFA, GFGFX, GGXYamide and MSFamide peptides; *Meloidogyne incognita/javanica* ESTs encoded GFGFX, GGXYamide, MSFamide, GGARAF, LQFamide and YGGWamides; *Onchocerca volvulus* and *Ostertagia ostertagia* ESTs encoded GFGFX peptides; *Strongyloides stercoralis* ESTs encoded FAFA, GFGFX and GGARAF peptides; *Toxocara canis* ESTs encoded FRPamide peptides.

The *afp-5* gene in *A. suum* encodes seven different peptides with C-terminal (D/S)R(D/N) F(M/L)(N/H/S)Famide signatures, but it is unclear if these are best annotated as FLPs or NLPs. Products from two other *A. suum nlps* have been identified by mass spectrometric methods.[67] These also include peptides identical to and structurally related to *C. elegans* NLP-12s (YRPLQFamides). *As-nlp-12* and *Trichostrongylus colubriformis* (*Tc-nlp-12*) transcripts were characterized and ESTs encoding NLP-12 peptides were identified from *Meloidogyne* spp., *Necator americanus, O. ostertagia* and *Wuchereria bancrofti*.[76] An additional and novel NLP signature (RWNamide) was predicted on two peptides, NRRRNAAARWNamide and NRRRNATARWNamide from an *A. suum* EST and a peptide corresponding to the former sequence was identified by mass spectrometry.[67] From the small amount of data available on NLPs from parasitic nematodes it is clear that they are highly divergent peptides that occur across Phylum Nematoda and, due to their dissimilarity to vertebrate peptides their signaling systems could make appealing drug targets. However, much work is needed to unravel the roles of these peptide signaling molecules.

Nematode NLP Distribution/Expression

As with *flp* and *ins* expression, the expression of *nlp-1* to *32* in *C. elegans* has been investigated using promoter sequences from *nlps* to drive GFP expression in transgenic animals.[42] Expression was not identified for *nlp-4, 17, 22, 25, 28* or *32* and with a small number of exceptions (*nlp-23, 26, 29, 30* and 31), most of the other *nlps* displayed neuronal expression. Those not expressed in neurons could have been wrongly designated as *nlps*. It is noteworthy that all of those not expressed in nerves are expressed to varying degrees in hypodermal tissue and two of these (*nlp-29* and *31*) are the aforementioned anti-microbial peptides. The other *nlps* were expressed in a wide variety of neurons including those of the head and tail, sensory neurons, circumpharyngeal nerve ring and associated ganglia, RVG, nerve cords, vulva and pharynx.[42] Just like the situation with *flps, nlps* appear to have distinct but overlapping distribution that presents a complex mosaic of expression across the nervous system of *C. elegans*. Remarkably, 9 different *nlp* genes are expressed in the ASI (amphidial) neurons (*nlp-1, 5, 6, 7, 9, 14, 18, 24* and *27*; note that *flp-10* and *21* are also expressed in these cells) revealing that individual neurons can possess a highly complex array of neuropeptide signaling molecules. The application of a microarray profiling technique to monitor elevated expression of mRNAs in embryonic motorneurons identified multiple *nlps*, including *nlp-3, 5, 7, 9, 10, 11, 15, 17, 18, 21, 28, 29, 30* and *31*.[63] A pan-neural dataset from larval *C. elegans* revealed enhanced expression of *nlp-1, 2, 3, 5, 6, 7, 8, 9, 10, 11, 12, 13, 14, 15, 17, 18, 20* and *21*.[64] As well as neuronal expression, multiple *nlps* are additionally expressed in the intestine (*nlp-1, 2, 3, 6, 8, 9, 13, 14, 15, 16, 18, 20, 21, 27* and *29*), spermatheca (*nlp-5, 13, 18, 19, 20, 24* and *27*), embryo (*nlp-9, 11, 21* and *31*), rectal gland (*nlp-18*) and vulval secretory cells (*nlp-2*).[42]

Reverse transcriptase (rt-)PCR indicated that *nlp-12* was expressed throughout the larval and adult stages of *C. elegans*; transcription was also identified in the L3 and adult stages of *T. colubriformis*. Curiously, ISH identified *Tc-nlp-12* expression in a single tail neuron, matching that reported for *C. elegans*.[76] However, rt-PCR indicated the expression of *As-nlp-12* in both head and tail tissue,[76] matching mass spectrometric data which identified masses identical to

NLP-12 predicted peptides in the nerve ring, RVG, ventral and dorsal nerve cords.[67] Furthermore, peptides matching the mass of the RWNamides were detected by mass spectrometry in the nerve ring, ventral ganglion and RVG of *A. suum*.[67] Therefore, it seems that the pattern of neuropeptide distribution/expression in nematodes differs across species and may reflect significant plasticity in the nervous systems of nematodes which could facilitate the various behaviors and life styles they adopt in spite of a rather structurally-simple nervous system. Clearly, the sophistication and huge success (in terms of biomass) displayed by nematodes may be, in part, due to their highly complex networks of peptide signaling molecules that appear to facilitate much diversity and, ultimately, subtlety in intercellular communication.

Platyhelminth Neuropeptides

Unlike the situation in nematodes, flatworms have nervous systems that commonly comprise several thousands of neurons and there appears to be no rigorous conformity in structure across the classes. Having said that, a bilobed anterior brain that supplies longitudinal cords which are connected by commissures to create a ladder like (=orthogonal) arrangement provides the most common structural theme to their central nervous system (reviewed by ref. 77). Disadvantaged by the unavailability of complete and readily accessible genomic information, our knowledge of flatworm neuropeptides is sparse, even though the first report of neuropeptide immunoreactivity in the nervous system of flatworms was in 1981 with a report on vertebrate neurohormone immunoreactivities in the free-living turbellarian, *Dugesia lugubris*.[78] Many subsequent studies have employed immunocytochemistry to map the distribution of immunoreactivities to a vast range of vertebrate and invertebrate neuropeptides in all four classes of flatworms (reviewed by refs. 45, 47, 49, 77, 79-81). This section will focus only on those peptides for which peptide or nucleotide sequence data are available. Taking this approach, two distinct neuropeptide families dominate current understanding of flatworm neuropeptides, the FLPs and NPF.

Platyhelminth Neuropeptide F (NPF)

In the late 1980s/early 1990s, indirect immunocytochemistry and confocal microscopy combined to record the widespread occurrence in flatworm parasites of immunostaining with antisera that were raised against the C-termini of vertebrate neuropeptide Y (NPY) family peptides.[10-13,82,83] Most commonly, these antisera had been raised against pancreatic polypeptide (PP) or peptide YY (PYY); less commonly, immunoreactivity was reported to NPY, a third member of this vertebrate peptide family. Within vertebrates, these three peptides were usually 36 amino acids long and had several structural features in common: a polyprolyl N-terminus; tyrosyl residues situated 10 and 17 amino acids from the C-terminus; and, an RXRYamide C-terminal signature. Note that relatively recent gene duplication events that have led to the rapid evolution of the duplicate genes have generated additional members of the NPY family that have quite distinct structural features and biological functions—these will not be considered further here.[84-86]

In the early 1990s, NPF from the cestode *Moniezia expansa* (mxNPF) became the first flatworm neuropeptide to be structurally characterized following acid ethanol extraction of whole worms and a series of chromatographic purification steps interfaced with radioimmunometric monitoring of the immunoreactive fraction.[30] This peptide comprised 39 amino acid residues and possessed a distinctive RPRFamide C-terminal signature and came to define a new class of peptides, designated NPF. Although this peptide was a little longer than any previously reported NPY superfamily member, it did display key assets that led the discoverers to designate this peptide as an invertebrate NPY. These features included the C-terminal arginyl residues and tyrosyl residues situated 10 and 17 amino acids from the C-terminus; these are invariant amongst vertebrate NPY superfamily peptides (Table 4). Further efforts structurally characterized a 36 amino acid NPF from the land planarian *Artioposthia triangulata* (subsequently redesignated *Arthurdendyus triangulatus*).[33] This peptide had an identical GRPRFamide C-terminus and the signature tyrosyl residues in identical relative positions.

One feature which was characteristic of vertebrate NPY superfamily peptides and yet absent from these worm NPFs was a polyprolyl N-terminus, leaving their relationship to the vertebrate peptides unclear. However, characterization of *Mx-npf*, the first known flatworm neuropeptide encoding gene, revealed another characteristic that indicated a relationship to NPY superfamily peptides. This was a Phase-2 intron within the penultimate R residue, a feature common to both *npy* superfamily genes in vertebrates and *Mx-npf*.[87] However, this trait was not observed in the *At-npf* gene or in the more recently characterized *S. mansoni* and *S. japonicum npf* genes.[88,89] Alignment of the prepropeptides for all known platyhelminth NPFs and a selection of other invertebrate NPFs and human NPY reveals the diversity in peptide sequence around the conserved C-terminal and fixed tyrosyl residues (see Table 4). Further evidence to support the relationship between helminth NPF and vertebrate NPYs was derived from examination of the solution structure of Mx-NPF using nuclear magnetic resonance. This work revealed a random structure for the N- (Pro^1 to Asn^{16}) and C- (Gly^{35} to Phe^{39}) termini either side of an alpha helix with a structure described as similar to that of porcine NPY.[90]

Platyhelminth NPF Distribution/Expression

Information on the distribution/expression of NPF in flatworms is restricted to immunocytochemical data generated using C-terminal or whole-molecule directed antisera and fluorescence or confocal scanning laser microscopy. However, it is highly likely that previous staining patterns obtained using antisera to vertebrate NPY superfamily members (pancreatic polypeptide, peptide YY and/or neuropeptide Y) were in fact due to their cross-reactivity with NPF. Studies in the early-mid 1990s began to employ specific NPF antisera to localize expression to the nervous systems of *M. expansa* and other platyhelminth parasites.[91-97] General observations included the fact that immunostaining was confined to neuronal elements and that expression was widespread in the nervous system with both central and peripheral nerves being immunopositive. The widespread distribution of NPF-immunoreactivity was evident in cestodes, monogeneans and trematodes with much staining localizing not only to the cerebral ganglia and associated nerve cords but also to innervation of muscular organs such as suckers and/or holdfasts, egg chambers (ootypes), uteri and pharynges. Comparative studies on the distribution of NPF, serotonin and acetylcholine expression indicated that peptidergic signaling systems more closely following those of the cholinergic systems and were distinct and clearly distinguishable from serotoninergic nerve pathways (for example, see reference 98). Further, immunogold labeling of immunogenic peptides for electron microscopic observation localized staining in dense-cored secretory vesicles in a number of different flatworm parasites and confirmed that these were associated with secretory pathways.[92,93,95,97,99,100] The broad range of immunocytochemical studies on NPF in flatworms confirms the abundance of this peptide in flatworm neuronal tissues, a situation similar to that seen for NPY in the vertebrate brain.

Platyhelminth FMRFamide-Like Peptides (FLPs)

FMRFamide-like peptides (FLPs) in flatworms conform to the most common FLP signature of a C-terminal tetrapeptide comprising an aromatic residue, a hydrophobic residue and an RFamide. The only published data on FLP sequences from flatworms have been derived from biochemical studies that employed ethanolic extraction and chromatographic purification procedures similar to those used for *M. expansa* NPF, except that FLP antisera were employed in radioimmunometric-based peptide monitoring. Again, the first success was seen with the large and widely available tapeworm, *M. expansa* with the structural characterization of the hexapeptide GNFFRFamide.[31] At this time, this peptide remains the only FLP that has been structurally characterized from a parasitic platyhelminth.

Three other FLPs have been structurally characterized from free-living turbellarians including: GYIRFamide (from *Bdelloura candida*, an ecto-commensal of the horseshoe crab; *Girardia tigrina*, a fresh water planarian; *Procerodes littoralis*, a marine planarian); YIRFamide (from *B. candida*); RYIRFamide (from the land planarian, *A. triangulatus*) (see Table 5) (R.N. Johnston,

Table 4. Neuropeptide F (NPF) prepropeptide precursor sequences. Selected sequelogs are shown for comparison

Peptide	Prepropeptide Neuropeptide F Amino Acid Sequences*
At-NPF	MSYLSFLFVSVIYLAIYLINFSSVTVDA**KVVHLRPRSSFSSEDEYQIYLRNVSKYIQLYGRPRFG**KRDLDRGQFEQYEK
Me-NPF	MDLNKEVMDEFSAYVKR**PDQDSIVNPSDLVLDNKAALRDYLRQINEYFAIIGRPRFG**
Sj-NPF	MHCHGYRQFSSKWNVKLMNSMINILLILLIRCV LFQSNIILCNAIELSSNSNLNGPDEQQR**AQALAKLMTLFYTSDAFNKYMENLDAYYMLRGRPRFG**KRNSYHRENNYDGVMQNHLLNSLLRQQLFRKFPTDNLDNDGDENFVI
Sm-NPF	MYSYMYGQYSFNWNVKSMNSMSILLILLLKFILFQ SNSIFINADELNLNTNNNLSNIPNDQQR**AQALAKLMSLFYTSDAFNKYMENLDAYYMLRGRPRFG**KRNYNPIKDNNDLMKNDLINNYLRQKLLQKYLIDNYNNDGDENFLRN
Ac-NPF	MQRVILVVLLLSCMAVLSVRADNSE**MLAPPPRPEEFTSAQQLRQYLAALNEYYSIMGRPRFG**KRGDSFRKREFFRTNGERYPEDAAAWTEFQ
Ls-NPF	MHKLLLVSLLVLSLAVMEVLCTEA**MLTPPQRPEEFKNPNELRKYLKALNEYYAIVGRPRFG**KRNGARVSDMFRPSGDDFGDYSANWGDF
Dm-NPF	MCQTMRCILVACVALALLAAGCRVEA**SNSRPPRKNDVNTMADAYKFLQDLDTYYGDRARVRFG**KRGSLMEILRNHEMDNINLGKNANNGGEFVSIE
Hs-NPY	MLGNKRLGLSGLTLALSLLVCLGALAEA**YPSKPDNPGEDAPAEDMARYYSALRHYINLITRQRYG**KRSSPETLISDLLMRESTENVPRTRLEDPAMW

*Single letter annotation for amino acids used throughout. Amino acids that are identical throughout the family members shown are highlighted in bold and common residues are underlined. At, *Arthurdendyus triangulatus* (turbellarian flatworm); Me, *Moniezia expansa* (cestode flatworm); Sj, *Schistosoma japonicum* (trematode flatworm); Sm, *Schistosoma mansoni*; Ac, *Aplysia californica* (mollusk); Ls, *Lymnaea stagnalis* (mollusk); Dm, *Drosophila melanogaster* (arthropod); Hs, *Homo sapiens*; NPY, neuropeptide Y. Note that all, except Me-NPF have dibasic cleavage sites (blocked in grey) following the glycyl amide donor at the C-terminus of the peptide (*M. expansa* possesses a stop codon immediately following the G).

Table 5. FMRFamide-like peptide (FLP) sequences from platyhelminths and a mollusk

Species (platyhelminth unless indicated)	Amino Acid Sequence*
Arthurdendyus triangulatus	R**YIRF**.NH_2
Bdelloura candida	G**YIRF**.NH_2
	YIRF.NH_2
Girardia tigrina	G**YIRF**.NH_2
Moniezia expansa	GN**FFRF**.NH_2
Procerodes littoralis	G**YIRF**.NH_2
Macrocallista nimbosa (mollusk)	**FMRF**.NH_2

*Single letter annotation for amino acids used throughout. The FLP signature comprising: aromatic amino acid, hydrophobic amino acid, arginine and phenylalaninamide, is shown in boldface text.

Queen's University Belfast, unpublished observations).[32,34,35] In all cases, only one FLP (and possibly two FLPs in the case of *B. candida*) has been identified in acid ethanol extractions of flatworms, a situation that contrasts markedly with the FLP diversity seen in nematodes. Although no trematode or monogenean FLPs have been structurally characterized, the chromatographic fractionation of acid ethanol extracts have only reported a single immunoreactive peak, consistent with the occurrence of one FLP or a small number of FLPs in these species.[34,95] More recent genomic and EST analyses have identified a raft of novel neuropeptide encoding transcripts, including some encoding FLPs, in *S. mansoni* and other free-living flatworms (P. McVeigh and G.R. Mair, Queen's University Belfast and T.A. Day, Iowa State University, unpublished data). It seems likely that our very limited knowledge of platyhelminth neuropeptides will be radically expanded soon.

Platyhelminth FLP Distribution/Expression

Many immunocytochemical studies document the widespread occurrence of FLP immunoreactivity in platyhelminths and provide a broad picture that is similar to that seen for NPF (reviewed by refs. 45,47,49,77). The first immunocytochemical study reporting FLP immunoreactivity in the nervous system of a parasite demonstrated anti-FMRFamide antisera cross-reaction with central and peripheral nerve elements in the tapeworm, *Diphyllobothrium dendriticum*.[7] Subsequent studies revealed that this observation was common to the other classes of flatworm parasite and confirmed a widespread distribution for this peptide family.[9,13] The data generated on FLP distribution displayed the same key features as that for NPF immunoreactivity in that both central and peripheral nerve elements were immunopositive. As with NPF, FLP distribution patterns are similar to those of acetylcholine but distinct from those for serotonin (for example see refs. 101-107). Furthermore, ultrastructural studies have localized immunoreactivity to dense-cored secretory vesicles in nerves of both the central and peripheral nervous systems of parasitic flatworms (reviewed by refs. 45,47,49,77,79-81,108). Of particular note, is the widespread abundance of FLP-immunoreactivity in the innervation of muscular structures associated with the body wall, attachment organs, reproductive systems and feeding organs such as the pharynx. One particularly interesting observation made using immunocytochemistry was the link between the expression of FLPs in the reproductive system of the monogenean parasite *Polystoma nearcticum* and spawning of the grey treefrog host, *Hyla versicolor*.[109] This work revealed a role for FLPs in the reproductive synchrony between the parasite and its host. All these observations have pointed to a role for flatworm FLPs in muscle modulation, hypotheses which were subsequently confirmed experimentally (see Chapter 5 in this book).

Targets for Parasite Control

Currently, much more is known about parasite neuropeptides than either the receptors they act upon or the associated signaling pathways (see Chapter 5 in this book). Indeed, only a small number of neuropeptides have been receptor-matched in *C. elegans* (see Chapter 5 in this book) and no neuropeptide receptors have been functionally characterized from a parasitic worm. Regardless, the peptides themselves are useful as tools to initiate receptor discovery efforts and also to validate signaling pathways as targets. The application of gene silencing through RNA interference (RNAi) as a mechanism to validate potential targets in parasitic worms has been widely discussed.[110-113] However, complications in the utility of RNAi abound from the lack of genomic data and optimized RNAi protocols for the vast majority of parasites, although the *S. mansoni* and *B. malayi* genome projects and ongoing efforts to address RNAi optimization are beginning to counter these weaknesses.[114-116] Further, the application of RNAi to parasitic nematodes and, indeed, to neuronal targets in *C. elegans*, has had mixed and moderate successes, respectively (for example see refs. 117-122).

Even where RNAi has proven successful, within the confines of target validation studies it can only really inform on the potential utility of antagonistic drugs and does not speak to the potential of agonists that act on the same pathways. Nevertheless, even before receptors and signaling pathways are determined and characterized, the application of RNAi for each neuropeptide-encoding gene could offer a rational approach to the selection of neuropeptide signaling pathways for further study. For example, null phenotypes associated with the silencing of a neuropeptide gene would effectively rule out that pathway from a target discovery program. In contrast, a lethal or incapacitating phenotype associated with silencing a neuropeptide gene would be extremely appealing; at time of writing a lethal phenotype associated with neuropeptide silencing has not been reported. However, the silencing of five different *flp* genes in *G. pallida* each generated worms with profound aberrant phenotypes which appear incompatible with survival in the host and which indicate the importance of neuropeptide signaling pathways to parasite behavior.[123] So even though only five neuropeptide genes have been examined in this way, already several offer appeal as targets for the control of plant parasitic nematodes. Clearly, interrogating the entire peptide complement in this way could provide a large set of validated target systems. Although there appear to be numerous hurdles to the application of RNAi in some animal parasites,[121] the success seen in the application of RNAi to interrogate gene function in plant parasitic nematodes (reviewed by refs. 120,124) and the potential for the application of plant-based RNAi strategies for parasite control are very promising.[125-128] Time will tell if the translation of these successes to animal and human parasites is a real possibility.

Conclusion

This chapter has focused on neuropeptide signaling molecules that play a major role in worm neurobiology and beyond. Due to the success of anthelmintics that have acted to compromise normal motor function in helminth parasites, peptides with associated roles have much appeal as conduits to target systems for parasite control. Although the peptides themselves do not provide useful chemotherapeutic targets, the various enzymes that contribute to generation of the mature peptide products, the receptors (ion channels or G-protein coupled receptors) they interact with, the enzymes that break down the peptides after signal initiation and components of the signaling pathways they trigger, all provide potential targets for parasite control (see Fig. 2).

Unfortunately, at this time, our knowledge of these facets of peptide signaling networks in parasitic worms is at best rudimentary, although data on neuropeptide degradation is starting to accumulate for *C. elegans*,[44,129,130] and an amidating enzyme has been characterized from *S. mansoni*.[131] At least some of these will have potential in mechanism-based drug discovery programs. The recent successes in the application of RNAi-based control measures for plant parasitic

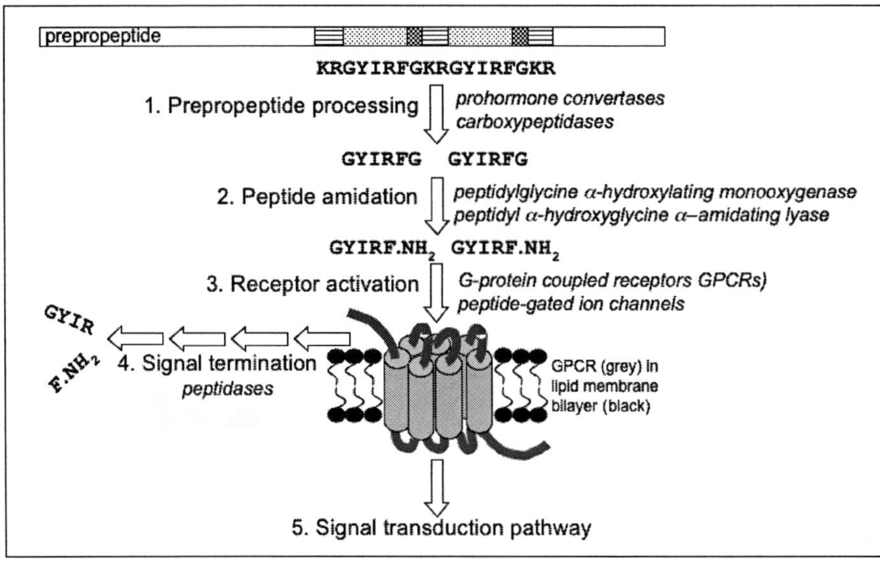

Figure 2. Within neuropeptide signaling systems in parasitic worms there are five obvious groups of targets that could be exploited for parasite control. Here, these are shown for the hypothetical GYIRFamide encoding gene. 1) The enzymes associated with preproprotein processing (prohormone convertases and carboxypeptidases) could provide targets that would compromise multiple neuropeptide signaling pathways by stopping the generation of mature peptide products. 2) A similar scenario would apply to the amidation process whereby peptidylglycine α-hydroxylating monooxygenase (PHM) and peptidyl α-hydroxyglycine α-amidating lyase (PAL) act sequentially to generate C-terminal amide moieties from glycyl residues. Most neuropeptides display C-terminal amidation and this is commonly essential to receptor binding and/or activation. Again, compromising the amidation process would prevent the generation of amidated peptide products and thus would broadly disrupt peptide signaling processes. 3) Receptors often receive the most attention with respect to drug target exploitation, mainly because high throughput screens involving receptors are well established. Critically, data on neuropeptide receptors in *C. elegans* and some parasites are now beginning to accumulate and appear to offer real potential for receptor exploitation. 4) Very little is known about neuropeptide signal termination in parasitic worms, but compromising the signal termination processes would have obvious merits in the disruption of peptide signaling pathways. Again, if the peptidases involved are widely conserved across peptide signaling networks then there is the potential for broad-scale disruption of these signaling pathways. 5) Least is known about the signal transduction pathways associated with neuropeptide action, but often these encompass various enzymatic steps which provide multiple opportunities for disruption of the associated signaling processes. All of these target groups may also be amenable to gene silencing through RNA interference (RNAi) which could provide a valuable tool for the validation of these targets in parasites. If RNAi approaches become feasible for parasite control (and they show promise for the control of plant parasitic nematodes) then the genes encoding neuropeptides could be added to the above-mentioned list of target groups.

nematodes underscore the potential of this approach and could provide a real opportunity to exploit neuropeptides for parasite control.

Note Added in Proof

A recent bioinformatic trawl of the available flatworm EST and genomic datasets identified ~60 distinct neuropeptide precursors encompassing 96 neuropeptides from 10 species of flatworm.[132] Although some of these peptides belong to peptide families previously recognised in flatworms (FLP- and NPF-like families) or other animal phyla (myomodulin-, buccalin- and neuropeptide FF (NPFF)-like peptide families) most are novel and, therefore, flatworm-specific. This is significant as ligand uniqueness underscores the potential for the cognate receptors to provide drug targets that are easily discriminated between parasite and host. For more details, readers are directed to the original manuscript available at http://www.sciencedirect.com/science?_ob=MImg&_imagekey=B6T7F-4W1BV9T-1-D&_cdi=5057&_user=126523&_orig=search&_coverDate=09%2F30%2F2009&_sk=999609988&view=c&wchp=dGLbVlz-zSkWA&md5=53005d80cea0568efb49434fe9ae1c8c&ie=/sdarticle.pdf.

Acknowledgements

The authors wish to acknowledge the support of the National Institutes of Health USA and the Department of Agriculture and Rural Development for Northern Ireland.

References

1. White JG, Southgate E, Thompson JN et al. The structure of the nervious system of the nematode Caenorhabditis elegans. Philos Trans R Soc Lond B Biol Sci 1986; 314(1165):1-340.
2. Davey KG. Neurosecretion and molting in some parasitic nematodes. Am Zool 1966; 6(2):243-9.
3. Rogers WP. Neurosecretory granules in the infective stage of Haemonchus contortus. Parasitology 1968; 58(3):657-62.
4. Leach L, Trudgill DL, Gahan PB. Immunocytochemical localization of neurosecretory amines and peptides in the free-living nematode, Goodeyus ulmi. Histochem J 1987; 19(9):471-5.
5. Atkinson HJ, Isaac IE, Harris PD et al. FMRFamide-like immunoreactivity within the nervous-system of the nematodes Panagrellus redivius, Caenorhabditis elegans and Heterodera glycines. J Zool 1988; 216:663-71.
6. Davenport TR, Lee DL, Isaac RE. Immunocytochemical demonstration of a neuropeptide in Ascaris suum (Nematoda) using an antiserum to FMRFamide. Parasitology 1988; 97(1):81-8.
7. Gustafsson MK, Wikgren MC, Karhi TJ et al. Immunocytochemical demonstration of neuropeptides and serotonin in the tapeworm Diphyllobothrium dendriticum. Cell Tissue Res 1985; 240(2):255-60.
8. Gustaffson MK, Lehtonen MA, Sundler F. Immunocytochemical evidence for the presence of "mammalian" neurohormonal peptides in neurones of the tapeworm Diphyllobothrium dendriticum. Cell Tissue Res 1986; 243(1):41-9.
9. Gustafsson MK. Immunocytochemical demonstration of neuropeptides and serotonin in the nervous systems of adult Schistosoma mansoni. Parasitol Res 1987; 74(2):168-74.
10. Fairweather I, MaCartney GA, Johnston CF et al. Immunocytochemical demonstration of 5-hydroxytryptamine (serotonin) and vertebrate neuropeptides in the nervous system of excysted cysticercoid larvae of the rat tapeworm, Hymenolepis diminuta (Cestoda, Cyclophyllidea). Parasitol Res 1988; 74(4):371-9.
11. Magee RM, Fairweather I, Johnston CF et al. Immunocytochemical demonstration of neuropeptides in the nervous system of the liver fluke, Fasciola hepatica (Trematoda, Digenea). Parasitology 1989; 98(2):227-38.
12. Maule AG, Shaw C, Halton DW et al. Localization, quantification and characterization of pancreatic polypeptide immunoreactivity in the parasitic flatworm Diclidophora merlangi and its fish host (Merlangius merlangus). Gen Comp Endocrinol 1989a; 74(1):50-6.
13. Maule AG, Halton DW, Johnston CF et al. Immunocytochemical demonstration of neuropeptides in the fish-gill parasite, Diclidophora merlangi (Monogenoidea). Int J Parasitol 1989b; 19(3):307-16.
14. Cowden C, Stretton AO, Davis RE. AF1, a sequenced bioactive neuropeptide isolated from the nematode Ascaris suum. Neuron 1989; 2(5):1465-73.
15. Geary TG, Price DA, Bowman JW et al. Two FMRFamide-like peptides from the free-living nematode Panagrellus redivivus. Peptides 1992; 13(2):209-14.
16. Cowden C, Stretton AO. AF2, an Ascaris neuropeptide: isolation, sequence and bioactivity. Peptides 1993; 14(3):423-30.
17. Cowden C, Stretton AO. Eight novel FMRFamide-like neuropeptides isolated from the nematode Ascaris suum. Peptides 1995; 16(3):491-500.
18. Rosoff ML, Doble KE, Price DA et al. The flp-1 propeptide is processed into multiple, highly similar FMRFamide-like peptides in Caenorhabditis elegans. Peptides 1993; 14(2):331-8.
19. Maule AG, Shaw C, Bowman JW et al. KSAYMRFamide: a novel FMRFamide-related heptapeptide from the free-living nematode, Panagrellus redivivus, which is myoactive in the parasitic nematode, Ascaris suum. Biochem Biophys Res Commun 1994a; 200(2):973-80.

20. Maule AG, Shaw C, Bowman JW et al. The FMRFamide-like neuropeptide AF2 (Ascaris suum) is present in the free-living nematode, Panagrellus redivivus (Nematoda, Rhabditida). Parasitology 1994b; 109(3):351-6.
21. Maule AG, Shaw C, Bowman JW et al. Isolation and preliminary biological characterization of KPNFIRFamide, a novel FMRFamide-related peptide from the free-living nematode, Panagrellus redivivus. Peptides 1995; 16(1):87-93.
22. Keating CD, Holden-Dye L, Thorndyke MC et al. The FMRFamide-like neuropeptide AF2 is present in the parasitic nematode Haemonchus contortus. Parasitology 1995; 111(4):515-21.
23. Marks NJ, Shaw C, Maule AG et al. Isolation of AF2 (KHEYLRFamide) from Caenorhabditis elegans: evidence for the presence of more than one FMRFamide-related peptide-encoding gene. Biochem Biophys Res Commun 1995a; 217(3):845-51.
24. Marks NJ, Maule AG, Geary TG et al. APEASPFIRFamide, a novel FMRFamide-related decapeptide from Caenorhabditis elegans: structure and myoactivity. Biochem Biophys Res Commun 1997; 231(3):591-5.
25. Marks NJ, Maule AG, Geary TG et al. KSAYMRFamide (PF3/AF8) is present in the free-living nematode, Caenorhabditis elegans. Biochem Biophys Res Commun 1998; 248(2):422-5.
26. Marks NJ, Maule AG, Li C et al. Isolation, pharmacology and gene organization of KPSFVRFamide: a neuropeptide from Caenorhabditis elegans. Biochem Biophys Res Commun 1999a; 254(1):222-30.
27. Marks NJ, Sangster NC, Maule AG et al. Structural characterisation and pharmacology of KHEYLRFamide (AF2) and KSAYMRFamide (PF3/AF8) from Haemonchus contortus. Mol Biochem Parasitol 1999b; 100(2):185-94.
28. Marks NJ, Shaw C, Halton DW et al. Isolation and preliminary biological assessment of AADGAPLIRFamide and SVPGVLRFamide from Caenorhabditis elegans. Biochem Biophys Res Commun 2001; 286(5):1170-6.
29. Davis RE, Stretton AO. The motornervous system of Ascaris: electrophysiology and anatomy of the neurons and their control by neuromodulators. Parasitology 1996; 113:S97-117.
30. Maule AG, Shaw C, Halton DW et al. Neuropeptide F: a novel parasitic flatworm regulatory peptide from Moniezia expansa (Cestoda: Cyclophyllidea). Parasitol 1991; 102:309-16.
31. Maule AG, Shaw C, Halton D et al. GNFFRFamide: a novel FMRFamide-immunoreactive peptide isolated from the sheep tapeworm, Moniezia expansa. Biochem Biophys Res Commun 1993a; 193(3):1054-60.
32. Maule AG, Shaw C, Halton DW et al. RYIRFamide: a turbellarian FMRFamide-related peptide (FaRP). Regul Pept 1994c; 50(1):37-43.
33. Curry WJ, Shaw C, Johnston CF et al. Neuropeptide F: primary structure from the tubellarian, Artioposthia triangulata. Comp Biochem Physiol C 1992; 101(2):269-74.
34. Johnston RN, Shaw C, Halton DW et al. GYIRFamide: a novel FMRFamide-related peptide (FaRP) from the triclad turbellarian, Dugesia tigrina. Biochem Biophys Res Commun 1995; 209(2):689-97.
35. Johnston RN, Shaw C, Halton DW et al. Isolation, localization and bioactivity of the FMRFamide-related neuropeptides GYIRFamide and YIRFamide from the marine turbellarian Bdelloura candida. J Neurochem 1996; 67(2):814-21.
36. Rosoff ML, Burglin TR, Li C. Alternatively spliced transcripts of the flp-1 gene encode distinct FMRFamide-like peptides in Caenorhabditis elegans. J Neurosci 1992; 12(6):2356-61.
37. Duret L, Guex N, Peitsch MC et al. New insulin-like proteins with atypical disulfide bond pattern characterized in Caenorhabditis elegans by comparative sequence analysis and homology modeling. Genome Res 1998; 8(4):348-53.
38. Gregoire FM, Chomiki N, Kachinskas D et al. Cloning and developmental regulation of a novel member of the insulin-like gene family in Caenorhabditis elegans. Biochem Biophys Res Commun 1998; 249(2):385-90.
39. Nelson LS, Kim K, Memmott JE et al. FMRFamide-related gene family in the nematode, Caenorhabditis elegans. Mol Brain Res 1998; 58(1-2):103-11.
40. Li C, Kim K, Nelson LS. FMRFamide-related neuropeptide gene family in Caenorhabditis elegans. Brain Res 1999a; 848:26-34.
41. Li C, Nelson LS, Kim K et al. Neuropeptide gene families in the nematode Caenorhabditis elegans. Ann N Y Acad Sci 1999b; 897:239-52.
42. Nathoo AN, Moeller RA, Westlund BA et al. Identification of neuropeptide-like protein gene families in Caenorhabditis elegans and other species. Proc Natl Acad Sci USA 2001; 98:14000-5.
43. McVeigh P, Leech S, Mair GR et al. Analysis of FMRFamide-like peptide (FLP) diversity in phylum Nematoda. Int J Parasitol 2005a; 35(10):1043-60.
44. McVeigh P, Geary TG, Marks NJ et al. The FLP-side of nematodes. Trends Parasitol 2006a; 22(8):385-96.
45. Maule AG, Mousley A, Marks NJ et al. Neuropeptide signaling systems—potential drug targets for parasite and pest control. Curr Top Med Chem 2002; 2(7):733-58.
46. Mousley A, Marks NJ, Maule AG. Neuropeptide signalling: a repository of targets for novel endectocides? Trends Parasitol 2004; 20(10):482-7.

47. Gustaffson MK, Halton DW, Kreshchenko ND et al. Neuropeptides in flatworms. Peptides 2002; 23(11):2053-61.
48. Li C. The ever-expanding neuropeptide gene families in the nematode Caenorhabditis elegans. Parasitol 2005; 131:S109-S27.
49. McVeigh P, Kimber MJ, Novozhilova et al. Neuropeptide signalling systems in flatworms. Parasitology 2005b; 131:S41-S55.
50. Husson SJ, Mertens I, Janssen T et al. Neuropeptidergic signaling in the nematode Caenorhabditis elegans. Prog Neurobiol 2007; 82(1):33-55.
51. Kim K, Li C. Expression and regulation of an FMRFamide-related neuropeptide gene family in Caenorhabditis elegans. J Comp Neurol 2004; 475:540-50.
52. Nelson LS, Rosoff ML, Li C. Disruption of a neuropeptide gene, flp-1, causes multiple behavioral defects in Caenorhabditis elegans. Science 1998; 281:1686-90.
53. Husson SJ, Clynen E, Baggerman G et al. Discovering neuropeptides in Caenorhabditis elegans by two dimensional liquid chromatography and mass spectrometry. Biochem Biophys Res Commun 2005; 335(1):76-86.
54. Yew JY, Davis R, Dikler S et al. Peptide products of the afp-6 gene of the nematode Ascaris suum have different biological actions. J Comp Neurol 2007; 502:872-82.
55. Sithigorngul P, Stretton AO, Cowden C. Neuropeptide diversity in Ascaris: an immunocytochemical study. J Comp Neurol 1990; 294(3):362-76.
56. Schinkmann K, Li C. Localization of FMRFamide-like peptides in Caenorhabditis elegans. J Comp Neurol 1992; 316(2):251-60.
57. Cowden C, Sithigorngul P, Brackley P et al. Localization and differential expression of FMRFamide-like immunoreactivity in the nematode Ascaris suum. J Comp Neurol 1993; 333(3):455-68.
58. Brownlee DJ, Fairweather I, Johnston CF. Immunocytochemical demonstration of neuropeptides in the peripheral nervous system of the roundworm Ascaris suum (Nematoda, Ascaroidea). Parasitol Res 1993; 79(4):302-8.
59. Brownlee DJ, Fairweather I, Johnston CF et al. Immunocytochemical demonstration of neuropeptides in the central nervous system of the roundworm, Ascaris suum (Nematoda: Ascaroidea). Parasitology 1993; 106(3):305-16.
60. Brownlee DJ, Brennan GP, Halton DW et al. Ultrastructural localization of pancreatic polypeptide- and FMRFamide immunoreactivities within the central nervous system of the nematode, Ascaris suum (Nematoda: Ascaroidea). Parasitology 1994; 108(5):587-93.
61. Brownlee DJ, Fairweather I, Johnston CF et al. Immunocytochemical demonstration of peptidergic and serotoninergic components in the enteric nervous system of the roundworm, Ascaris suum (Nematoda, Ascaroidea). Parasitology 1994; 108(1):89-103.
62. Rogers C, Reale V, Kim K et al. Inhibition of Caenorhabditis elegans social feeding by FMRFamide-related peptide activation of NPR-1. Nat Neurosci 2003; 6:1178-85.
63. Fox RM, Von Stetina SE, Barlow SJ et al. A gene expression fingerprint of C. elegans embryonic motor neurons. BMC Genomics 2005; 6(1):42-65.
64. Von Stetina SE, Watson JD, Fox RM et al. Cell-specific microarray profiling experiments reveal a comprehensive picture of gene expression in the C. elegans nervous system. Genome Biol 2007; 8(7):R135 [Epub ahead of print].
65. Kimber MJ, Fleming CC, Prior A et al. Localisation of Globodera pallida FMRFamide-related peptide encoding genes using in situ hybridization. Int J Parasitol 2002; 32:1095-105.
66. Yew JY, Dikler S, Stretton AO. De novo sequencing of novel neuropeptides directly from Ascaris suum tissue using matrix-assisted laser desorption/ionization time-of-flight/time-of-flight. Rapid Commun Mass Spectrom 2003; 17(24):2693-8.
67. Yew JY, Kimberly KK, Dikler S et al. Mass spectrometric map of neuropeptide expression in Ascaris suum. J Comp Neurol 2005; 488:396-413.
68. Eastman C, Horvitz HR, Yin Y. Coordinated transcriptional regulation of the unc-25 glutamic acid decarboxylase and the unc-47 GABA vesicular transporter by the Caenorhabditis elegans UNC-30 homeodomain protein. J Neurosci 1999; 19(15):6225-34.
69. Hart AC, Sims S, Kaplan JM. Synaptic code for sensory modalities revealed by C. elegans GLR-1 glutamate receptor. Nature 1995; 378:82-5.
70. Tsalik EL, Niacaris T, Wenick AS et al. LIM homeobox gene-dependent expression of biogenic amine receptors in restricted regions of the C. elegans nervous system. Dev Biol 2003; 263(1):81-102.
71. Wilson R, Ainscough R, Anderson K et al 2.2 Mb of contiguous nucleotide sequence from chromosome III of C. elegans. Nature 1994; 368:32-8.
72. Kawano T, Ito Y, Ishiguro M et al. Molecular cloning and characterization of a new insulin/IGF-like peptide of the nematode Caenorhabditis elegans. Biochem Biophys Res Commun 2000; 273:431-6.

73. Pierce SB, Costa M, Wisotzkey R et al. Regulation of DAF-2 receptor signaling by human insulin and ins-1, a member of the unusually large and diverse C. elegans insulin gene family. Genes Dev 2001; 15:672-86.
74. Li W, Kennedy SG, Ruvkun R. daf-28 encodes a C. elegans insulin superfamily member that is regulated by environmental cues and acts in the DAF-2 signaling pathway. Genes Dev 2003; 17(7):844-58.
75. Couillault C, Pujol N, Reboul J et al. TLR-independent control of innate immunity in Caenorhabditis elegans by the TIR domain adaptor protein TIR-1, an ortholog of human SARM. Nat Immunol 2004; 5(5):488-94.
76. McVeigh P, Leech S, Marks NJ et al. Gene expression and pharmacology of nematode NLP-12 neuropeptides. Int J Parasitol 2006b; 36(6):633-40.
77. Halton DW, Maule AG. Flatworm nerve and muscle: structural and functional analysis. Canadian J Zool 2004; 82:316-33.
78. Schilt J, Richoux JP, Dubois MP. Demonstration of peptides immunologically related to vertebrate neurohormones in Dugesia lugubris (Turbellaria, Tricladida). Gen Comp Endocrinol 1981; 43(3):331-5.
79. Halton DW, Fairweather I, Shaw C et al. Regulatory peptides in parasitic platyhelminths. Parasitol Today 1990; 6(9):284-90.
80. Halton DW, Shaw C, Maule AG et al. Regulatory peptides in helminth parasites. Adv Parasitol 1994a; 34:163-227.
81. Halton DW, Maule AG, Mair GR et al. Monogenean neuromusculature: some structural and functional correlates. Int J Parasitol 1998; 28(10):1609-23.
82. Skuce PJ, Johnston CF, Fairweather I et al. Immunoreactivity to the pancreatic polypeptide family in the nervous system of the adult human blood fluke, Schistosoma mansoni. Cell Tissue Res 1990; 261(3):573-81.
83. Skuce PJ, Johnston CF, Fairweather I et al. A confocal scanning laser microscope study of the peptidergic and serotoninergic components of the nervous system in larval Schistosoma mansoni. Parasitology 1990; 101(2):227-34.
84. Herzog H, Hort Y, Schneider R et al. Seminalplasmin: recent evolution of another member of the neuropeptide Y gene family. Proc Natl Acad Sci USA 1995; 92(2):594-8.
85. Couzens M, Liu M, Tüchler C et al. Peptide YY-2 (PYY2) and pancreatic polypeptide-2 (PPY2): species-specific evolution of novel members of the neuropeptide Y gene family. Genomics 2000; 64(3):318-23.
86. Conlon JM, Larhammar D. The evolution of neuroendocrine peptides. Gen Comp Endocrinol 2005; 142(1-2):53-9.
87. Mair GR, Halton DW, Shaw C et al. The neuropeptide F (NPF) encoding gene from the cestode, Moniezia expansa. Parasitology 2000; 120(1):71-7.
88. Dougan PM, Mair GR, Halton DW et al. Gene organization and expression of a neuropeptide Y homolog from the land planarian Arthurdendyus triangulatus. J Comp Neurol 2002; 454(1):58-64.
89. Humphries JE, Kimber MJ, Barton YW et al. Structure and bioactivity of neuropeptide F from the human parasites Schistosoma mansoni and Schistosoma japonicum. J Biol Chem 2004; 279(38):39880-5.
90. Miskolzie M, Kotovych G. The NMR-derived conformation of neuropeptide F from Moniezia expansa. J Biomol Struct Dynamics 2002; 19:991-8.
91. Maule AG, Shaw C, Halton DW et al. Neuropeptide F (Moniezia expansa): localization and characterization using specific antisera. Parasitology 1992a; 105(3):505-12.
92. Maule AG, Brennan GP, Halton DW et al. Neuropeptide F-immunoreactivity in the monogenean parasite Diclidophora merlangi. Parasitol Res 1992b; 78(8):655-60.
93. Hrckova G, Halton DW, Maule AG et al. Neuropeptide F-immunoreactivity in the tetrathyridium of Mesocestoides corti (Cestoda: Cyclophyllidea). Parasitol Res 1993; 79(8):690-5.
94. Marks NJ, Maule AG, Halton DW et al. Distribution and immunochemical characteristics of neuropeptide F (NPF) (Moniezia expansa)—immunoreactivity in Proteocephalus pollanicola (Cestoda: Proteocephalidea). Comp Biochem Physiol C 1993; 104(3):381-6.
95. Marks NJ, Halton DW, Maule AG et al. Comparative analyses of the neuropeptide F (NPF)- and FMRFamide-related peptide (FaRP)-immunoreactivities in Fasciola hepatica and Schistosoma spp. Parasitology 1995b; 110(4):371-81.
96. Gustafsson MK, Fagerholm HP, Halton DW et al. Neuropeptides and serotonin in the cestode, Proteocephalus exiguus: an immunocytochemical study. Int J Parasitol 1995; 25(6):673-82.
97. Brennan GP, Ramasamy P. Ultrastructure of the surface structures and electron immunogold labeling of peptide immunoreactivity in the nervous system of Pseudothoracocotyla indica (Polyopisthocotylea: Monogenea). Parasitol Res 1996; 82(7):638-46.
98. Maule AG, Halton DW, Shaw C et al. The cholinergic, serotoninergic and peptidergic components of the nervous system of Moniezia expansa (Cestoda, Cyclophyllidea). Parasitology 1993b; 106(4):429-40.

99. Brennan GP, Halton DW, Maule AG et al. Electron immunogold labeling of regulatory peptide immunoreactivity in the nervous system of Moniezia expansa (Cestoda: Cyclophyllidea). Parasitol Res 1993; 79(5):409-15.
100. Halton DW, Maule AG, Brennan GP et al. Grillotia erinaceus (Cestoda, Trypanorhyncha): localization of neuroactive substances in the plerocercoid, using confocal and electron-microscopic immunocytochemistry. Exp Parasitol 1994b; 79(3):410-23.
101. Maule AG, Halton DW, Johnston CF et al. The serotoninergic, cholinergic and peptidergic components of the nervous system in the monogenean parasite, Diclidophora merlangi: a cytochemical study. Parasitology 1990; 100(2):255-73.
102. Biserova NM, Dudicheva VA, Terenina NB et al. The nervous system of Amphilina foliacea (Platyhelminthes, Amphilinidea) an immunocytochemical, ultrastructural and spectrofluorometrical study. Parasitology 2000; 121(4):441-53.
103. Zurawski T, Mousley A, Mair GR et al. Immunomicroscopical observations on the nervous system of adult Eudiplozoon nipponicum (Monogenea: Diplozoidae). Int J Parasitol 2001; 31(8):783-92.
104. Kotikova EA, Raikova OI, Reuter M et al. The nervous and muscular systems in the free-living flatworm Castrella truncata (Rhabdocoela): an immunocytochemical and phalloidin fluorescence study. Tissue Cell 2002; 34(5):365-74.
105. Stewart MT, Marks NJ, Halton DW. Neuroactive substances and associated major muscle systems in Bucephaloides gracilescens (Trematoda: Digenea) metacercaria and adult. Parasitol Res 2003; 91(1):12-21.
106. Stewart MT, Mousley A, Koubkova B et al. Development in vitro of the neuromusculature of two strigeid trematodes, Apatemon cobitidis proterorhini and Cotylurus erraticus. Int J Parasitol 2003; 33(4):413-24.
107. Stewart MT, Mousley A, Koubkova B et al. Gross anatomy of the muscle systems and associated innervation of Apatemon cobitidis proterorhini metacercaria (Trematoda: Strigeidea), as visualized by confocal microscopy. Parasitology 2003; 126(3):273-82.
108. Shaw C, Maule AG, Halton DW. Platyhelminth FMRFamide-related peptides. Int J Parasitol 1996; 26(4):335-45.
109. Armstrong EP, Halton DW, Tinsley RC et al. Immunocytochemical evidence for the involvement of an FMRFamide-related peptide in egg production in the flatworm parasite Polystoma nearcticum. J Comp Neurol 1997; 377(1):41-8.
110. Behm CA, Bendig MM, McCarter JP et al. RNAi-based discovery and validation of new drug targets in filarial nematodes. Trends Parasitol 2005; 21(3):97-100.
111. Jones AK, Buckingham SD, Sattelle DB. Chemistry-to-gene screens in Caenorhabditis elegans. Nat Rev Drug Discov 2005; 4(4):321-30.
112. Lochnit G, Bongaarts R, Geyer R. Searching new targets for anthelminthic strategies: Interference with glycosphingolipid biosynthesis and phosphorylcholine metabolism affects development of Caenorhabditis elegans. Int J Parasitol 2005; 35(8):911-23.
113. Zawadzki JL, Presidente PJ, Meeusen EN et al. RNAi in Haemonchus contortus: a potential method for target validation. Trends Parasitol 2006; 22(11):495-9.
114. Aboobaker AA, Blaxter M. Use of RNA interference to investigate gene function in the human filarial nematode parasite Brugia malayi. Mol Biochem Parasitol 2003; 129(1):41-51.
115. Issa Z, Grant WN, Stasiuk S et al. Development of methods for RNA interference in the sheep gastrointestinal parasite, Trichostrongylus colubriformis. Int J Parasitol 2005; 35(9):935-40.
116. Krautz-Peterson G, Radwanska M, Ndegwa D et al. Optimizing gene suppression in schistosomes using RNA interference. Mol Biochem Parasitol 2007; 153(2):194-202.
117. Simmer F, Tijsterman M, Parrish S et al. Loss of the putative RNA-directed RNA polymerase RRF-3 makes C. elegans hypersensitive to RNAi. Curr Biol 2002; 12(15):1317-9.
118. Kennedy S, Wang D, Ruvkun G. A conserved siRNA-degrading RNase negatively regulates RNA interference in C. elegans. Nature 2004; 427(6975):645-9.
119. Asikainen S, Vartiainen S, Lakso M et al. Selective sensitivity of Caenorhabditis elegans neurons to RNA interference. Neuroreport 2005; 16(18):1995-9.
120. Fleming CC, McKinney S, McMaster S et al. Getting to the root of neuronal signalling in plant parasitic nematodes using RNA interference. Nematology 2007; [in press].
121. Geldhof P, Visser A, Clark D et al. RNA interference in parasitic helminths: current situation, potential pitfalls and future prospects. Parasitology 2007; 134(5):609-19.
122. Knox DP, Geldhof P, Visser A et al. RNA interference in parasitic nematodes of animals: a reality check? Trends Parasitol 2007; 23(3):105-7.
123. Kimber MJ, McKinney S, McMaster S et al. flp gene disruption in a parasitic nematode reveals motor dysfunction and unusual neuronal sensitivity to RNA interference. FASEB J 2007; 21(4):1233-43.
124. Bakhetia M, Charlton WL, Urwin PE et al. RNA interference and plant parasitic nematodes. Trends Plant Sci 2005; 10(8):362-7.

125. Huang G, Allen R, Davis EL et al. Engineering broad root-knot resistance in transgenic plants by RNAi silencing of a conserved and essential root-knot nematode parasitism gene. Proc Natl Acad Sci USA 2006; 103(39):14302-6.
126. Yadav BC, Veluthambi K, Subramaniam K. Host-generated double stranded RNA induces RNAi in plant-parasitic nematodes and protects the host from infection. Mol Biochem Parasitol 2006; 148(2):219-22.
127. Steeves RM, Todd TC, Essig JS et al. Transgenic soybeans expressing siRNAs specific to a major sperm protein gene suppress Heterodera glycines reproduction. Funct Plant Biol 2006; 33:991-999.
128. Fairbairn DJ, Cavallaro AS, Bernard M et al. Host-delivered RNAi: an effective strategy to silence genes in plant parasitic nematodes. Planta 2007; [Epub ahead of print].
129. Husson SJ, Clynen E, Baggerman G et al. Defective processing of neuropeptide precursors in Caenorhabditis elegans lacking proprotein convertase 2 (KPC-2/EGL-3): mutant analysis by mass spectrometry. J Neurochem 2006; 98(6):1999-2012.
130. Husson SJ, Janssen T, Baggerman G et al. Impaired processing of FLP and NLP peptides in carboxypeptidase E (EGL-21)-deficient Caenorhabditis elegans as analyzed by mass spectrometry. J Neurochem 2007; 102(1):246-60.
131. Mair GR, Niciu MJ, Stewart MT et al. A functionally atypical amidating enzyme from the human parasite Schistosoma mansoni. FASEB J 2004; 18(1):114-21.
132. McVeigh P, Mair GR, Atkinson L et al. Discovery of multiple neuropeptide families in the phylum Platyhelminthes. Int J Parasitol 2009; 39(11):1243-52.

Chapter 5

Neuropeptide Physiology in Helminths

Angela Mousley, Ekaterina Novozhilova, Michael J. Kimber and Tim A. Day*

Abstract

Parasitic worms come from two distinct, distant phyla, *Nematoda* (roundworms) and *Platyhelminthes* (flatworms). The nervous systems of worms from both phyla are replete with neuropeptides and there is ample physiological evidence that these neuropeptides control vital aspects of worm biology. In each phyla, the physiological evidence for critical roles for helminth neuropeptides is derived from both parasitic and free-living members. In the nematodes, the intestinal parasite *Ascaris suum* and the free-living *Caenorhabditis elegans* have yielded most of the data; in the platyhelminths, the most physiological data has come from the blood fluke *Schistosoma mansoni*. FMRFamide-like peptides (FLPs) have many varied effects (excitation, relaxation, or a combination) on somatic musculature, reproductive musculature, the pharynx and motor neurons in nematodes. Insulin-like peptides (INSs) play an essential role in nematode dauer formation and other developmental processes. There is also some evidence for a role in somatic muscle control for the somewhat heterogeneous grouping of peptides known as neuropeptide-like proteins (NLPs). In platyhelminths, as in nematodes, FLPs have a central role in somatic muscle function. Reports of FLP physiological action in platyhelminths are limited to a potent excitation of the somatic musculature. Platyhelminths are also abundantly endowed with neuropeptide Fs (NPFs), which appear absent from nematodes. There is not yet any data linking platyhelminth NPF to any particular physiological outcome, but this neuropeptide does potently and specifically inhibit cAMP accumulation in schistosomes. In nematodes and platyhelminths, there is an abundance of physiological evidence demonstrating that neuropeptides play critical roles in the biology of both free-living and parasitic helminths. While it is certainly true that there remains a great deal to learn about the biology of neuropeptides in both phyla, physiological evidence presently available points to neuropeptidergic signaling as a very promising field from which to harvest future drug targets.

Introduction

Although there have been considerable efforts to discern specific physiological roles for selected neuropeptides in parasitic nematodes and platyhelminths, these are largely disjointed and incomplete, often prescribed by the limitations inherent in the various experimental species. However, there has already emerged a very clear picture that neuropeptide signaling systems play an absolutely central role in the biology of both nematodes and platyhelminths.

In nematodes, the majority of our understanding of neuropeptide function and physiology is derived from the porcine intestinal parasite *Ascaris suum* and the free-living nematode *Caenorhabditis elegans*. *A. suum* is comparatively huge and is very amenable to physiological examination, including recording of muscle tension responses in the body wall, the ovijector and the pharynx, as well as a host of eletrophysiological techniques. *C. elegans*, in contrast, offers a wealth of genomic data

*Corresponding Author: Tim A. Day—Neuroscience Program and Department of Biomedical Sciences, 2008 Veterinary Medicine Building, Iowa State University, Ames, Iowa 50011-1250 USA. Email: day@iastate.edu

Neuropeptide Systems as Targets for Parasite and Pest Control, edited by Timothy G. Geary and Aaron G. Maule. ©2010 Landes Bioscience and Springer Science+Business Media.

and vast amenability to genetic manipulation. For both species, there is a great deal of structural information regarding the neuropeptide complement present in the worms. Although these data account for the lion's share, significant information has been gathered from other species.

In platyhelminths, our knowledge of neuropeptide function and physiology is likewise acquired from both free-living and parasitic members of the phylum. Much of the physiological data are drawn directly from the most important parasitic species, including *Schistosoma* and *Fasciola*. Although schistosomes were historically noted to be difficult to culture and hard to work with experimentally, those limitations are somewhat obviated by the long history of schistosome biology which provides a number of useful bioassays for insight into function. The impending refinement of the schistosome genome also brings focus and experimental opportunity to this important parasite. There is a great deal of promise in the model flatworm *Schmidtea mediterranea*, which is very tractable to both genetic and experimental manipulation and for which genome compilation is also quite advanced.

As noted in the previous chapter, the array of neuropeptides present in nematodes is startling. The data reviewed here will address FMRFamide-like peptides (FLPs); there is a great deal of data derived from *A. suum*, *C. elegans* and plant parasitic nematodes demonstrating a central role for FLP signaling in nematode biology. Nematode FLPs have many varied effects on somatic musculature, reproductive musculature, the pharynx and motor neurons in nematodes. These FLP effects on worm tissues are potent and quite specific. Further, injection of synthetic FLPs or genetic manipulations of *flp*-gene expression produce important behavioral effects in worms. The data reviewed here will also address insulin-like peptides (INSs), for which there is comparatively less physiological data. However, the data do demonstrate a vital role for INSs in dauer formation and other critical developmental processes. Lastly, the data reviewed here will also address the neuropeptide-like proteins (NLPs) of nematodes. There has only been a small bit of physiological experimentation targeting NLP biology in nematodes and it suggests a role in somatic muscle control.

The present data do not show platyhelminths to be endowed with an array of neuropeptides as broad as the nematodes, but there is ample evidence for neuropeptide centrality in platyhelminth physiology. In platyhelminths, FLPs are potently myoexcitatory in a range of flatworms which includes endoparasitic flukes and tapeworms, ectoparasitic monogeneans and free-living species. FLP myoexcitation in flatworms is very potent and very specific. Platyhelminths are also abundantly endowed with neuropeptide Fs (NPFs), which appear absent from nematodes. There is not yet any data directly linking platyhelminth NPF to any particular physiological outcome, but this neuropeptide does potently and specifically inhibit cAMP accumulation in schistosomes.

Neuropeptide Function in Nematodes

FLP Function in Control and Modulation of Somatic Body Wall Muscle—Ascaris suum, Ascaridida galli *and* Haemonchus contortus

Through published works, approximately 25 nematode and 10 arthropod-derived FLPs with structurally diverse carboxy-terminal motifs have been shown to induce a variety of effects on somatic body wall muscle strips of *Ascaris*. A recent summary of their effects identified four key response types defined as body wall response Type 1 (bwRT1) through to body wall response Type 4 (bwRT4). The characteristics of these responses are as follows: bwRT1 (slow and prolonged inhibition), bwRT2 (fast and transient inhibition), bwRT3 (sustained contraction) and bwRT4 (biphasic activity: transient relaxation/sustained contraction).

Nearly 20 years ago, Cowden et al[1] were the first to examine the effects of an RFamide peptide (AF1) on *Ascaris* muscle activity. Subsequently, both AF1 and AF2 were shown to induce analogous biphasic activities on body wall muscle strips comprising transient, rapid relaxations followed by a prolonged phase of increased contractility that have since been described as defining bwRT4.[2-7] Different receptors have been implicated in the AF1/AF2-response as a result of studies by Bowman et al[2] and Maule et al[5] who showed that the biphasic response is abolished upon denervation of the muscle strip. Consequently, a muscle-based receptor mediating the transient inhibitory phase and a neuronal receptor conveying the eventual excitatory response was proposed. To date, these speculations have not been confirmed and the receptor(s) responsible for relaying the AF1 and/or AF2 signal remain(s) to be identified and localized. Using electrophysiological techniques, Trailovic et al[8] demonstrated that AF2 enhanced muscle cell membrane potential responses to ACh which is in contrast to the study by Maule et al[5] who saw no excitation following nerve cord removal. More recently, AF2 was shown to modulate voltage-activated calcium currents in *A. suum*.[9] Structure-activity studies employing an alanine scan series and N-terminally truncated analogues have shed light on the complexity of the AF1 and AF2 responses.[2] Substitution of alanine for any amino acid in AF1 has profound effects on activity; Ala^1, Ala^2 and Ala^7 analogues are inactive (where the number indicates amino acid position from the N-terminus), Ala^3AF1 and Ala^5AF1 relayed only inhibitory responses and substitutions with alanine at position 4 displayed only the excitatory phase of the AF1 response. Ala^6AF1 was found to antagonise the AF1 effect while having no effect on the AF2 response indicating that AF1 and AF2 may act at different receptors despite identical physiology. In addition, AF1/AF2 hybrids (KHEFIRFa, KNEYIRFa and KNEFLRFa) induce either similar or less potent effects than AF1.[2] In contrast to AF1, all alanine-substituted AF2 derivatives are active on *A. suum* muscle strip preparations, although with reduced potencies. In addition, Ala^6AF2 does not display the antagonistic activities that were noted for Ala^6AF1.[7] The effects of AF2 have also been characterized on the body wall muscle of the sheep nematode *Haemonchus contortus*. While injection of AF2 into the pseudocoelomic cavity had no significant effect on spontaneous contractions, it significantly inhibited ACh-induced contractions.[10]

Inhibitory peptides have also been well-characterised on the body wall muscle. The first bwRT1 RFamides to be examined were the *Panagrellus* FLPs PF1 and PF2 which induced inhibitory effects on dorsal, ventral and denervated *A. suum* muscle strip preparations.[5,11-13] The PF1-induced relaxation is defined by a slow onset resulting in flaccid paralysis and appears to be insensitive to external Cl^- and K^+, whilst Ca^{2+}-free medium reduces its action.[5] An important role for nitric oxide (NO) has been implicated in the PF1-induced relaxation on the body wall muscle and is discussed later in this chapter (see section on second messenger systems). In addition, PF1 has been shown to antagonise the excitatory effects induced by levamisole, acetylcholine and the biphasic bwRT4-inducing RFamides (AF1 and AF2). Structure activity studies using an alanine scan series and N-terminal deletions revealed that the two N-terminal residues (SD) were not necessary for activity and that the Phe^5 and the C-terminal RFamide are essential for receptor binding and activity. In addition to the *Panagrellus* FLPs, eight other nematode RFamides (AF6, AF11, AF19, AF21, AF22, AF23 and the two *flp*-13 peptides, AADGAPLIRFa and APEASPFIRFa) and eight arthropod FLPs (pQDVDHVFLRa, PDVDHVFLRFa, VFLRFa, TNRNFLRFa, SDRNFLRFa, DPSFLRFa, KPNQDFMRFa and GNSFLRFa) have been shown to induce a bwRT1-like response.[14-18] The effects of AADGAPLIRFa and APEASFIRFa could be distinguished from each other by their requirement for external ions. AADGAPLIRFa-induced inhibition was insensitive to external K^+ or Ca^{2+} but the relaxation was abolished in Cl^- free media, in contrast the inhibitory effect produced by APEASFIRFamide was abolished in high K^+ but not in Ca^{2+}- or Cl^- free media. These sensitivities are in contrast to the requirements of PF1 (Ca^{2+}) as described above and highlights further divisions amongst the bwRT1-inducing peptides.

An additional inhibitory response type (bwRT2) has been described and is currently exclusive to PF4. This involves a rapid relaxation of the body wall muscle in *A. suum*.[19,20] Structure activity studies performed by Kubiak et al[21] using analogues with either alanine or glycine substitutions in position 2 or 5, have revealed the importance of the proline residue in position 2 to the biological activity and metabolic stablility of PF4. It was proposed that Pro^2 confers resistance to endogenous aminopeptidases. The inhibitory action of PF4 is abolished when Cl^- is absent from the bathing media, suggesting that its actions are mediated via chloride channels; GABA and PF4 have similar activity profiles on body wall muscle.[20,22] Subsequent studies showed that the nematode GABA channel blocker, NCS 281-93, failed to antagonise PF4-induced relaxation, indicating that PF4 acts on a Cl^- channel which is not GABA-gated.[20] More recently, through electrophysiology studies, Purcell et al[23] showed that the time course of PF4-induced hyperpolarization was consistent with a ligand-gated channel and could be distinguished from the time course of the PF1 inhibition (characteristic of transduction pathway activation) and subsequently they provided further evidence to show that the PF4 activates low-conductance Cl^- channels in *Ascaris*.[24]

Interestingly, most of the peptides examined on the body wall muscle exerted excitatory effects described by bwRT3. AF3 and AF4 induce concentration-dependent contractions of dorsal somatic muscle strips.[1,25] These two peptides have also been examined on dorsal muscle strip preparations from the domestic fowl parasite, *A. galli*[25] where they were shown to increase muscle contraction. Further examination of the mechanism underlying the excitatory effects of AF3 revealed that its actions were independent of the cholinergic system.[26] Four additional *Ascaris* FLPs (AF10, AF13, AF14, AF20) which share carboxy-terminal homology with AF3 and AF4 also display excitatory activities (bwRT3), as do AF5, AF9, AF17 and the *C. elegans flp*-18-encoded peptide SVPGVLRFa.[1,14,16]

AF8 displays a unique differential activity of nerve-cord dependent excitatory effects on ventral (bwRT3) and inhibitory effects on dorsal (bwRT1) muscle strip preparations.[5,14,27] The PF3 inhibition was not abolished in Cl^- free media or by pretreatment with NOS inhibitors but was reduced in K^+ and Ca^{2+} free media. The mechanism of action of this peptide is unclear, however its nerve cord dependence suggests that PF3 influences neurotransmitter release.[20] A more recent study has shown that PF3 modulates the cholinergic system in the sheep nematode *H. contortus*. In this study the effects of PF3 and AF2 were examined on two different *Haemonchus* isolates—an MH isolate, which was susceptible to ACh and the cholinomimetic levamisole and a resistant Lawes isolate, displaying reduced sensitivity to cholinergic drugs. The Lawes isolate was found to be significantly less sensitive to AF8 than MH isolates; AF1 effects were identical in both isolates,[10] suggesting that AF8 modulates cholinergic transmission.

Two *Ascaris* FLPs (AF7 and AF16) exerted little or no effect on body wall muscle[14] as did the *C. elegans* RFamides, KPSFVRFa and VPSAADMMIRFa.[10]

FLP Function in Reproductive Muscle Control—Ascaris suum

The effects of 30 nematode FLPs and two FLP chimeras have been examined on reproductive muscle in *Ascaris*. The highly prolific nature of *A. suum* (with egg-laying studies reporting output of up to 2 million eggs per day[28]) reflects a well developed reproductive system; the female system consists of paired ovaries and tubular uteri which unite to form a highly muscular ovijector unit comprising the vagina vera and vagina uteri.[29] A bioassay system to measure the activity of the ovijector was developed by Fellowes et al[30] who showed that the ovijector displayed an inherent rhythmical activity, characterised by circular muscle contraction. This pilot study reported the effects of three FLPs (AF1, AF2 and AF8) and was subsequently expanded to include the activities of a further six FLPs (PF1, PF2, PF4, AF3, AF4 and KPSFVRFa).[10,31,32]

More recently, Moffett et al,[33] following examination and analysis of the activities of 21 FLPs (representing 17 *C. elegansflp*-encoding genes) and including the data from the studies by Fellowes et al[30,31] and Marks et al[32] for comparison, reported five different categories of effect on the ovijector muscle, defined as ovijector response Type 1 (ovRT1) to ovijector response Type 5 (ovRT5). Significantly, more than half of the FLPs examined (PF1, PF2, PF3, PF4, AF2, SPREPIRFa, SPLGTMRFa, APSPSFIRFa, SPMERSAMVRFa, KPSFVRFa, QPKARSGYIRFa, NGAPQPFVRFa, RNKFEFIRFa, ASSAPLIRFa, GPSGPLRFa, GQTFVRFa, KSQYIRFa and AMMRFa; encoded by multiple *flp*-genes) modulated the ovijector by causing qualitatively indistinguishable circular muscle relaxation and resultant flaccid paralysis (ovRT1). The excitatory effects, as displayed by circular muscle contraction with increases in contraction frequencies, were evident upon addition of seven FLPs (SPREPIRFa, SVPGVLRFa, WANQVRFa, ASWASSVRFa, GLGPRPLRFa, AVPGVLRFa, GDVPGVLRFa; encoded on four *C. elegans flp*-genes and two *Ascaris flp* genes) and define the second response type, ovRT2. The third response type (ovRT3), induced by only one peptide (LRGEPIRFa; *flp*-2) was similar to ovRT2 in that it comprised a transient contraction of circular muscle however the absence of increases in contractility meant that it was placed into a distinct response type. Four RFamides (AF1, AGAKFIRFa, APKPKFIRFa, SPSAKWMRFa) induced an ovRT4 response which comprised a transient contraction followed by a period of inactivity. The last response type (ovRT5) was attributable to the *flp*-11 peptide AMRNALVRFamide and involved only an increase in contraction frequency.

In addition to nematode-derived FLPs, a recent study explored the activity of 10 arthropod FLPs on the ovijector; eight of the peptides examined induced an ovRT1 like response (pQDVDHVFL-RFa, PDVDHVFLRFa, VFLRFa, DPSFLRFa, TNRNFLRFa, SDRNFLRFa, KPNQDFMRFa, EQFDDYGHMRFa) whilst two (HVLRFamide and GNSFLRFa) induced a response that was comparable to ovRT4.[17]

A direct relationship between response type and FLP receptor diversity on the ovijector was proposed by the Moffett et al[33] suggesting the presence of at least five receptors, however no such correlation was noted between peptide structure and resultant response type. Indeed the receptor responsible for conveying the ovRT1 response is not selective with respect to ligand structure (the ovRT1-inducing peptides comprise 12 different C-terminal tetrapeptide motifs). Interestingly, of those ovRT1-inducing peptides that were similar in structure, there appeared to be differences in potency; potent inhibitory peptides included the *Panagrellus* FLPS (PF1, PF2 and PF4) however a peptide that shares the C-terminal motif (FIRFamide) with PF4, RNKFEFIRFa was one of the least potent ovRT1-inducing peptides such that a relationship between structure and potency of the ovRT1 response was not apparent.

Unfortunately, most of the peptides examined, are the only representatives of the *flp*-gene on which they are encoded making it difficult to determine a relationship between *flp*-gene and ovijector response type, as was possible for the pharyngeal studies (see later), however at least two peptides representing each of *flp*-1, -2, -5, -11 and -19 were examined. Of these, both peptides on *flp*-1 induced the same effect (ovRT1) as did the two *flp*-19 and *flp*-5 peptides (ovRT2 and ovRT4 respectively), however the three peptides encoded on *flp*-2 each induced a different response type (ovRT1, ovRT2 and ovRT3) and the two *flp*-11 peptides also induced different response types (ovRT1 and ovRT5).

It is possible that the response type (representing either local or peripheral action) is being driven by receptor/ligand locality rather than structural attributes and stoichiometry of the interacting peptides. To date, the only *flp*-gene that has been shown to be expressed in the female reproductive system is *flp*-19—Kim and Li[34] showed the expression of *flp*-19 in the HSN neuron which synapses onto vulva muscle and is believed to function in egg-laying.[35] The *flp*-19 peptides, WANQVRFamide and ASWASSVRFamide induced similar potent ovRT2 actions.

The most potent FLP examined on the ovijector was the *flp*-22 peptide, SPSAKWMRFa. The characteristics of this nonapeptide that could account for its potency include its length and the presence of the proline residue at position 2 from the N-terminus (previously shown to reduce rate of metabolism of FLPs—see Kubiak et al[21]). However, it should be noted that, amongst the other less potent peptides examined, more than 10 were at least nine amino acids in length and nine possessed a proline at position 2 from the N-terminus. The significance of these studies will not be fully realised until the activity profiles of FLPs on the ovijector are matched with the expression patterns of the encoding genes and the receptors for these peptides are characterised and localised.

*FLP Function in Pharyngeal Muscle Control—*Ascaris suum *and* Caenorhabditis elegans

The effects of FLPs on pharyngeal muscle have been examined in two nematode species, *C. elegans* and *A. suum*. In the free-living worm extracellular and intracellular recordings were made from a semi-intact pharynx where the nervous connections were not severed.[36,37] In *A. suum*, pharyngeal pumping activity, measured by changes in intrapharyngeal pressure was monitored using a pressure transducer system[38,39] and extracellular recordings, performed by Yew et al[18], were made by placing suction electrodes anywhere on the surface of the pharynx; the *Ascaris* pharyngeal preparation also possessed nervous input. In many of these studies, pharyngeal pumping was stimulated and maintained with a constant perfusion of 5-HT (serotonin) to examine the effects of potentially inhibitory peptides; the effects of excitatory peptides were determined in nonstimulated tissue preparations.[36-39] In contrast, Yew et al[18] employed macrophoretic application of both 5-HT and FLPs. The pharyngeal studies are significant as they are one of the few physiology studies that, through virtually technique-matched methodologies (intracellular recordings), enable comparisons of FLP activities between free-living and parasitic nematode species (*A. suum* and *C. elegans*).

The effects of 13 FLPs and the molluscan peptide FMRFamide were investigated on pharyngeal pumping behaviour in *A. suum*.[18,38,39] The majority of FLPs examined (AF2, AF3, AF4, AF6, AF16, PF1, PF2, PF4, FMRFa and the flatworm-dervied FLP, GNFFRFa) had no effect on 5-HT-induced pumping, but five FLPs (AF1, AF8, AF21, AF22 and AF23) induced an overall inhibition of pharyngeal activity; AF1 induced an inhibition that culminated in a hyper-relaxed pharynx, whereas AF8/PF3 induced a biphasic inhibition that resulted in hyper-contraction of the pharyngeal tissue.[38,39] In addition to the differences in effect, the AF8-induced inhibition lasted ~1.7 times longer than that of AF1.[39] The differences noted could either be due to their interaction with separate populations of receptors on the pharyngeal tissue or their action via different second messenger pathways; however, no FLP receptors have yet been identified on the *A. suum* pharynx.

Rogers et al[37] initially carried out examination of FLP activities on the pharynx from *C. elegans*, using an intracellular recording from the pharyngeal muscle. A total of 10 FLPs were examined in these preliminary experiments; four FLPs (AF1, AF2, AF8 and GAKFIRFa) were excitatory increasing pharyngeal action potential in a similar manner to 5-HT, whilst six FLPs (PF1, PF2, SAEPFGTMRFa, KPSFVRFa, APEASPFIRFa and AQTVFVRFa) were inhibitory causing an octopamine-like inhibition on 5-HT-stimulated tissue. It is interesting to note that AF1 was the most potent peptide inducing an increase in action potential that was comparable in potency to 5-HT-induced excitation. The AF1-induced response was reduced in mutants displaying defects in both neurotransmission (*snb*-1) and presynaptic inhibition (*goa*-1) indicating that the action of AF1 is mediated indirectly via the pharyngeal neuronal circuit, however the activity of AF1 remained unaltered in 5-HT synthesis-defective mutants (*tph*-1) and biogenic amine release defective mutants (*cat*-1, *cat*-4) highlighting a FLP-directed pathway that is distinct from that attributed to 5-HT and octopamine.

Of the inhibition-inducing FLPs examined, APEASPFIRFa was most potent. The effects of APEASPFIRFa were not altered in *snb*-1 or *goa*-1 mutants pointing towards direct effects on pharyngeal muscle. Further evidence for a pharyngeal muscle-based APEASPFIRFamide receptor is highlighted by *flp*-gene expression studies[40]; *flp*-13 (encoding APEASPFIRFa) is expressed in the pharyngeal motorneurons, M3 and M5 and in I5, an interneuron with synaptic output to the pharyngeal muscle—M3 and I5 are believed to play a role in the timing of the relaxation of the pharynx.

The study carried out by Rogers et al[37] was recently expanded and the effects of an additional 17 FLPs were examined, as well as those tested in the preliminary experiments described above (27 in total; encoded by some 23 *flp* genes) using electropharyngeogram (EPG) recordings of extracellular voltage.[36] With respect to the EPG recordings, 12 FLPs (APEASPFIRFa, AMRNALVRFa, SAEPFGTMRFa, PF2, EMPGVLRFa, SLLDYRFa, PF1, GGPQGPLRFa, KPSFVRFa, AF15, WANQVRFa, AF9; encoded by 11 *flp*-genes) were shown to inhibit pharyngeal activity whilst nine FLPs (AF1, AF2, AF8, KSQYIRFa, KSAFVRFa, SPSAKWMRFa, GAKFIRFa, PTFIRFa and SPREPIRFa; encoded by 8 *flp* genes) were excitatory confirming the earlier work of Rogers et al (2001) using intracellular recordings. Six of the FLPs examined were inactive on the EPG recorded from the pharyngeal muscle (SPMQRSSMVRFa, QPKARSGYIRFa, RNKFEFIRFa, EIVFHQISPIFFRFa, AMMVRFa, TKFQDFLRFa). Of significance, AF1, KSAFVRFa and KSQYIRFa were the most potent excitatory peptides; AMRNALVRFa and APEASPFIRFa were the most potent inhibitory peptides.

In the study by Papaioannou et al[36] the authors compared the FLP-induced activities to the expression of the encoding genes, as described by Kim and Li,[34] facilitating the evaluation of a relationship between expression and activity. Significantly, the majority of active peptides (inhibitory and excitatory) examined were encoded on genes that were found to be expressed in the pharyngeal system. For example, the *flp*-17 excitatory peptides (KSAFVRFa and KSQYIRFa) are expressed in the pharyngeal motorneuron M5 and the *flp*-13 excitatory peptide (APEASPFIRFa) is expressed in I5, M3 and M5 as described above. Furthermore, four of the inactive peptides (SPMQRSSMVRFa, QPKARSGYIRFa, RNKFEFIRFa and AMMVRFa) are encoded by *flp* genes that are not expressed in the pharyngeal system. On the other hand, the *flp*-1 peptides, PF1 and PF2, were not very potent and yet *flp* -1 is expressed in the pharyngeal motorneuron M5. Moreover, one of the most potent inhibitory peptides (AMRNALVRFa) was not expressed in the pharynx. The exact relationship between expression and activity remains to be determined and will only be uncovered when the expression profiles of the full complement of *flp*-genes have been examined and compared to the biological activities of the encoded peptides in pharyngeal tissue. What is clear is that FLP-gene expression in the pharyngeal system is not a prerequisite for activity highlighting the likelihood of alternative routes of modulating pharyngeal muscle for some of the FLPs.

In addition to the extracellular recordings, the study by Papaioannou et al,[36] also employed intracellular recordings [similar to those described by Rogers et al[37] to examine the structure activity relationships of peptides encoded on three *flp* genes (*flp*-3, -13 and -18) in both wild type animals and *snb*-1 mutants. All of the peptides examined induced qualitatively different inhibitory effects that remained unaltered in the mutant animals.

Of interest are the differences between the FLP-induced pharyngeal activities (based on the intracellular recordings) between the free-living and parasitic nematode species despite similarities in the 5-HT response. For example, AF1 and AF8 inhibited pharyngeal activity in *A. suum* but stimulated pumping in *C. elegans*; PF1, PF2 and AF2 also had conflicting effects (PF1 and PF2 were inhibitory in *C. elegans* but had no effect in *A. suum*, AF2 was excitatory in *C. elegans* but was inactive in *A. suum*). Although there were some variations in the intracellular recording methodologies used for the free-living and parasitic nematodes, these minor differences are unlikely to account for the disparity between activities. This is the first indicator of functional differences of the same FLP on the same tissue between nematode species.

FLP Effects on Motorneuron Activity—Ascaris suum

In addition to the postsynaptic FLP action described above, FLPs have also been shown to have direct effects on neuronal activities in *Ascaris*.[4,14,41] *A. suum* comprises 298 neurons of which approximately 75 synapse onto either the ventral or dorsal components of the somatic body wall muscle and are termed motorneurons.[42,43] *Ascaris* motorneurons can be divided into seven classes based on their distribution (axonal and dendritic position) and physiology: ventral excitatory (V1, V2), ventral inhibitory (VI), dorsal excitatory (DE1, DE2, DE3) and dorsal inhibitory (DI).

Using electrophysiological techniques Cowden et al[4] were the first to demonstrate FLP-induced neuronal effects in *Ascaris*. In this study AF1 was shown to exert inhibitory effects on the electrical properties of inhibitory motorneurons (VI and DI) in addition to decreases in their input resistance (AF1 had no effect on dorsal excitatory motorneurons). These effects were unaltered following synaptic isolation of VI and DI suggesting that the AF1 receptor is located on inhibitory motorneurons,[14] however this has not been confirmed and will ultimately depend upon the characterisation and localisation of the elusive AF1 receptor.

Some six years later, fuelled by the identification of numerous other *Ascaris* FLPs, Davis and Stretton[14] described the neuronal effects of an additional three RFamides (AF5, AF7 and AF11) on all motorneuron classes and, more recently, in a separate study,[41] the activities of another 14 FLPs (AF2, AF3, AF4, AF6, AF8, AF9, AF10, AF13, AF14, AF15, AF16, AF17, AF19, AF20) were examined on the membrane potential and input resistance of DE2 and DI. In addition, Yew et al[18] have examined the effects of a further three FLPs (AF21, AF22 and AF23) on DE1, DE2 and DI bringing the total number of FLPs examined on *Ascaris* motorneurons to 21. Despite complex physiological effects, a general pattern emerged that enabled Davis and Stretton[41] to delineate five major response types associated with the effects of FLPs on DE2 and/or DI motorneurons, thought to be attributable to distinct FLP receptor subtypes; the recent data of Yew et al[18] were not included.

The first response type is induced by the addition of the FIRFamide peptides (AF1, AF5, AF6 and AF7) and AF15 and it is characterised by depolarizing effects on DE2 motorneurons and hyperpolarizing effects on DI motorneurons. AF1 produced strong and consistent DE2 neuron depolarization with sporadic weak and transient hyperpolarizations. Significantly, the strongest decrease in DE2 input resistance was produced by AF1. AF1 superfusions also resulted in weak DI neuron hyperpolarizations that occurred early, transiently and inconsistently. AF5 also induced strong and constant DE2 neuron depolarizations that were consistent with decreases in input resistance. In addition, AF5 superfusions resulted in weak DI neuron hyperpolarizations. AF15 had similar effects to AF5 including DE2 depolarization and DI hyperpolarization, perhaps reflecting their sequence similarities at the C-terminus (TFV/IRFamide). In contrast to the effects of AF1, AF5 and AF15, superfusion of AF6 and AF7 produced weak DE2 neuron depolarizations. AF6 and AF7 also produced weak DI neuron hyperpolarizations.

The unique biphasic effect of AF2 on the input resistance of the DE2 motorneurons comprises the second response type; AF2 induced an early, transient increase followed by a sustained decrease in input resistance (in contrast to the decrease in input resistance noted for AF1). AF2-superfusion also resulted in the strongest depolarization of DE2 neurons that was noted for any of the 18 *Ascaris* FLPs examined; it had no significant effect on DI neurons.

AF9-induced DE2 neuron depolarization represents the third response type with superfusion of AF9 producing a weak, transient DE2 neuron hyperpolarization followed by a sustained depolarization with no significant effect on the input resistance. An early transient, weak DI neuron hyperpolarization was also noted for AF9.

The fourth response type is based on the potency of the AF8-induced depolarization of DI motorneurons. Superfusion with AF8 also produced a small hyperpolarization of DE2 neurons. Together these effects could account for the relaxation of dorsal muscle strips noted by Maule et al[5] (see section on somatic muscle above), although it was suggested that, since DI has a presynaptic output to DE2, the hyperpolarization of DE2 may, in fact, reflect an indirect interaction mediated by DI highlighting the potential for diverse circuit interactions.

The final response was observed on superfusion of either AF17 or AF19, both of which induced identical strong DE2 and DI neuron hyperpolarizations and decreases in DE2 and DI input resistance. AF19 induced that largest DE2 neuron hyperpolarization that was noted for all 18 *Ascaris* FLPs.

The effects of AF21, AF22 and AF23 were only recently examined on DE1, DE2 and DI neurons and therefore were not included in the response analysis performed by Davis and Stretton.[41] Whilst it is difficult to assign the responses of AF21, AF22 and AF23 to one of the five established responses, AF22 induced DE1, DE2 and DI hyperpolarizations with decreases in input resistance that were comparable to the responses induced by AF17 and AF19. AF21 and AF23 produced depolarizations of all three dorsal motorneurons and large decreases in input resistance that were not similar to any of the previously established response types.

Of the other peptides examined, the six PGVLRFamides (AF3, AF4, AF10, AF13, AF14 and AF20) induced either weak DE2-depolarizations and decreases in DE2 input resistance or negligible effects. AF11 produced weak DE2 depolarization and a slight increase in the input resistance relative to control. AF16 was the only FLP examined that had no effect.

FLP Effects on Second Messenger Systems—Ascaris suum *and* Ascaridida galli

Our limited knowledge of endogenous neuropeptide signal transduction pathways in nematodes has been largely derived from experiments on the effects of selected FLPs [AF1, AF2, AF3, AF8, AF10, AF17, PF4, ASPSFIRFa (*flp*-4), SPMQRSMVRFa (*flp*-7), KPSFVRFa (*flp*-9) and APEASPFIRFa (*flp*-13)] on endogenous levels of one signaling molecule, cAMP, following either treatment of the muscle strip (*A. suum* and *A. galli*) or whole worm injections (*A. suum*). An additional study used *Ascaris* body wall muscle strip physiology to investigate the role of another second messenger, nitric oxide (NO), in the FLP (PF1) signal transduction pathway.[11] Whilst your attention is drawn to a recent review by McVeigh et al[44] who provide a useful illustrative summary of the proposed nematode FLP signaling pathways, published details on the role of second messengers in FLP-signaling pathways will be subsequently discussed.

cAMP levels (generated by adenylate cyclase following G-protein activation) have been shown to increase following the treatment of *A. suum* muscle strips with either AF1 or AF2[7,45,46] or following whole worm AF1/AF2 injections.[47] Moreover, similar increases in cAMP levels are noted upon exposure of denervated muscle preparations to AF2, such that the increase in cAMP is thought to be attributable to the inhibitory phase of the AF2-induced, biphasic response.[7] In contrast, AF3 and AF17, which cause distinct excitatory effects in *A. suum* somatic muscle, decrease cAMP levels in *A. suum* as do peptides encoded on *flp*-4, -7, -9 and -13;[26,47] AF3 also decreases cAMP in *A. galli*.[26] AF10 and AF8 had no effect on intracellular cAMP levels.[7,47]

The studies by Thompson et al[7] and Geary et al[45] signified a link between increases in cAMP levels and muscle relaxation; decreases in cAMP levels appear to be associated with muscle contraction. However this inverse relationship was not confirmed by Reinitz et al[47] when they compared the effects of FLP-induced cAMP levels to changes in worm length (decreases in cAMP were associated with increases in body length, reminiscent of muscle relaxation) leaving us with a somewhat confused and complex picture of the FLP-induced cAMP pathway.

With respect to NO, the story is a lot less complicated with only one study implicating its involvement in FLP signaling. Bowman et al[11] underlined NO involvement in the PF1-induced response by showing that an NO donor (sodium nitroprusside) mimicked the effect of PF1 on the body wall muscle of *A. suum* (slow muscle relaxation) and that blockers of NO synthesis (*N*-nitro-*L*—arginine) reduced its activity. The revelation that NOS is most abundant in the hypodermal tissues of *A. suum* suggests a site of action for PF1 receptors. In this respect, activation of the PF1 receptor would stimulate Ca^{2+}-dependent NOS inducing the release of NO from the hypodermis, which diffuses readily to the muscle and causes relaxation.

FLP Function in Behaviour, Locomotion and Body Posture—Ascaris suum

The specific in vitro-assessed FLP-induced effects on somatic body wall, ovijector and pharyngeal muscles, motorneuron activities and second messenger systems described above highlight the key role played by FLP peptides in controlling neuromuscular function in nematodes. This section considers the correlation between the in vitro findings and the FLP-effects in vivo, characterised by behavioural assays.

In 1989 Cowden et al[4] showed that AF1 abolished localised locomotory waves following injection into *A. suum*. Since then, a total of 21 *Ascaris* FLPs and a further seven *C. elegans flp* gene-encoded peptides (not represented by the *Ascaris* FLPs) have been examined on general qualitative behaviour of intact adult *A. suum*.[18,41,47] In these technique-matched studies, FLPs were directly injected into the pseudocoelomic cavity of large female worms then monitored for behavioural effects. Reinitz et al[47] initiated this work by examining the effects of six *Ascaris* FLPs (AF1, AF2, AF8, AF10, AF11 and AF17) and seven *C. elegans flp*-encoded peptides (*flp*-3, -4, -7, -9, -12, -13 and -15) on locomotory behaviour and body waveforms; Davis and Stretton[41] expanded this study to investigate a further 10 *Ascaris* FLPs using more detailed behavioural attributes for scoring the effects, including modulation of locomotion (increased, decreased or abolished), body posture characteristics (linear, coiled, corkscrew, hairpin, loop or figure-8 postures) and head searching activities (decreased or head tremor/jerkiness).

With respect to the *Ascaris* FLPs, injection of AF1 inhibited the propagation of locomotory waves, reduced the number of waveforms and decreased body length. In addition, worms injected with AF1 showed a more linear body posture and a marked decrease in locomotion. Peptides inducing similar effects to AF1 are AF5 and AF7, which can be further characterised by an anterior head tremor or jerkiness. In contrast, the only other FIRFamide examined, AF6, exhibited inconsistent, short-lived effects that were nondescript.

The most striking effect was noted on addition of AF11 which resulted in a complete cessation of locomotion that was immediately apparent upon injection and the most complex picture of behaviours was painted by the effects of the PGVLRFamides (the six peptides encoded on the

Ascaris FLP gene, *afp*-1; AF3, 4, 10, 13, 14 and 20) all of which, upon injection, induced similar exaggerated postures, multiple body waves, head thrashing activities, transient coiling and impaired head searching behaviours[41]; AF10 also induced a decrease in body length.[47] AF9 exhibited a variety of effects on all of the behavioural attributes examined including decreased locomotion, linear anterior body posture and variable posterior posture between worms examined (coiled, corkscrew, loop, hairpin and figure 8 postures) and a reduction in head-searching activities.

The MRFamides, AF8 and AF16, induced very different effects with AF16 representing one of the least active FLPs examined (limited effects were weak and transient); AF8-injection resulted in behavioural characteristics that included abolished movement, anterior ventral coils and predominantly corkscrew postures as well as head tremor activity.

AF15, 17 and 19 had similar effects on the behaviour of *A. suum* adults—locomotion and head searching activities were both abolished upon injection of each peptide resulting in inactive worms. Regarding the body posture all three peptides induced a posture that was somewhere between linear and coiled, described as curved and incorporating bends at the points of injection.

With respect to the *C. elegans flp*-encoding peptides examined by Reinitz et al (2001), the *flp*-4, -7, -9 and -13 peptides inhibited locomotion, decreased waveforms and increased body length. In contrast, the *flp*-3 and -15 peptides had no effect on behaviour or body length and the *flp*-12 peptide, whilst decreasing waveform did not affect the length of the body.

Most of the results generated in the two studies are comparable with the only disparity relating to the most abundant nematode FLP, AF2. Whilst Reinitz et al[47] describe an AF2-induced behavioural effect similar to what is noted on injection of AF1 (reduced waveform, decreased body length), the observations of Davis and Stretton[41] are somewhat opposite comprising abrupt thrashing and anterior tremor; the reasons for these differences remain unclear. It should also be noted that, in general, FLPs with marked behavioural effects in *Ascaris* also have prominent effects of DE2 and/or DI motorneurons as described in the section above. Moreover, FLPs with no effect on behaviour, locomotion and/or body posture (e.g., AF6 and AF16) were also inactive or induced weak effects on DE2 and DI motorneurons. This is not surprising given the role of DE2 and DI in locomotion. Exceptions to these generalisations include the PGVLRFamides (pronounced behavioural effect versus weak effects on DE2 and DI) and AF11 (strongest behavioural effect versus weak effects on DE2 and DI). The reasons for these deviations from the general correlation between FLP-induced behavioural effect and motorneuron activity remain to be determined, however it is possible that the dramatic behavioural effects noted on addition of the PGVLRFamides and AF11 do not involve the DE2 and DI motorneurons.

More recently, Yew et al[18] examined the effects of AF21, AF22 and AF23; injection of each peptide resulted in an inhibition of locomotion, decrease in the number of waveforms and reduction in head searching activities. Significantly, the effects of AF22 persisted more than 11 hours post-injection whereas the effects of AF23 wore off after 6-9 hours and worms returned to normal activity approximately 2-3 hours post AF21-injection.

FLP Function Determined by Gene Knockout Studies—Caenorhabditis elegans

Manipulation of gene expression in *C. elegans* through inactivation or over-expression of target *flp* genes, with the resulting generation and subsequent phenotypic analysis of *flp*-gene-mutant animals, has provided insights into FLP function in nematodes (see Li, 2005[40] for review). To date, *C. elegans* knockouts for 11 *flp* genes (*flp*-1, -3, -4, -6, -8, -9, -10, -12, -19, -20 and -21) have been characterized and, although most of the mutants examined did not display any obvious phenotype (*flp*-3, -4, -6, -10, -12, -19 and -20), evaluation of the phenotypes of others (*flp*-1, -8, -9 and -21) revealed a range of behavioral defects.[40,48,49] For example, disruption of *flp*-1 results in a number of motor and sensory function abnormalities including inability to sense regions of high osmolarity, insensitivity to nose touch stimuli, loopy and uncoordinated movement, wandering and hyperactivity behaviors and absence of 5-HT-induced inhibition of locomotion,[48] in addition to defects in egg laying abilities (timing between and numbers of, eggs)[50] and fat storage.[40] Animals lacking *flp*-8 display defecation defects and *flp*-9 mutants show slight sluggishness and have compromised swimming abilities. With respect to *flp*-21, mutants exhibit mild aggregation behavior during feeding,[51] a phenotype which is consistent with the *flp*-21 receptor (NPR-1).

It is interesting to note that the majority of the peptides encoded on those genes, that show no phenotype when knocked out in *C. elegans*, have profound effects on neuromuscular function in other nematode species (*Ascaris*, *Ascaridida*, *Haemonchus*, *Globodera*). Most significantly, the *flp*-6 peptide, AF8, which is represented in both free-living and plant/animal-parasite species throughout the phylum, has been shown to be potently active on a range of muscle activities (somatic body wall, pharynx, ovijector) in several parasite species (see section on FLP effects on muscle for details); as discussed, *flp*-6 mutant *C. elegans* showed no observable phenotype. Whilst these inconsistencies could signify true functional differences of individual FLPs between nematodes species, perhaps substantiated somewhat by the recently emerging data by Yew et al,[52] demonstrating differences in expression patterns between sequelogous *flp*-genes in *Ascaris* and *C. elegans*, they could also be explained by functional redundancy where different FLPs fulfil the role of their knocked out counterparts. However it is more likely that the differences noted reflect the inability of the reverse genetic methodology and phenotype scoring technique used by Nelson et al[48] and Li et al[49] to delineate subtle or complex phenotypes. Li discusses this drawback and indicates that novel assays have been developed that address the limitations of the current reverse genetic methodology and that most of the *flp*-mutants are currently undergoing rescreening.[40]

FLP Function Determined by RNA Interference Studies—Globodera pallida *and* Caenorhabditis elegans

RNA interference (RNAi) provides a novel mechanism to knock down selected genes and offers a rapid and efficient method to determine gene function. This has particular relevance to nematode parasites, many of which have remained unsuited to the reverse genetics approach described above for *C. elegans* (due to the limitations in the transgenic methodologies for nematode parasites and the difficulties associated with in vitro laboratory maintenance that are necessary to establish mutant animals), in addition to the lack of alternative bioassays. Since the discovery of RNAi in *C. elegans*,[53] RNAi techniques have been developed for a number of nematode parasites including the animal parasites, *Nippostrongylus brasiliensis*[54], *Haemonchus contortus*[55-58] and *Brugia malayi*[59,60] and the plant parasites, *Heterodera glycines*,[61-63] *Meloidogyne incognita*[64-68] and *Globodera pallida*,[61,69] amongst others, albeit with variable success (see Knox et al[70]). In large scale RNAi screens in *C. elegans*, several phenotypes have been noted for only four *flp*-genes including embryonic lethal (*flp*-2, -18 and -24), slow growth (*flp*-6 and -2) and larval arrest (*flp*-2).[71-75] This may not be a true reflection of *flp*-gene function as it has been reported that some genes (especially those that are neuronally-based) are less sensitive to RNAi;[76-80] indeed in the RNAi screen, phenotypes were not noted for *flp*-1, -8, -9 and -21 despite the abnormalities determined in knock-out studies.[40] It should be noted that there were some differences in the types of behavioural assays used to determine phenotype. Several studies have demonstrated that use of mutant strains (*rrf*-3, *eri*-1 and *unc*-13 knockouts) enhances the sensitivity of neuronal-expressed transcripts to RNAi,[74,78,81-83] however these mutants have not been used in RNAi screens to investigate *flp*-gene knock-down. More recently, Esposito et al[84] described a transgene approach to silencing specific neuronal genes in *C. elegans* highlighting an additional route of overcoming RNAi insensitivity.

In spite of the problems encountered, the technique of RNAi has been successfully adapted to the study of plant parasitic nematode neuropeptide function. Using simple soaking methodologies to introduce the dsRNA, Kimber et al[69] showed the susceptibility of five *flp*-genes to RNAi in the infective stage (J2) of the potato cyst nematode, *G. pallida*, confirming the presence of a fully functional RNAi pathway in this parasite. Silencing of *Gp-flp*-1, -6, -12, -14 and -18 resulted in profound motor dysfunction that was assessed by a reduction in the ability of the J2s, soaked in the appropriate dsRNA construct, to migrate a sand column relative to water-soaked control worms. In addition, silencing resulted in a number of phenotypes that were distinct from the normal sinusoidal body form and movement associated with the plant parasitic nematodes, including slow movement, twitching, coiled, poker-straight and stationary phenotypes. The results observed in this study are consistent with the gene expression data in *Globodera,* [85] where three of the genes examined were found to be expressed in motorneurons (*Gp-flp*-12, -14 and -18) and one in sensory neurons (*Gp-flp*-1) that may correspond to potential loss of geotactic behaviour impairing migration ability in silenced worms and available physiology data in other nematode species. Inconsistencies were noted with the transgenic studies of the *flp*-gene knockouts in *C. elegans* for *flp*-6 and -12 where no phenotype was observed in mutant animals; the reasons for these differences remain unclear but may point towards real functional differences between nematode species as noted for the differences in FLP-induced pharyngeal pumping activities between *A. suum* and *C. elegans.*

NLP Function in Nematodes: Introduction

Despite their diversity in terms of number and structure in *C. elegans* (see Chapter 4 for details), which appears to be reflected across the nematode phyla (McVeigh, Marks, Day and Maule; personal communication), we know very little about the function of NLPs in nematodes. Apart from data that have arisen from large scale RNAi screens, the limited information we have is based on three other publications; two probing the effects of NLPs on locomotory behavior, cAMP levels and body wall muscle activities in *Ascaris*[47,86] and another implicating an anti-microbial role for some NLPs in *C. elegans,*[87] in addition to some unpublished material.

NLP Function in Control and Modulation of Somatic Body Wall and Reproductive Muscle—Ascaris suum

McVeigh et al[86] showed that DYRPLQFa induced a nerve-cord independent excitatory effect (reminiscent of bwRT3) on both dorsal and ventral muscle strips that appeared to be insensitive to external high K^+ and Ca^{2+}; DYRPLQFa was inactive on the *Ascaris* ovijector. The activities of three additional NLPs (AIPFNGGMYa, AFAAGWNRa and GGWa representing *nlp*-10, -23 and -29) have also been examined on *Ascaris* somatic and ovijector muscle; only AFAAGWNRa was active on the body wall muscle inducing a slow-onset inhibition that was comparable to bwRT1—all of the other peptides examined were inactive on the body wall and ovijector (Mousley, Marks, Maule; unpublished observations). It is interesting to note that *nlp*-15 and *nlp*-3 have been shown to be expressed in the egg-regulating neuron, HSN.

NLP Function in Behaviour, Locomotion and Body Posture—Ascaris suum

In their study on the effects of FLPs on locomotory behaviour and cAMP levels in *Ascaris* Reinitz et al[47] also examined the activities of two NLPs, DYRPLQF (*nlp*-12) and NDFSRDIMSFa (*nlp*-13). Whilst NDFSRDIMSFa was inactive, injection of DYRPLQFa induced ventral coiling but had no effect on cAMP levels.

NLP Role as Anti-Microbial Peptides—Caenorhabditis elegans

A number of the *nlp* genes have been shown have anti-microbial properties. Couillalt et al,[87] following a cDNA microarray analysis of 8,000 genes between noninfected *C. elegans* and worms that had been exposed to both fungi (*Drechmeria coniospora*) and bacteria (*Serratia marcescens*), reported upregulation of three *nlp*-genes in infected animals; *nlp*-29 was upregulated following either fungal or bacterial exposure whereas *nlp*-31 and *nlp*-33 were only unregulated following fungal infection. The mature peptide encoded on *nlp*-31 was shown to have anti-fungal activity against *D. coniospora, Neurospora crassa* and *Aspergillus fumigatus.*[87]

FLP Function Determined by RNA Interference Studies—Caenorhabditis elegans

Knock-down of five *C. elegansnlp* genes (*nlp*-11, -17, -20, -39 and -40) resulted in a range of different phenotypes. Notably, silencing *nlp*-20 and *nlp*-39 produced an embryonic lethal phenotype[71-73,76,88] and suppression of *nlp*-11 and *nlp*-17 resulted in fat-content increases and abnormalities in morphology respectively. Silencing *nlp*-40 expression produced a variety of phenotypes including sterile progeny, sick and slow growth animals.[71]

INS Function Determined by Gene Knockout Studies—Caenorhabditis elegans

To date, more than 38 genes (*ins*-1 to *ins*-37 and *daf*-28) encoding insulin-like peptides have been identified in *C. elegans*. With respect to function, insulin-related peptides have been implicated mainly in reproductive development and, in particular, dauer formation. Kimura et al[89] and Riddle and Albert[90] showed that an insulin-like peptide receptor (DAF-2) mediates a pathway determining dauer formation; in addition, *daf*-28 knockout mutants displayed short-lived dauer formation.[91] *Ins*-1 and-9 loss-of function mutants showed no dauer phenotype, but overexpression of *ins*-1 caused a low level of dauer arrest in wildtype animals that was further enhanced in *daf*-2 and *daf*-7 mutants.[92] Larval arrest resulted following overexpression of *ins*-9 in a *daf*-2 mutant but not in wildtype *C. elegans*. These findings indicate that *ins*-1 and *ins*-9 peptides can modulate the DAF-2 signaling pathway thereby promoting dauer formation; similar results were reported for *ins*-18.[92] Insulin-like peptides also appear to play a role in determining lifespan and limiting body size as determined by studies on the DAF-2 receptor by Kenyon et al[93] and McCulloch and Gems.[94]

Neuropeptide Function in Platyhelminths

Although relatively less is known about the function of neuropeptides in flatworms, that which is known points to neuropeptidergic systems as attractive sources for novel drug targets within the phylum. The peptide families which are known to be present, the FLPs and the NPFs, both produce potent effects in critical aspects of worm biology.

Neuropeptidergic transmission is very important in platyhelminths, even though there is presently less evidence for the broad diversity of neuropeptides in flatworms as compared to nematodes and arthoropods. It is not clear if this lack of diversity of known neuropeptides in flatworms actually reflects the presence of a more limited repertoire, or is simply a function of our limited knowledge. Platyhelminths have, as a phylum, been less tractable to some of the approaches that have yielded knowledge regarding the diversity of neuropeptides in other phyla. There is, however, more than enough structural and functional data to be sure that neuropeptides play a central role in the biology of flatworms as they do in other invertebrate phyla.

As in nematodes, our understanding of the role of neuropeptides in flatworm physiology is derived from both parasitic and free-living species. The functional data is derived from a range of parasitic flatworms including flukes, tapeworms and ectoparasitic mongeneans. These parasitic species provide the most direct insight into the possibility of neuropeptidergic systems yielding useful drug targets, but it is often difficult to obtain large quantities of the parasites and they remain difficult to manipulate genetically and otherwise. In contrast, the free-living flatworms are relatively simple to culture, it is possible to obtain larger numbers of worms and they are markedly more amenable to genetic manipulation. As related below, the functional data suggest a conservation of neuropeptide function spanning the parasitic and free-living flatworms, indicating that the free-living worms will be useful models for neuropeptide function in parasitic worms.

FLP Function in Platyhelminths

FMRFamide-like peptides (FLPs) are known to be myoexcitatory in a number of different flatworm preparations. FLPs are abundantly distributed throughout the nervous systems of every flatworm that has been examined (for review, see Chapter 4) and this includes marked representation in the peripheral nervous elements serving the somatic musculature. This distribution suggests a role for FLPs in flatworm muscle control and physiological data confirm this.

Physiological data demonstrate the myoexcitatory activity of flatworm FLPs in varied flatworm muscle preparations; no inhibitory effects of FLPs have yet to be reported in platyheminths, which is in contrast to nematodes. In schistosomes, FLPs are potently myoexcitatory when applied to individual, dispersed muscle fibers.[95] Specifically, the FLPs GYIRFamide, YIRFamide and RYIRFamide elicited contractions at concentrations as low as 0.1 nM with half- effective concentrations in the 1-10 nM range.[96] In *Fasciola*, FLPs are potently myoexcitatory when applied to muscle strips. Specifically, GYIRFamide and RYIRFamide were both myoexcitatory, with GYIRFamide causing an increase in both the amplitude and frequency of contractions at a threshold of 50 nM.[97] In the free-living flatworm *Procerodes littoralis*, FLPs are potently myoexcitatory when applied to individual muscle fibers. Both YIRFamide and GYIRFamide showed thresholds at 0.1 nM and GYIRFamide was most potent with a half-effective concentration in the nanomolar range.[98] In the mongenean *Diclidophora merlangi*, these YIRFamide-containing FLPs are also myoexcitatory. Using worms with the lateral edges trimmed for peptide access, YIRFamide, RYIRFamide and GYIRFamide all contracted the worms, with GYIRFamide providing the most dramatic effects.[99] Lastly, in the cestode *Mesocestoides corti* the motility of tetrahyridial larvae was increased in the presence of the platyhelminth FLPs.[100] The cestode-derived neuropeptide GNFFRFamide was the most potent, with a threshold of 100 nM. In summary, potent myoexcitation is the uniform physiological effect for FLPs in flatworms.

A couple of patterns are worth noting. For all the muscle preparations examined from monogeneans, trematodes and turbellarians, all of the YIRFamide-containing motifs had efficacy and GYIRFamide tended to be the most potent and effective. However, the YIRFamide-containing FLPs have all been discovered in turbellarians and, as of time of this publication, the presence of these motifs in trematode FLPs has not been confirmed. The potent effects of FLPs derived from free-living flatworms on the important parasites *S. mansoni* and *F. hepatica* support the idea that there is sufficient conservation of FLP structure within the phylum such that free-living worms could prove to be very useful models for uncovering FLP function in flatworm parasites.

The only FLP thus far identified in parasitic flatworm is from a cestode, GNFFRFamide from the giant sheep tapeworm *Moneizia expansa*.[101] The GNFFRFamide structure is notably distinct from the YIRFamide-containing turbellarian-derived FLPs discussed above. However, in most of the preparations discussed above, GNFFRFamide proved to have some excitatory activity, although it was significantly lower than the turbellarian FLPs. GNFFRFamide was about 100 times less potent than RYIRFamide on schistosome muscles and on the *Procerodes* muscle fibers it was 10 times less potent. On the only cestode preparation examined, GNFFRFamide was significantly more potent than the YIRFamide-containing neuropeptides. This supports the supposition that GNFFRFamide is a somewhat structurally-unique cestode equivalent of the more classical FLPs described in other flatworms.

There is evidence that the FLP-induced myoexcitation in *Fasciola* muscle strips is mediated by a phosphoplipase C (PLC)—protein kinase C (PKC) pathway.[102] Specifically, the PLC inhibitor neomycin sulfate did not inhibit baseline contractility in muscle strips, but it did inhibit the GYIRFamide response. Likewise, 10 μM of the PKC inhibitor chelerythrine chloride was without effect on basal contractility, but it almost abolished the GYIRFamide response. These data suggest that the FLP-induced myoexcitation in flatworms is PKC-mediated.

In summary, the single function associated with platyhelminth FLPs to date is myoexcitation. As discussed in the previous chapter, the immunocytochemical distribution of flatworm FLPs suggests a role for these neuropeptides in somatic muscle control. Taken together, the location and function of FLPs in flatworms advance the hypothesis that these are important excitatory neuromuscular transmitters in platyhelminths.

NPF Function in Platyhelminths

NPFs are linked to cAMP inhibition in flatworms, although few of the specifics of NPF function in flatworms are known. NPFs were first discovered in the cestode *M. expansa* in 1991[103] and that discovery hinged on the flatworm neuropeptide cross reacting with antisera targeting veterbrate NPY family peptides. Subsequent resolution of the primary sequence revealed that the tapeworm peptide not only shared immunoreactiivty with vertebrate NPYs, but it also shared some important, conserved amino acid structures. Specifically, the first flatworm NPF proved similar in size (39 amino acids, while NPYs are almost always 36 amino acids), it featured a conserved carboxy-terminal structure of R-X-R-Y/Famide and it featured conserved tyrosine residues at postions 10 and 17 relative to the carboxy terminus. Other NPFs subsequently discovered in the flatworms[104-106] shared these structural elements of vertebrate NPY family peptides. These now include known structures from the cestode *M. expansa*, the trematodes *S. mansoni* and *S. japonicum* and the terrestrial turbellarian *Arthurdendyus trinagulatus*, as well as a large number of NPFs from other invertebrate phyla.[107-112] Later, identification of the NPF-encoding gene in *M. expansa* revealed striking similarity in gene structure between the flatworm NPF and and vertebrate NPYs.[106]

Immunocytochemistry, conserved primary structural elements and conserved gene structure strongly suggest orthology between invertebrate NPFs and vertebrate NPY family peptides. This produced the hypothesis that flatworm NPFs would intitate the same singaling pathway utilized by vertebrate NPYs, which are overwhelmingly coupled to the inhibition of cAMP. Inhibition of cAMP is so closely associated with NPYs that it has been referred to as the universal signaling mechanism associated with NPY family peptides.[113,114] Indeed, schistosome NPF does potently inhibit cAMP accumulation in schistosome homogenates,[105] further linking invertebrate NPFs to vertebrate NPYs.[115]

Schistosome NPF's inhibition of cAMP in schistosome homogenates is very potent, with significant effects evident at 10^{-11} M and a half-inhibitory concentration of 170 pM.[105] The inhibition could also be produced by structurally similar and closely-related peptides, but these were not nearly as potent. For example, *Moniezia* NPF and porcine NPY also produce concentration-dependent inhibition, but are about 1000-fold less potent. The data clearly link flatworm NPF to cAMP inhibition; the potency and specificity of NPF inhibition of cAMP in schistosomes leaves little uncertainty that this is the associated endogenous biochemical pathway for NPF in the worms. However, exactly which physiological responses are elicited remains unknown.

There are also data suggesting a role for NPF in platyhelminth muscle control. When truncated *Moniezia* NPF was applied to *Fasciola* muscle strips, it proved myoexcitatory.[116] However, the truncated NPF had a threshold of 10 μM, which was 10,000 times less potent than the flatworm FLP RYIRFamide in the same assay. *Moniezia* NPF also proved to be myoexcitatory when applied to intact *Mesocestoides vogae* larvae.[117] Again, the NPF myoexcitation in the tapeworm larvae required 10 μM, a concentration significantly higher than is needed for FLP (GNFFRFamide) myoexcitation. Since very high concentrations of NPF are required for myoexcitation and since NPFs share a RFamide carboxy-terminal motif with more potent myoexcitatory FLPs, NPF myoexcitation could be attributed to a somewhat nonspecific interaction of the NPF with a FLP receptor. However, in *M. vogae* larvae, the FLP effect was blocked by GNFFR(d)Famide, but it did not block NPF myoexcitation.[117] This would indicate that NPF is triggering myoexcitation through a receptor distinct from the one used by GNFFRFamide.

In summary, platyhelminth NPFs are abundantly distributed throughout the nervous system of every flatworm examined and NPFs have potent control over cAMP, one of the most important intracellular signaling molecules throughout all animal phyla. However, outside of some slender indication that NPFs may have a role in somatic muscle control, there is not yet direct evidence attributing NPF to any particular biological function.

Conclusion

The focus of this book is the consideration of neuropetide signaling as a potential target for novel chemotherapeutics. In both nematodes and platyhelminths, there is an abundance of physiological evidence demonstrating that neuropeptides play critical roles in the biology of helminths, both free-living and parasitic.

One very substantial example is the obvious role for FLPs in somatic muscle function in both phyla. In the important nematode parasite *Ascaris*, a broad range of FLPs have potent and varied actions on the somatic musculature. In the important platyhelminth parasites, *Schistosoma* and *Fasciola*, FLPs are potently myoexcitatory, with specific effects below the nanomolar range. There is compelling physiological evidence that FLPs have a critical role in endogenous somatic muscle control. Since helminth parasites require directed muscle function for survival in their hosts and since so many effective anthelmintics work at the level of the neuromusculature, such a potent action on the somatic musculature must present itself as a reasonable candidate for chemotherapeutic interference. Adding to their appeal is the fact that similar neuropeptides have no analogous role in the mammalian hosts of these parasites.

One alluring point raised by the physiological studies is the cross-phyla action of some FLPs; some platyhelminth FLPs elicit physiological responses in nematodes and some nematode FLPs are active in platyhleminths. The physiological actions of these peptides across these phyla boundaries suggest that there could be sufficient conservation of structure amongst these neuropeptides and their receptors such that drugs targeting them could have effect against worms from both phyla.

It is certainly true that we have a great deal to learn about the biology of FLPs in both phyla, but the physiological evidence presently available points to FLP signaling as a very rationale field from which to harvest future drug targets.

There is markedly less known about the physiological roles of INSs and NLPs in nematodes and NPFs in platyhelminths. However, their abundant distribution and the initial physiological findings suggest that these neuropeptide families could have potential to yield new drug targets. In all cases, it is reasonable to expect more physiology studies on the role of these peptide families. Elucidation of their biological function(s) will address their validity as sources of drug targets.

Acknowledgements

The authors wish to acknowledge the support of the National Institutes of Health (R01 AI49162) and the Department of Agriculture and Rural Development for Northern Ireland.

References

1. Cowden C, Stretton AO. Eight novel FMRFamide-like neuropeptides isolated from the nematode Ascaris suum. Peptides 1995; 16:491-500.
2. Bowman JW, Friedman AR, Thompson DP et al. Structure-activity relationships of KNEFIRFamide (AF1), a nematode FMRFamide-related peptide, on Ascaris suum muscle. Peptides 1996; 17:381-387.
3. Cowden C, Stretton AO. AF2, an Ascaris neuropeptide: isolation, sequence and bioactivity. Peptides 1993; 14:423-430.
4. Cowden C, Stretton AO, Davis RE. AF1, a sequenced bioactive neuropeptide isolated from the nematode Ascaris suum. Neuron 1989; 2:1465-73.
5. Maule AG, Geary TG, Bowman JW et al. Inhibitory effects of nematode FMRFamide-related peptides (FaRPs) on muscle strips from Ascaris suum. Invert Neurosci 1995; 1:255-265.
6. Pang FY, Mason J, Holden-Dye L et al. The effects of the nematode peptide, KHEYLRFamide (AF2), on the somatic musculature of the parasitic nematode Ascaris suum. Parasitol 1995; 110:353-362.
7. Thompson DP, Davis JP, Larsen MJ et al. Effects of KHEYLRFamide and KNEFIRFamide on cyclic adenosine monophosphate levels in Ascaris suum somatic muscle. Int J Parasitol 2003; 33:199-208.
8. Trailovic SM, Clark CL, Robertson AP et al. Brief application of AF2 produces long lasting potentiation of nAChR responses in Ascaris suum. Mol Biochem Parasitol 2005; 139:51-64.
9. Verma S, Robertson AP, Martin RJ. The nematode neuropeptide, AF2 (KHEYLRF-NH2), increases voltage-activated calcium currents in Ascaris suum muscle. Br J Pharmacol 2007; 151:888-899.
10. Marks NJ, Sangster NC, Maule AG et al. Structural characterisation and pharmacology of KHEYLR-Famide (AF2) and KSAYMRFamide (PF3/AF8) from Haemonchus contortus. Mol Biochem Parasitol 1999; 100:185-94.
11. Bowman JW, Winterrowd CA, Friedman AR et al. Nitric oxide mediates the inhibitory effects of SDPNFLRFamide, a nematode FMRFamide-related neuropeptide, in Ascaris suum. J Neurophysiol 1995; 74:1880-1888.
12. Franks CJ, Holden-Dye L, Williams RG et al. A nematode FMRFamide-like peptide, SDPNFLRFamide (PF1), relaxes the dorsal muscle strip preparation of Ascaris suum. Parasitol 1994; 108:229-236.
13. Holden-Dye L, Franks CJ, Williams RG et al. The effect of the nematode peptides SDPNFLRFamide (PF1) and SADPNFLRFamide (PF2) on synaptic transmission in the parasitic nematode Ascaris suum. Parasitol 1995; 110:449-455.
14. Davis RE, Stretton AO. The motornervous system of Ascaris: electrophysiology and anatomy of the neurons and their control by neuromodulators. Parasitol 1996; 113:S97-117.
15. Marks NJ, Maule AG, Geary TG et al. APEASPFIRFamide, a novel FMRFamide-related decapeptide from Caenorhabditis elegans: structure and myoactivity. Biochem Biophys Res Commun 1997; 231:591-5.
16. Marks NJ, Shaw C, Halton DW et al. Isolation and preliminary biological assessment of AADGAPLIR-Famide and SVPGVLRFamide from Caenorhabditis elegans. Biochem Biophys Res Commun 2001; 286:1170-6.

17. Mousley A, Marks NJ, Halton DW et al. Arthropod FMRFamide-related peptides modulate muscle activity in helminths. Int J Parasitol 2004; 34:755-68.
18. Yew JY, Davis R, Dikler S et al. Peptide products of the afp-6 gene of the nematode Ascaris suum have different biological actions. The Journal of Comparative Neurology 2007; 502:872-882.
19. Holden-Dye L, Brownlee DJ, Walker RJ. The effects of the peptide KPNFIRFamide (PF4) on the somatic muscle cells of the parasitic nematode Ascaris suum. Br J Pharmacol 1997; 120:379-386.
20. Maule AG, Shaw C, Bowman JW et al. Isolation and preliminary biological characterization of KPNFIR-Famide, a novel FMRFamide-related peptide from the free-living nematode, Panagrellus redivivus. Peptides 1995; 16:87-93.
21. Kubiak TM, Maule AG, Marks NJ et al. Importance of the proline residue to the functional activity and metabolic stability of the nematode FMRFamide-related peptide, KPNFIRFamide (PF4). Peptides 1996; 17:1267-1277.
22. Geary TG, Bowman JW, Friedman AR et al. The pharmacology of FMRFamide-related neuropeptides in nematodes: new opportunities for rational anthelmintic discovery? Int J Parasitol 1995; 25:1273-1280.
23. Purcell J, Robertson AP, Thompson DP et al. The time-course of the response to the FMRFamide-related peptide PF4 in Ascaris suum muscle cells indicates direct gating of a chloride ion-channel. Parasitol 2002; 124:649-656.
24. Purcell J, Robertson AP, Thompson DP et al. PF4, a FMRFamide-related peptide, gates low-conductance Cl- channels in Ascaris suum. Eur J Pharmacol 2002; 456:11-17.
25. Trim N, Holden-Dye L, Ruddell R et al. The effects of the peptides AF3 (AVPGVLRFamide) and AF4 (GDVPGVLRFamide) on the somatic muscle of the parasitic nematodes Ascaris suum and Ascaridia galli. Parasitol 1997; 115:213-222.
26. Trim N, Brooman JE, Holden-Dye L et al. The role of cAMP in the actions of the peptide AF3 in the parasitic nematodes Ascaris suum and Ascaridia galli. Mol Biochem Parasitol 1998; 93:263-71.
27. Maule AG, Shaw C, Bowman JW et al. KSAYMRFamide: a novel FMRFamide-related heptapeptide from the free-living nematode, Panagrellus redivivus, which is myoactive in the parasitic nematode, Ascaris suum. Biochem Biophys Res Commun 1994; 200:973-980.
28. Olsen LS, Kelley GW, Sen HG. Longevity and egg-production of Ascaris suum. Trans Am Microsc Soc 1958; 77:380-383.
29. Fellowes RA, Dougan PM, Maule AG et al. Neuromusculature of the ovijector of Ascaris suum (Ascaroidea, Nematoda): an ultrastructural and immunocytochemical study. J Comp Neurol 1999; 415:518-28.
30. Fellowes RA, Maule AG, Marks NJ et al. Modulation of the motility of the vagina vera of Ascaris suum in vitro by FMRFamide-related peptides. Parasitol 1998; 116:277-287.
31. Fellowes RA, Maule AG, Marks NJ et al. Nematode neuropeptide modulation of the vagina vera of Ascaris suum: in vitro effects of PF1, PF2, PF4, AF3 and AF4. Parasitol 2000; 120:79-89.
32. Marks NJ, Maule AG, Li C et al. Isolation, pharmacology and gene organization of KPSFVRFamide: a neuropeptide from Caenorhabditis elegans. Biochem Biophys Res Commun 1999; 254:222-30.
33. Moffett CL, Beckett AM, Mousley A et al. The ovijector of Ascaris suum: multiple response types revealed by Caenorhabditis elegans FMRFamide-related peptides. Int J Parasitol 2003; 33:859-876.
34. Kim K, Li C. Expression and regulation of an FMRFamide-related neuropeptide gene family in Caenorhabditis elegans. J Comp Neurol 2004; 475:540-50.
35. Trent C, Tsuing N, Horvitz HR. Egg-laying defective mutants of the nematode Caenorhabditis elegans. Genetics 1983; 104:619-647.
36. Papaioannou S, Marsden D, Franks CJ et al. Role of a FMRFamide-like family of neuropeptides in the pharyngeal nervous system of Caenorhabditis elegans. J Neurobiol 2005; 65:304-319.
37. Rogers CM, Franks CJ, Walker RJ et al. Regulation of the pharynx of Caenorhabditis elegans by 5-HT, octopamine and FMRFamide-like neuropeptides. J Neurobiol 2001; 49:235-244.
38. Brownlee DJ, Holden-Dye L, Fairweather I et al. The action of serotonin and the nematode neuropeptide KSAYMRFamide on the pharyngeal muscle of the parasitic nematode, Ascaris suum. Parasitol 1995; 111:379-384.
39. Brownlee DJ, Walker RJ. Actions of nematode FMRFamide-related peptides on the pharyngeal muscle of the parasitic nematode, Ascaris suum. Ann N Y Acad Sci 1999; 897:228-238.
40. Li C. The ever-expanding neuropeptide gene families in the nematode Caenorhabditis elegans. Parasitol 2005; 131:S109-27.
41. Davis RE, Stretton AO. Structure-activity relationships of 18 endogenous neuropeptides on the motor nervous system of the nematode Ascaris suum. Peptides 2001; 22:7-23.
42. Angstadt JD, Donmoyer JE, Stretton AO. The number of morphological synapses between neurons does not predict the strength of their physiological synaptic interactions: a study of dendrites in the nematode Ascaris suum. The Journal of Comparative Neurology 2001; 432:512-527.

43. Stretton AO, Fishpool RM, Southgate E et al. Structure and physiological activity of the motoneurons of the nematode Ascaris. Proc Natl Acad Sci USA 1978; 75:3493-3497.
44. McVeigh P, Geary TG, Marks NJ et al. The FLP-side of nematodes. Trends Parasitol 2006; 22:385-396.
45. Geary TG, Marks NJ, Maule AG et al. Pharmacology of FMRFamide-related peptides in helminths. Ann N Y Acad Sci 1999; 897:212-27.
46. Rex E, Harmych S, Puckett T et al. Regulation of carbohydrate metabolism in Ascaris suum body wall muscle: a role for the FMRFamide AF2, not serotonin. Mol Biochem Parasitol 2004; 133:311-313.
47. Reinitz CA, Herfel HG, Messinger LA et al. Changes in locomotory behavior and cAMP produced in Ascaris suum by neuropeptides from Ascaris suum or Caenorhabditis elegans. Mol Biochem Parasitol 2000; 111:185-97.
48. Nelson LS, Rosoff ML, Li C. Disruption of a neuropeptide gene, flp-1, causes multiple behavioral defects in Caenorhabditis elegans. Science 1998; 281:1686-90.
49. Li C, Nelson LS, Kim K et al. Neuropeptide gene families in the nematode Caenorhabditis elegans. Ann N Y Acad Sci 1999; 897:239-52.
50. Waggoner LE, Hardaker LA, Golik S et al. Effect of a neuropeptide gene on behavioral states in Caenorhabditis elegans egg-laying. Genetics 2000; 154:1181-1192.
51. Rogers C, Reale V, Kim K et al. Inhibition of Caenorhabditis elegans social feeding by FMRFamide-related peptide activation of NPR-1. Nat Neurosci 2003; 6:1178-85.
52. Yew JY, Kutz KK, Dikler S et al. Mass spectrometric map of neuropeptide expression in Ascaris suum. J Comp Neurol 2005; 488:396-413.
53. Fire A, Xu S, Montgomery MK et al. Potent and specific genetic interference by double-stranded RNA in Caenorhabditis elegans. Nature 1998; 391:806-11.
54. Hussein AS, Kichenin K, Selkirk ME. Suppression of secreted acetylcholinesterase expression in Nippostrongylus brasiliensis by RNA interference. Mol Biochem Parasitol 2002; 122:91-94.
55. Geldhof P, Murray L, Couthier A et al. Testing the efficacy of RNA interference in Haemonchus contortus. Int J Parasitol 2006; 36:801-810.
56. Geldhof P, Whitton C, Gregory WF et al. Characterisation of the two most abundant genes in the Haemonchus contortus expressed sequence tag dataset. Int J Parasitol 2005; 35:513-522.
57. Kotze AC, Bagnall NH. RNA interference in Haemonchus contortus: suppression of beta-tubulin gene expression in L3, L4 and adult worms in vitro. Mol Biochem Parasitol 2006; 145:101-110.
58. Zawadzki JL, Presidente PJ, Meeusen EN et al. RNAi in Haemonchus contortus: a potential method for target validation. Trends Parasitol 2006; 22:495-499.
59. Aboobaker AA, Blaxter ML. Use of RNA interference to investigate gene function in the human filarial nematode parasite Brugia malayi. Mol Biochem Parasitol 2003; 129:41-51.
60. Behm CA, Bendig MM, McCarter JP et al. RNAi-based discovery and validation of new drug targets in filarial nematodes. Trends Parasitol 2005; 21:97-100.
61. Urwin PE, Lilley CJ, Atkinson HJ. Ingestion of double-stranded RNA by preparasitic juvenile cyst nematodes leads to RNA interference. Mol Plant Microbe Interact 2002; 15:747-752.
62. Lilley CJ, Goodchild SA, Atkinson HJ et al. Cloning and characterisation of a Heterodera glycines aminopeptidase cDNA. Int J Parasitol 2005; 35:1577-1585.
63. Alkharouf NW, Klink VP, Matthews BF. Identification of Heterodera glycines (soybean cyst nematode [SCN]) cDNA sequences with high identity to those of Caenorhabditis elegans having lethal mutant or RNAi phenotypes. Exp Parasitol 2007; 115:247-258.
64. Rosso MN, Dubrana MP, Cimbolini N et al. Application of RNA interference to root-knot nematode genes encoding esophageal gland proteins. Mol Plant Microbe Interact 2005; 18:615-620.
65. Bakhetia M, Charlton W, Atkinson HJ et al. RNA interference of dual oxidase in the plant nematode Meloidogyne incognita. Mol Plant Microbe Interact 2005; 18:1099-1106.
66. Shingles J, Lilley CJ, Atkinson HJ et al. Meloidogyne incognita: molecular and biochemical characterisation of a cathepsin L cysteine proteinase and the effect on parasitism following RNAi. Exp Parasitol 2007; 115:114-120.
67. Huang G, Allen R, Davis EL et al. Engineering broad root-knot resistance in transgenic plants by RNAi silencing of a conserved and essential root-knot nematode parasitism gene. Proc Natl Acad Sci USA 2006; 103:14302-14306.
68. Fanelli E, Di Vito M, Jones JT et al. Analysis of chitin synthase function in a plant parasitic nematode, Meloidogyne artiellia, using RNAi. Gene 2005; 349:87-95.
69. Kimber MJ, McKinney S, McMaster S et al. flp gene disruption in a parasitic nematode reveals motor dysfunction and unusual neuronal sensitivity to RNA interference. FASEB J 2007; 21:1233-1243.
70. Knox DP, Geldhof P, Visser A et al. RNA interference in parasitic nematodes of animals: a reality check? Trends Parasitol 2007; 23:105-107.

71. Maeda I, Kohara Y, Yamamoto M et al. Large-scale analysis of gene function in Caenorhabditis elegans by high-throughput RNAi. Curr Biol 2001; 11:171-176.
72. Kamath RS, Ahringer J. Genome-wide RNAi screening in Caenorhabditis elegans. Methods 2003; 30:313-321.
73. Sonnichsen B, Koski LB, Walsh A et al. Full-genome RNAi profiling of early embryogenesis in Caenorhabditis elegans. Nature 2005; 434:462-469.
74. Simmer F, Moorman C, van der Linden AM et al. Genome-wide RNAi of C. elegans using the hypersensitive rrf-3 strain reveals novel gene functions. PLoS Biology 2003; 1:E12.
75. Fisk Green R, Lorson M, Walhout AJ et al. Identification of critical domains and putative partners for the Caenorhabditis elegans spindle component LIN-5. Molecular Genetics and Genomics 2004; 271:532-544.
76. Fraser AG, Kamath RS, Zipperlen P et al. Functional genomic analysis of C. elegans chromosome I by systematic RNA interference. Nature 2000; 408:325-330.
77. Asikainen S, Vartiainen S, Lakso M et al. Selective sensitivity of Caenorhabditis elegans neurons to RNA interference. Neuroreport 2005; 16:1995-1999.
78. Kennedy S, Wang D, Ruvkun G. A conserved siRNA-degrading RNase negatively regulates RNA interference in C. elegans. Nature 2004; 427:645-649.
79. Timmons L, Court DL, Fire A. Ingestion of bacterially expressed dsRNAs can produce specific and potent genetic interference in Caenorhabditis elegans. Gene 2001; 263:103-112.
80. Kamath RS, Martinez-Campos M, Zipperlen P et al. Effectiveness of specific RNA-mediated interference through ingested double-stranded RNA in Caenorhabditis elegans. Genome Biology 2000; 2:RESEARCH0002.1-0002.10.
81. Sijen T, Fleenor J, Simmer F et al. On the role of RNA amplification in dsRNA-triggered gene silencing. Cell 2001; 107:465-476.
82. Simmer F, Tijsterman M, Parrish S et al. Loss of the putative RNA-directed RNA polymerase RRF-3 makes C. elegans hypersensitive to RNAi. Curr Biol 2002; 12:1317-1319.
83. Chapin A, Correa P, Maguire M et al. Synaptic neurotransmission protein UNC-13 affects RNA interference in neurons. Biochem Biophys Res Commun 2007; 354:1040-1044.
84. Esposito G, Di Schiavi E, Bergamasco C et al. Efficient and cell specific knock-down of gene function in targeted C. elegans neurons. Gene 2007; 395:170-176.
85. Kimber MJ, Fleming CC, Prior A et al. Localisation of Globodera pallida FMRFamide-related peptide encoding genes using in situ hybridisation. Int J Parasitol 2002; 32:1095-105.
86. McVeigh P, Leech S, Marks NJ et al. Gene expression and pharmacology of nematode NLP-12 neuropeptides. Int J Parasitol 2006; 36:633-640.
87. Couillault C, Pujol N, Reboul J et al. TLR-independent control of innate immunity in Caenorhabditis elegans by the TIR domain adaptor protein TIR-1, an ortholog of human SARM. Nature Immunology 2004; 5:488-494.
88. Rual JF, Ceron J, Koreth J et al. Toward improving Caenorhabditis elegans phenome mapping with an ORFeome-based RNAi library. Genome Res 2004; 14:2162-2168.
89. Kimura KD, Tissenbaum HA, Liu Y et al. daf-2, an insulin receptor-like gene that regulates longevity and diapause in Caenorhabditis elegans. Science 1997; 277:942-946.
90. Riddle DL, Albert PS. Genetic and environmental regulation of dauer larva development. C elegans II 1997:739–768.
91. Malone EA, Thomas JH. A screen for nonconditional dauer-constitutive mutations in Caenorhabditis elegans. Genetics 1994; 136:879-886.
92. Pierce SB, Costa M, Wisotzkey R et al. Regulation of DAF-2 receptor signaling by human insulin and ins-1, a member of the unusually large and diverse C. elegans insulin gene family. Genes Dev 2001; 15:672-86.
93. Kenyon C, Chang J, Gensch E et al. A C. elegans mutant that lives twice as long as wild type. Nature 1993; 366:461-464.
94. McCulloch D, Gems D. Body size, insulin/IGF signaling and aging in the nematode Caenorhabditis elegans. Exp Gerontol 2003; 38:129-136.
95. Day TA, Maule AG, Shaw C et al. Platyhelminth FMRFamide-related peptides (FaRPs) contract Schistosoma mansoni (Trematoda:Digenea) muscle fibres in vitro. Parasitol 1994; 109:455-9.
96. Day TA, Maule AG, Shaw C et al. Structure-activity relationships of FMRFamide-related peptides contracting Schistosoma mansoni muscle. Peptides 1997; 18:917-21.
97. Graham MK, Fairweather I, McGeown JG. The effects of FaRPs on the motility of isolated muscle strips from the liver fluke, Fasciola hepatica. Parasitol 1997; 114:455-65.
98. Moneypenny CG, Kreshchenko N, Moffett CL et al. Physiological effects of FMRFamide-related peptides and classical transmitters on dispersed muscle fibres of the turbellarian, Procerodes littoralis. Parasitol 2001; 122:447-55.

99. Moneypenny CG, Maule AG, Shaw C et al. Physiological effects of platyhelminth FMRF amide-related peptides (FaRPs) on the motility of the monogenean Diclidophora merlangi. Parasitol 1997; 115:281-8.
100. Hrckova G, Velenbny S, Halton DW et al. Mesocestoides corti (syn. M. vogae): modulation of larval motility by neuropeptides, serotonin and acetylcholine. Parasitol 2002; 124:409-21.
101. Maule AG, Shaw C, Halton DW et al. GNFFRFamide: a novel FMRFamide-immunoreactive peptide isolated from the sheep tapeworm, Moniezia expansa. Biochem Biophys Res Commun 1993; 193:1054-60.
102. Graham MK, Fairweather I, McGeown JG. Second messengers mediating mechanical responses to the FaRP GYIRFamide in the fluke Fasciola hepatica. Am J Physiol Regul Integr Comp Physiol 2000; 279: R2089-94.
103. Maule AG, Shaw C, Halton DW et al. Neuropeptide F: a novel parasitic flatworm regulatory peptide from Moniezia expansa (Cestoda: Cyclophyllidea). Parasitol 1991; 102:309-316.
104. Dougan PM, Mair GR, Halton DW et al. Gene organization and expression of a neuropeptide Y homolog from the land planarian Arthurdendyus triangulatus. J Comp Neurol 2002; 454:58-64.
105. Humphries JE, Kimber MJ, Barton YW et al. Structure and bioactivity of neuropeptide F from the human parasites Schistosoma mansoni and Schistosoma japonicum. J Biol Chem 2004; 279:39880-39885.
106. Mair GR, Halton DW, Shaw C et al. The neuropeptide F (NPF) encoding gene from the cestode, Moniezia expansa. Parasitol 2000; 120:71-7.
107. Brown MR, Crim JW, Arata RC et al. Identification of a Drosophila brain-gut peptide related to the neuropeptide Y family. Peptides 1999; 20:1035-42.
108. Huang Y, Brown MR, Lee TD et al. RF-amide peptides isolated from the midgut of the corn earworm, Helicoverpa zea, resemble pancreatic polypeptide. Insect Biochem Mol Biol 1998; 28:345-56.
109. Leung PS, Shaw C, Maule AG et al. The primary structure of neuropeptide F (NPF) from the garden snail, Helix aspersa. Regul Pept 1992; 41:71-81.
110. Stanek DM, Pohl J, Crim JW et al. Neuropeptide F and its expression in the yellow fever mosquito, Aedes aegypti. Peptides 2002; 23:1367-78.
111. Vanden Broeck J. Neuropeptides and their precursors in the fruitfly, Drosophila melanogaster. Peptides 2001; 22:241-54.
112. Rajpara SM, Garcia PD, Roberts R et al. Identification and molecular cloning of a neuropeptide Y homolog that produces prolonged inhibition in Aplysia neurons. Neuron 1992; 9:505-13.
113. Michel MC, Beck-Sickinger A, Cox H et al. XVI International Union of Pharmacology recommendations for the nomenclature of neuropeptide Y, peptide YY and pancreatic polypeptide receptors. Pharmacol Rev 1998; 50:143-50.
114. Michel MC. Receptors for neuropeptide Y: multiple subtypes and multiple second messengers. Trends Pharmacol Sci 1991; 12:389-94.
115. McVeigh P, Kimber MJ, Novozhilova E et al. Neuropeptide signalling systems in flatworms. Parasitol 2006; 131:S41-S55.
116. Marks NJ, Johnson S, Maule AG et al. Physiological effects of platyhelminth RFamide peptides on muscle-strip preparations of Fasciola hepatica (Trematoda: Digenea). Parasitol 1996; 113:393-401.
117. Hrckova G, Velebny S, Halton DW et al. Pharmacological characterisation of neuropeptide F (NPF)-induced effects on the motility of Mesocestoides corti (syn. Mesocestoides vogae) larvae. Int J Parasitol 2004; 34:83-93.

CHAPTER 6

Neuropeptide Gene Families in *Caenorhabditis elegans*

Chris Li* and Kyuhyung Kim

Abstract

Neuropeptides are short sequences of amino acids that function in all multicellular organisms to communicate information between cells. The first sequence of a neuropeptide was reported in 1970[1] and the number of identified neuropeptides remained relatively small until the 1990s when the DNA sequence of multiple genomes revealed treasure troves of information. By blasting away at the genome, gene families, the sizes of which were previously unknown, could now be determined. This information has led to an exponential increase in the number of putative neuropeptides and their respective gene families.

The molecular biology age greatly benefited the neuropeptide field in the nematode *Caenorhabditis elegans*. Its genome was among the first to be sequenced[2] and this allowed us the opportunity to screen the genome for neuropeptide genes. Initially, the screening was slow, as the Genefinder and BLAST programs had difficulty identifying small genes and peptides. However, as the bioinformatics programs improved, the extent of the neuropeptide gene families in *C. elegans* gradually emerged.

Identification of Neuropeptide Genes in *C. elegans*

The identification of neuropeptide genes has been driven by the interests of different researchers. Scans based on similarity to specific peptides or neuropeptide precursor molecules were performed to sample the genome. Because pathways dependent on an insulin-like receptor DAF-2 are involved in metabolism and longevity in *C. elegans*, there was an impetus to identify the DAF-2 insulin ligand. This work led to the identification of a surprisingly large gene family encoding insulin-related peptides; with the exception of *daf-28*, these genes are referred to as *ins* genes[3-7] (Table 1). Our lab started to identify genes encoding FMRFamide (Phe-Met-Arg-Phe-amide)-related peptides, which were initially isolated as a cardioactive peptide[8] but later found to be involved in feeding and pain modulation (for review see ref. 9); these genes were named the *flp* genes.[10-13] Anne Hart examined the *C. elegans* genome for other neuropeptide genes and identified a large number of genes, most of which do not belong to any family;[11,13,14] these genes are collectively referred to as the neuropeptide-like peptide or *nlp* genes. Other groups have added to the continually growing collections of gene families. At present, 119 neuropeptide genes, including 40 insulin-encoding genes, 31 *flp* genes, and 48 *nlp* genes, have been identified in *C. elegans* (Tables 1-3). These genes encode over 250 putative neuropeptides. These numbers, however, are likely to be an underestimate of the actual numbers of genes and neuropeptides in *C. elegans* (see below).

Many of the neuropeptide genes, particularly ones that are in the same family, are clustered on a chromosome; this genomic organization may arise by recent gene duplications. For instance, chromosome I has seven *ins* genes within 25,000 bp; these genes do not appear to be part of an

*Corresponding Author: Chris Li—Department of Biology, City College of the City University of New York. Email: cli@sci.ccny.cuny.edu

Neuropeptide Systems as Targets for Parasite and Pest Control, edited by Timothy G. Geary and Aaron G. Maule. ©2010 Landes Bioscience and Springer Science+Business Media.

***Table 1. Neuropeptide genes encoding insulin-like peptides in* C. elegans**

Gene[#] Cosmid LG	Putative Peptides[@]	Expression Pattern[+]	Function or Phenotype	Receptor	Ref[#]
daf-28 Y116F11B.1 V	VPGVAVRACGRRLVPYVWSVCGDACEPQ EGIDIATQCCTYQCTAEYIQTACCPRLL	ASI, ASJ, PQR, other neurons, hindgut, pharyngeal muscle, hypodermis	promotes reproductive growth	DAF-2	6
ins-1 F13B12.5 IV	SIRLCGSRLTTTLLAVCRNQLCTGLTAF GGIATECCEKRCSFAYLKTFCCNQDDN	ASI, ASJ, ASH, NSM, other neurons, intestine, vulval muscles	DAF-2 antagonist?	DAF-2?	7, 94
ins-2 ZK75.2 II	VQKRLCGRRLILFMLATCGECDTD SSEDLSHICCIKQCDVQDIIRVCCPNSFRK	Amphidial, labial, ventral cord and tail neurons, pharynx, vulva			7
ins-3 ZK75.3 II	GDKVKICGTKVLKMVMVMCGGECSS TNENIATECCEKMCTMEDITTKCCPSR	Amphidial, labial, lateral, ventral cord and dorsal projecting neurons			7
ins-4 ZK75.1 II	VPAGEVRACGRRLLLFVWSTCGEPCTPQ EDMDIATVCCTTQCTPSYIKQACCPEK	Amphidial, labial, ventral cord, dorsal projecting and tail neurons, hypodermis		DAF-2	7, 81
ins-5 ZK84.3 II	ADRHTNYRSCALRLIPHVWSVCGDACQPQ NGIDVAQKCCSTDCSSDYIKETCCPFD	Amphidial, labial, ventral cord, lateral projecting and tail neurons, vulva			7
ins-6 ZK84.6 II	VPAPGETRACGRKLISLVMAVCGDLCNPQ EGKDIATECCGNQCSDDYIRSACCP	Amphidial, labial, ventral cord and tail neurons		DAF-2	7

continued on next page

Table 1. Continued

Gene[#] Cosmid LG	Putative Peptides[@]	Expression Pattern[+]	Function or Phenotype	Receptor	Ref[#]
ins-7 ZK1251.2 IV	VPDEKKIYRCGRRIHSYVFAVCGKACESN TEVNIASKCCREECTDDFIRKQCCP	Amphidial, labial, ventral cord and tail neurons			7
ins-8 ZK1251.11 IV	VPEQKNKLCGKQVLSYVMALCEKACDSN TKVDIATKCCRDACSDEFIRHQCCP	Amphidial, labial, ventral cord and tail neurons, vulva			7
ins-9 C06E2.8 X	TLETEKIYRCGRKLYTDVLSACNGPCEPG TEQDLSKICCGNQCTFVIRKACCADKL	ASI, ASJ	Overexpression causes embryonic and larval arrest	DAF-2?	7
ins-10 T08G5.12 V	AFPFQICVKKMEKMCRIINPEQCAQVNKITEI GALTDCCTGLCSWEEIRISCCSVL				7
ins-11 C17C3.4 II	APHHDKRHTACVLKIFKALNVMCNHEGDAD VLRRTASDCCRESCSLTEMLASCTLTSSEESTRDI	Labial, ventral cord and tail neurons			7
ins-12 C17C3.19 II	APSHEKTHKKCSDKLYLAMKSLCSYRGYSE FLRNSATKCCQDNCEISEMMALCVVAPNFDDDLLH				7
ins-13 C17C3.18 II	NKCOYSKKKYKICGVRALKHMKVYCTRGMTRD YGKLLVTCCSKGCNAIDIQRICL				7

continued on next page

Table 1. Continued

Gene[#] Cosmid LG	Putative Peptides[@]	Expression Pattern[+]	Function or Phenotype	Receptor	Ref[#]
ins-14 F41G3.16 II	SEDIKCDAKFISRITKLCIHGITED KLVRLLTRCCTSHCSKAHLKMFCTLKPHEEEPHHEI				7
ins-15 F41G3.17 II	GNDFQPRDNKHHSYRSCGESLSRRVAFLCNGGAIQT EILRALDCCSTGCTDKQIFSWCDFQI				7
ins-16 Y39A3A.5 III	RELKRCSVKLFDILSVICGTESDAE ILQKVAVKCCQEQCGFEEMCQHANLKIDKI				7
ins-17 F56F3.6 III	GSLKLCPPGGASFLDAFNLICPMRRRRR SVSENYNDGGGSLLGRTMNMCCETGCEFTDIFAICNPFG				7
ins-18 T28B8.2 I	ISLQQADGRMKMCPPGGSTFTMAWSMSCSMRR KRALIAPSIRQLQTICCQVGCNVEDLLAYCAPI	Amphidial, ventral cord, tail and pharyngeal neurons	DAF-2 antagonist?	DAF-2?	7
ins-19 T10D4.13 II	YIIDSSESYEVLMLFGYKRTCGRRLMNRINRVCVKDID PADIDPKIKLSEHCCIKGCTDGWIKKHICSEEVLNFGFFEN		Overexpres-sion causes larval arrest	DAF-2?	7
ins-20 ZK84.7 II	KEPKHHHHHHRHKGYCGVKAVKKLKQICPDLCSNVDD NLLMEMCSKNLTDDDILQRCCPE				7

continued on next page

Table 1. Continued

Gene[#] Cosmid LG	Putative Peptides[@]	Expression Pattern[+]	Function or Phenotype	Receptor	Ref[#]
ins-21 M04D8.1 III	SKSHSKKHVRFLCATKAVKHIRKVCPDMCLTGE EVEVNEFCRMGYSDSQIKYICCPE	Amphidial, ventral cord and tail neurons			7
ins-22 M04D8.2 III	MDAHTDKYVRTLCGKTAIRNIANLCPPKPEMKGICSTGE YPSITEYCSMGFSDSQIKFMCCDNQ	Amphidial, labial, ventral cord, lateral process projecting and tail neurons	Modulates acetylcholine signaling		7, 27
ins-23 M04D8.3 III	QVTDAHSELHVRRVCGTAIIKNIMRLCPGVPACENGE VPSPTEYCSMGYSDSQVKYLCCPTSQ	Amphidial, labial and ventral cord neurons			7
ins-24 ZC334.3 I	MGLIRANQGPQKACGRSMMMKVQKLCAGGCTIQNDD LTIKSCSTGYTDAGFISACCPSGFVF				7
ins-25 ZC334.8 I	KPEAQRRCGRYLIRFLGELCNGPCSGVSSVD IATIACATAVPIEDLKNMCCPNL				7
ins-26 ZC334.1 I	IGNHHHGTKAGLTCGMNIIERVDQLCNGQCTRNYDA LVIKSCHRGVSDMEFMVACCPTMKLFIH				7
ins-27 ZC334.11 I	FLAPSTAAKRRCGRRLIPYVYSICGGPCENGD IIIEHCFSGTTPTIAEVQKACCPELSEDPTFSS				7

continued on next page

Table 1. Continued

Gene[#] Cosmid LG	Putative Peptides[@]	Expression Pattern[+]	Function or Phenotype	Receptor	Ref[#]
ins-28 ZC334.9 I	ASPTCGRALLHRIQSVCGLCTIDAHHE LIAIACSRGLGDKEIIEMCCPI				7
ins-29 ZC334.10 I	DFGAQRRCGRHLVNFLEGLCGGPCSEAPTVE LASWACSSAVSIQDLEKLCCPSNLA				7
ins-30 ZC334.2 I	REPVVAAQGAKKTCGRSLLIKIQQLCHGICTVHADD LHETACMKGLTDSQLINSCCPPIPQTPFVF				7
ins-31 a T10D4.4 II	FVHHFDHSMFARPEKTCGGLLIRRVDRICPNLNY TYKIEWELMDNCCEVVCEDQWIKETFCRAPRFNFFGPSF		Overexpres-sion causes larval arrest[$]; modulates acetylcholine signaling	DAF-2?	7[$],27
ins-31 b	KALERSCGPKLFTRVKTVCGE DINVDNKVKISDHCCTPEGGCTDDWIKENVCKQTRFNFFRQFL				7[$]
ins-31 c	DSPQRSCGPQLFKRVNTLCNE NINVENNVSVSKSCCESAAGCTDDWIKKNVCTQHKPFVFRPGFY				7[$]
ins-32 Y8A9A.6 II	RSRRELICGRRLSKTVTNLCVEMN PQKEEDIATKCCKNKGCSREYIKSIMCPDE				7, OST
ins-33 W09C5.4 I	HGQKHCGTKIVRKLQMLCPKMCTISDD TLLTEMCSHSLFDDEIQLRCCPKEDE				7, OST

continued on next page

Table 1. Continued

Gene[#] Cosmid LG	Putative Peptides[@]	Expression Pattern[+]	Function or Phenotype	Receptor	Ref[#]
ins-34 F52B11.6 IV	KTTAAPLAQVNPQCLRRLTLLARGVCRQPCQPSDKPK TSAQQLLQLACSARRPTNEQIISYCCPEKSG				7
ins-35 K02E2.4 V	KMDENAFGINNRHCQRALKVYSFAICGAICQNYEK ILMEGCGSTVMLTMQRTKLICCPEPVDSDELFN				7, OST
ins-36 Y53H1A.4 I	IRKRHPEGKLVIRDCKRYLIMYSRTICKEKCEKFD ERNDITFSINLQFIFTDLLVEGCHSNQTLSNERTRELCCPNAGSN				7
ins-37 F08G2.6 II	NPIHPVPNAAFLPYRSCGSHLVHRAFEACSGKKD RSSDVDLWKMCCKDECTDLDIKESLCKYASQGYGV				7, OST
ins-38 C17C3.20 II	SQRSHVFSYKKHCGRRIVSLVQACD IDHDLSIDCCTQNCSSEFVSQRSHVFSY				
ins-39 F21E9.4 X	IFRMTISFCSQVECQNMEAMQKICNTTTPTI HVGELCCPEFFEQVKDDFVTLL				EST

[#]Genes for which ESTs, ORFeomes (OST), or cDNAs have been isolated are in bold; encoded peptides are based on sequence homologies to other insulin-like peptides. Cosmid and LG data only indicated for *ins-31 a; ins-31 b* and *ins-31 c* are the same. [$]Unclear which *ins-31* construct was used for overexpression and functional data. See References. [#]References are as follows: EST or OST in EST or ORFeome databases indicated only if not identified in canonical reference and sequence spans at least one intron. Modified from Li[10].

operon as some of the genes are transcribed on different strands. Some of the *nlp* genes that encode similar peptides are also physically close on the chromosome. The largest *flp* cluster is on the X chromosome, although with the exception of three *flp* genes, most of the *flp* genes on X are not as closely clustered as the *ins* genes.

Because peptides encoded by different *flp* genes were tested for bioactivity (see below), a nomenclature was developed to differentiate among the different peptides encoded by a single gene; each FLP peptide was assigned an individual gene name and number unless the gene encodes only one peptide sequence. For example, the *flp-1* gene encodes eight distinct peptides designated as FLP-1-1, FLP-1-2, etc. (Table 2). As more of the NLP and INS peptides are isolated and their sequences confirmed, the peptides encoded by these genes may also receive specific designations.

Expression of Neuropeptide Genes

The Genefinder programs are fairly robust at predicting genes, but the genomic organization of the predicted genes is frequently inaccurate. In particular, because *C. elegans* genes are generally trans-spliced to an SLI leader sequence,[15] the predictions of the 5' coding regions are often incorrect. Hence, to determine whether a candidate neuropeptide gene is transcribed and to confirm the genomic organization of the gene, researchers have isolated cDNAs through two basic strategies. One is to isolate cDNAs using reverse transcription (RT)-polymerase chain reaction (PCR); the second is to take advantage of the *C. elegans* EST and ORFeome databases and scan the databases for the relevant cDNAs. Through these approaches, the genomic organization of 37 of the 40 insulin-encoding genes,[4-7] 28 of the 31 *flp* genes[12,16-18] (I. Miskelly, N.J. Marks and A. Maule, pers comm; unpublished results) and 39 of the 48 *nlp* genes[14,19] have been determined. Based on these data, we predict that most, if not all, of the candidate neuropeptide genes are expressed. The genes have several common characteristics. First, most of the neuropeptide genes are relatively small and often have only a few exons. For instance, the coding and intronic regions of sixteen *flp* genes are less than 1 kbp.[16,17] Second, the transcripts are very short in length. Among the *nlp* genes, for example, a cluster of six *nlp* genes (*nlp-27, 28, 29, 30, 31* and *34*) on chromosome V encode very similar transcripts and have coding and intronic regions that range from 190 to 344 bp, which means that their transcripts are even smaller.[14] Genefinder programs, although improving, are generally not optimized to find small genes with a small number of exons. Hence, new peptide genes are still being identified and the entire complement of neuropeptide genes in *C. elegans* remains unknown.

Many of the neuropeptide genes encode more than one peptide. However, because the peptides encoded by different family members are very similar (for instance, the insulin-like peptides share common A and B domains and the FLPs all contain a common C-terminal Arg-Phe-NH_2), it has been difficult to generate reagents that specifically recognize individual peptides. For instance, many of the anti-FMRFamide antibodies recognize the C-terminal Arg-Phe-NH_2 and should theoretically cross-react with all FLPs. To determine the cellular expression pattern of the different neuropeptide genes, *C. elegans* researchers have taken advantage of the ease with which transgenic animals can be generated. Constructs containing gene fusions of the promoter region of a neuropeptide gene with a reporter gene, such as *lacZ* or green fluorescent protein (GFP), are used for germline transformation to generate transgenic animals. Despite the inherent caveats with reporter expression patterns, the expression patterns of 15 *ins*, 19 *flp* and 27 *nlp* genes have been determined by examining transgenic animals carrying reporter fusions (Tables 1-3, which include references); of these, one GFP expression pattern, that of *flp-8*, has been confirmed with an anti-FLP-8 monoclonal antibody,[20,21] suggesting that the GFP expression patterns represent protein localization.

From these expression patterns, a number of observations can be made. First, each gene has a unique expression pattern. Second, over half the cells in the nervous system express at least one neuropeptide gene. Third, there is considerable overlap in the expression pattern of different genes and a single cell may express multiple neuropeptide genes (Fig. 1). For instance, ASI is a chemosensory neuron that expresses *daf-28, ins-1* and *9, nlp-1, 5, 6, 9, 14, 18, 24* and *27* and *flp-2, 10* and *21*;[6,7,14,16] although some of these expression patterns may be an artifactual, ASI has the potential to release a plethora of neuropeptides to modulate neuronal activity. Fourth, as in

Table 2. Neuropeptide genes encoding FMRFamide-related peptides (FLPs) in **C. elegans**

Gene[#] Cosmid LG	Putative Peptides[@]	Peptide Name	Name in Other Species	Expression Pattern[+]	Function or Phenotype	Receptor^	Ref[##]
flp-1 F23B2.5 IV	*SADPNFLRFG *SQPNFLRFG *ASGDPNFLRFG *SDPNFLRFG *AAADPNFLRFG **(K)PNFLRFG AGSDPNFLRFG **(K)PNFMRYG**	FLP-1-1 FLP-1-2 FLP-1-3 FLP-1-4 FLP-1-5 FLP-1-6 FLP-1-7 FLP-1-8	PF2 PF1 AF26	AIA, AIY, AVA, AVE, AVK, RIG, RMG, M5	involved in locomotion, egg laying, and fat deposition; FLP-1-1 and FLP-1-4 inhibit frequency of pharyngeal action potentials; modulates acetylcholine signaling	(C25G6.5,Y58A8 a.1,C16D6.2, Y59H11AL.1)	16,18,27,28, 36,63,96, 101,128,116, 119,127, KA
flp-2 W07E11.3 X	LRGEPIRFG *SPREPIRFG	FLP-2-1 FLP-2-2		AIA, RID, PVW, I5, MC (ASI, M4, head muscles, an extra pair of cells in the head)	FLP-2-2 increases the frequency of pharyngeal action potentials	T19F4.1a/b	12,16,36,119, 127
flp-3 W07E11.2 X	SPLGTMRFG *TPLGTMRFG *SAEPFGTMRFG *NPENDTPFGTMRFG *ASEDALFGTMRFG EDGNAPFGTMRFG *EAEEPLGTMRFG *SADDSAPFGTMRFG NPLGTMRFG	FLP-3-1 FLP-3-2 FLP-3-3 FLP-3-4 FLP-3-5 FLP-3-6 FLP-3-7 FLP-3-8 FLP-3-9		IL1, PQR; SP, CP9	All peptides inhibit frequency of pharyngeal action potentials	C53C7.1a (Y58A8a.1, C16D6.2)	12,16,28,36, 116,127,128

continued on next page

Table 2. Continued

Gene[#] Cosmid LG	Putative Peptides[@]	Peptide Name	Name in Other Species	Expression Pattern[+]	Function or Phenotype	Receptor^	Ref[##]
flp-4 C18D1.3 II	PTFIRFG ASPSFIRFG	FLP-4-1 FLP-4-2		ADL, ASEL, AVM, AWC, FLP, PHA, PHB, PVD, I5, I6, NSM	FLP-4-1 increases the frequency of pharyngeal action potentials	C16D6.2	12,16,116,127
flp-5 C03G5.7 X	APKPKFIRFG AGAKFIRFG *GAKFIRFG	FLP-5-1 FLP-5-2 FLP-5-3		PVT, RMG, I4, M4, pharyngeal muscle, amphidial neuron, (PB, I2); rays 1, 5, 7, HOB	FLP-5-3 increases frequency of pharyngeal action potentials	(C25G6.5)	12,16,28,116, 127,128,OH
flp-6 F07D3.2 V	x6 *KSAYMRFG ***pQQDSEVEREMM**	FLP-6-1 FLP-6-2	AF8/PF3	ASE, AFD, ASG, PVT, I1 (one or two pairs of head cells); rays 2, 5, 6, 7	FLP-6-1 increases frequency of pharyngeal action potentials		12,16,28,117, 127,128
flp-7 F49E10.3 X	x2 *TPMQRSSMVRFG x3 *SPMQRSSMVRFG SPMERSAMVRFG SPMDRSKMVRFG	FLP-7-1 FLP-7-2 FLP-7-3 FLP-7-4		ALA, AVG, PHB, PDA, PVW, RIC, SAA (RMDV/SMDV& PHA)		Y59H11AL.1, C26F1.6	12,16,36,118, 119
flp-8 F31F6.4 X	x3 *KNEFIRFG	FLP-8	AF1	AUA, PVM, URX (RMG/ ADA, an extra pair of cells in the head); CP9	increases frequency of pharyngeal action potentials; overexpression causes defecation defects		12,16,127,128, UP, AS

continued on next page

Table 2. Continued

Gene[#] Cosmid LG	Putative Peptides[@]	Peptide Name	Name in Other Species	Expression Pattern[+]	Function or Phenotype	Receptor^	Ref[##]
flp-9 C36H8.3 IV	x2 *KPSFVRFG	FLP-9			inhibits frequency of pharyngeal action potentials; knockout shows slight sluggishness	(Y59H11AL.1)	12,28,31,36, 127,128, UP
flp-10 T06C10.4 IV	QPKARSGYIRFG	FLP-10		AIM, ASI, AUA, BAG, BDU, DVB, PQR, PVR, URX, vulD			12,16
flp-11 K02G10.4 X	*AMRNALVRFG *ASGGMRNALVRFG *NGAPQPFVRFG ***SPLDEEDFAPESPLQG**	FLP-11-1 FLP-11-2 FLP-11-3 FLP-11-4	AF21	AUA, BAG, DA, DD, DVB, LUA, PHC, PVC, SAB, URX, VD, uv1, head muscle (socket cells); ray 4	FLP-11-1 strongly inhibits frequency of pharyngeal action potentions	Y59H11AL.1, C26F1.6 (C16D6.2)	12,16,28,36, 116,118,127
flp-12 C05E11.8 X	RNKFEFIRFG	FLP-12	AF24	AVH/AVJ, BAG, PDA, PVR, SAA, SDQ, SMB (BDU); rays 1, 4, 5, 7, CP9			12,16

continued on next page

Table 2. Continued

Gene[#] Cosmid LG	Putative Peptides@	Peptide Name	Name in Other Species	Expression Pattern+	Function or Phenotype	Receptor^	Ref[##]
flp-13 F33D4.3 IV	*AMDSPFIRFG *AADGAPFIRFG x2 *APEASPFIRFG AADGAPLIRFG *ASPSAPFIRFG *SPSAVPFIRFG *SAAAPLIRFG ASSAPFIRFG	FLP-13-1 FLP-13-2 FLP-13-3 FLP-13-4 FLP-13-5 FLP-13-6 FLP-13-7 FLP-13-8		ASE, ASG, ASK, BAG, DD, I5, M3, M5 (an extra pair of cells in the head); VSP	All peptides inhibit frequency of pharyngeal action potentials	(Y59H11AL.1)	12,16,28,29, 32,36,127,128
flp-14 Y37D8A.15 III	x4 *KHEYLRFG	FLP-14	AF2/PF51		Increases frequency of pharyngeal action potentials	(C25G6.5, C16D6.2)	11,33,36,116, 127,128
flp-15 ZK525.1 III	*GGPQGPLRFG *RGPSGPLRFG	FLP-15-1 FLP-15-2		PHA, I2, socket/sheath cells (pharyngeal muscle, several cells in the head)	FLP-15-1 inhibits frequency of pharyngeal action potentials	C10C6.2 C16D6.2	11,16,36,115, 116,127
flp-16 F15D4.8 II	x2 *AQTFVRFG *GQTFVRFG	FLP-16-1 FLP-16-2	AF15		FLP-16-1 inhibits frequency of pharyngeal action potentials		11,28,127,128
flp-17 C52D10.11 IV	x2 KSAFVRFG KSQYIRFG	FLP-17-1 FLP-17-2		BAG, M5 (an extra pair of cells in the head); rays 1, 5, 7	All peptides are potent stimulators of the frequency of pharyngeal action potentials		11,16,127

continued on next page

Table 2. Continued

Gene[#] Cosmid LG	Putative Peptides[@]	Peptide Name	Name in Other Species	Expression Pattern[+]	Function or Phenotype	Receptor^	Ref[##]
flp-18 Y48D7A.2 X	**(DFD)GAMPGVLRFG *EMPGVLRFG x3 **(SYFDEKK)-SVPGVLRFG *EIPGVLRFG *SEVPGVLRFG *DVPGVLRFG	FLP-18-1 FLP-18-2 FLP-18-3 FLP-18-4 FLP-18-5 FLP-18-6	*(afp-1)*	AVA, AIY, RIG, RIM, M2 (M3, two extra pairs of cells in the head); rays 2, 6	Modulates chemosensation, fat metabolism, foraging, and dauer formation; FLP-18-2 inhibits the frequency of pharyngeal action potentials	NPR-4/C16D6.2 Y58A8a.1 C53C7.1a NPR-1/C39E6.6, NPR-5/Y58G8A.4 (C25G6.5, F41E7.3)	11,16,32,36, 101a,114,116, 118,120,125, 127, UP
flp-19 M79.4 X	*WANQVRFG *ASWASSVRFG	FLP-19-1 GLP-19-2		AIN, AWA, BAG, HSN, URX (an extra pair of cells in the tail); rays 5, 7, 9, CEM	FLP-19-1 inhibits the frequency of pharyngeal action potentials		11,16,28,36, 127
flp-20 E01H11.3 X	x2 AMMRFG	FLP-20		ALM, ASEL, AVM, LUA, PLM, PVC, PVM, PVR, RIB/AIB (PVT)			11,16
flp-21 C26F1.10 V	GLGPRPLRFG	FLP-21	AF9	ADL, ASI, ASH, ASJ, FLP, URA, MC, M4, M2; CP6–9, SP, DVF	mutation causes mild aggregation behavior inhibits the frequency of pharyngeal action potentials	NPR-1 C25G6.5 Y58A8a.1	16,114,116, 120,125,127, OH

continued on next page

Table 2. Continued

Gene[#] Cosmid LG	Putative Peptides[@]	Peptide Name	Name in Other Species	Expression Pattern[+]	Function or Phenotype	Receptor^	Ref[##]
flp-22 F39H2.1 I	x3 *SPSAKWMRFG	FLP-22		AIM, ASG, AVA, AVG, AVL, CEP, PVD, PVW, RIC/AIZ, RIV, SMD, URA, uv1; 6 out of 9 CP	Increases the frequency of pharyngeal action potentials	(Y59H11AL.1)	11,16,28,36,127
flp-23 F22B7.2 III	VVGQQDFLRG (TKFQDFLRFG)	FLP-23					16, AM
flp-24 C24A1.1 III	*VPSAGDMMVRFG	FLP-24					17,36
flp-25 K04H4.7 III	DYDFVRFG	FLP-25-1					36,120
	*ASYDYIRFG	FLP-25-2					
flp-26 R173.4 X	**(E)FNADDLTLRFG	FLP-26-1					17,36
	* GGAGEPLAFSP-DMLSLRFG	FLP-26-2					
	* **FRLPFQFFGANED-FNSGLT**	FLP-26-3					
	***NYYESKPY**	FLP-26-4					

continued on next page

Table 2. Continued

Gene[#] Cosmid LG	Putative Peptides[@]	Peptide Name	Name in Other Species	Expression Pattern[+]	Function or Phenotype	Receptor^	Ref[##]
flp-27 C25H3.5 II	(EASAFGDIIGELKGK)- GLGGRMRFG	FLP-27-1					17,36
	***pQPIDEERPIFME**	FLP-27-2					
flp-28 W07E11.4 X	*APNRVLMRFG	FLP-28					36, UP
flp-32 R03A10.2 X	AMRNSLVRFG	FLP-32					AM
flp-33 T07D10.6 I	*APLEGFEDMSGFL- RTIDGIQKPRFG	FLP-33					64
flp-34 R09A1.5 V	ALNRDSLVASLN- NAERLRFG	FLP-34-1					SH, LS
	ADISTFASAIN- NAGRLRYG	FTP-34-2					

[#]Genes for which ESTs, ORFeomes (OST), cDNAs, or encoded peptides have been isolated are in bold. [@]Common sequences among peptides encoded by the same gene are underlined. No. of copies of peptide encoded by gene indicated. A C-terminal glycine donates an amide group during amidation. [+]Based on colocalization with new markers, some expression patterns have been revised from published data. Cells in parentheses are variably expressed and/or tentative identifications. Cells after semi-colons are male-specific. ^Receptors with an $EC_{50} < 1.5\ \mu M$ are indicated; receptors with an $EC_{50} > 1.5\ \mu M$ are in parentheses. *Peptides have been biochemically isolated; *peptides including residues in parentheses have been isolated; non-FLP peptides are indicated in **bold**. [1]FLP-14/AF2 has also been found in *Haemonchus contortus* and *Panagrellus redivius*. [##]References are as follows: pers comm: KA, K. Ashrafi; AH, Anne Hart; OH, Oliver Hobert; SH, Steven Husson; AM, A. Maule; LS, Lilianne Schoofs; AS, Antony Stretton; UP, unpublished results; EST or OST in EST or ORFeome databases indicated only if not identified in canonical reference and sequence spans at least one intron. AF, *Ascaris suum*; PF, *Panagrellus redivius*. Modified from Li[10].

Table 3. Neuropeptide genes encoding non-insulin, non-FLP peptides in C. elegans

Gene[#] Cosmid LG	Putative Peptides[@]	Expression Pattern[+]	Function or Phenotype	Ref[##]
nlp-1 C01C4.1 X	x3 *MDANAFRMSFG *MDPNAFRMSFG *VNLDPNSFRMSFG	ASI, AWC, PHB, BDU, 4 head neurons, intestine		13,14, 28,36,
nlp-2 T24D8.5 X	SIALGRSGFRPG SMAMGRLGLRPG x3 SMAYGRQGFRPG	1 head neuron, secretory cells near vulva, intestine		13,14, OST
nlp-3 F48C11.3 X	AINPFLDSMG AVNPFLDSIG YFDSLAGQSLG	ADF, ASE, ASH, AWB, ASJ, BAG, HSN, I1, I2, I3, I4, MI, M3, NSMR, 3 head neurons, VNC, occ. I6, M2, pm1VL, intestine		13,14
nlp-4 F59C6.6 I	SLILFVILLVAFAAARPVSEEVDRV DYDPRTEAPRRLPADDDEVDGEDRV DYDPRTDAPIRVPVDPEAEGEDRV			13,14
nlp-5 F35C11.1 II	SVSQLNQYAGFDTLGGMGLG ALSTFDSLGGMGLG ALQHFSSLDTLGGMGFG	ASI, 2 head neurons, spermatheca; 1 male tail neuron		13,14
nlp-6 T23E7.4 X	*(MA)APKQMVFGFG *YKPRSFAMGFG *AAMRSFNMGFG LIMGLG	ASI, IL1, 2 head neurons, 1 tail neuron, intestine		13,14, 28,36
nlp-7 F18E9.2 X	*LYLKQADFDDPRMFTSSFG *SMDDLDDPRLMTMSFG *MILPSLADLHRYTMYD	Modulates lifespan extension via dietary restriction. ADL, AFD, ASE, ASI, PHA, VNC, 4 head neurons, 2 RVG neurons		13,14, 28,36, 101c

continued on next page

Table 3. Continued

Gene[#] Cosmid LG	Putative Peptides[@]	Expression Pattern[+]	Function or Phenotype	Ref[##]
nlp-8 D2005.2 I	*AFDRFDNSGVFSFGA *AFDRMDNSDFFGA *SFDRMGGTEFGLM *YPYLIFPASPSSGDSRRLV	ASK, ADL, 6 head neurons, 2 tail neurons, I2, g1D, pm5L, pm5R, 2 RVG, processes in pharynx, intestine; HOB		13,14, 28,36
nlp-9 E03D2.2 V	GGARAFYGFYNAGNS GGGRAFNHNANLFRFD GGGRAFAGSWSPYLE *TPIAEAQGAPEDVDDRRELE	ASI, AWB, 4 head neurons, 1 tail neuron, VNC, spermatheca, vulval muscles, intestine		13,14, 28,36
nlp-10 F37A8.4 III	AIPFNGGMYG STMPFSGGMYG AAIPFSGGMYG GAMPFSGGMYG	ASK, ADL, CAN, 2 lateral neurons, 1 tail neuron, 2 ant. pharyngeal neurons; 1 male tail neuron		13,14
nlp-11 ZK1320.10 II	*HISPSYDVEIDAGNMRNLLDIG *SAPMASDYGNQFQMYNRLIDAG *SPAISPAYQFENAFGLSEALERAG	IL1, 2 head neurons, VNC, PVD, 3 tail neurons, precomma embryos		13,14, 28,36, EST, OST
nlp-12 M01D7.5 I	x2 *DYRPLQFG *DGYRPLQFG	DVA, coeloomocytes	Regulates fat storage; Modulates acetylcholine signaling; Receptors: CKR-1/T23B3.4, CKR-2/Y39A3B.5	13,14,27, 36,101b

continued on next page

Table 3. Continued

Gene# Cosmid LG	Putative Peptides@	Expression Pattern+	Function or Phenotype	Ref##
nlp-13 E03D2.1 V	*NDFSRDIMSFG *SGNTADLYDRRIMAFG QPSYDRDIMSFG *SAPSDFSRDIMSFG *SSSMYDRDIMSFG *SPVDYDRPIMAFG *AEDYERQIMAFG	3 head neurons, NSM, M2, I4, spermatheca, LUA, 1 tail cell, dorsal and ventral hypoderm, intestine		13,14, 28,36
nlp-14 D1009.4 X	x2 ALDGLDGSGFGFD x5 ALNSLDGAGFGFE x3 ALDGLDGAGFGFD *ALNSLDGQGFGFE x3 ALNSLDGNGFGFD	ASI, ASK and another amphidial neuron, PHA, VNC, 2 RVG neurons, intestine		13,14, 28, EST
nlp-15 CC4.2 I	*AFDSLAGSGFDNGFN x2 AFDSLAGSGFGAFN AFDSLAGSGFSGFD AFDSLAGQGFTGFE AFDTVSTSGFDDFKL	ASH, CAN, HSN, BDU, 5 head neurons, VNC, 3 RVG neurons, 1 tail neuron, intestine		13,14, 28
nlp-16 T13A10.5 IV	STEHHRV SEGHPHE ATHSPEGHIVAKDDHHGHE SSDSHHGHQ *SVDEHHGHQ *NAEDHHEHQ SEHVEHQAEMHEHQ STQEVSGHPEHHLV	7 head neurons, 1 lateral neuron, intestine		13,14, 36

continued on next page

Table 3. Continued

Gene[#] Cosmid LG	Putative Peptides[@]	Expression Pattern[+]	Function or Phenotype	Ref[##]
nlp-17 Y45F10A.5 IV	*GSLSNMMRIG QQEYVQFPNEGVVPCESCNLGTLMRIG			13,14, 28
nlp-18 F33A8.2 II	*SPYRAFAFA ARYGFA *SPYRTFAFA ASPYGFAFA *SDEENLDFLE	ASI, 4 head neurons, 2 tail neurons, spermatheca, NSM, 2 ant. pharyngeal neurons, rectal gland, intestine		13,14, 28,36
nlp-19 K09C8.6 X	IAGLRLPNFL IGLRLPNML MGMRLPNIIFL	4 head neurons, VNC in males, NSM, 4 post. pharyngeal neurons, spermatheca		13,14, EST, OST
nlp-20 F45E4.8 IV	FAFAFA *SGPQAHEGAGMRFAFA APKEFARFARASFA	4 head neurons, 4 tail neurons, spermatheca, intestine, 1 ant. pharyngeal neuron		13,14,36
nlp-21 Y47D3B.2 III	GGARAFSADVGDDY *GGARAFYDE *GGARAFLTEM *GGARVFQGFEDE GGARAFMMD GGGRAFGDMM GGARAFVENS (GGGRSFPVKPGRLDD) *pQYTSELEEDE	AFD, 5 head neurons, VNC, 1 ant. pharyngeal neuron, 1 tail neuron, embryo, intestine		13,14, 28,36

continued on next page

Table 3. Continued

Gene[#] Cosmid LG	Putative Peptides[@]	Expression Pattern[+]	Function or Phenotype	Ref[##]
nlp-22 T24D8.3 X	SIAIGRAGFRPG			14
nlp-23 T24D8.4 X	LYISRQGFRPA SMAIGRAGMRPG AFAAGWNRG	Tail, dorsal and ventral hypoderm		14
nlp-24 F35B12.7 V	QWGGGPYGGYGPRGYGGGYGGG YGGYGGRGPYGGYGGRGPYGYGG GPYGGGGLVGALLG	ASI, spermatheca, 1 pharyngeal neuron	Anti-microbial?	14,19
nlp-25 Y43F8C.1 V	*pQWGGGYGNPYGGYG GGGYGGGYGGGFGAQQAYNVQNAA		Anti-microbial?	14,19, 36
nlp-26 Y43F8C.2 V	QFGFGGQQSFGGRGGQFGGMQRGGFNGN *GGFGQQSQFGG GGSQFNGRGGNQFGG	Hypoderm		14,36
nlp-27 B0213.2 V	QWGYGGMPYGGYGGMGGYGMGGYGMGY MWGSPYGGYGGYGGYGGWG	ASI, 3 head neurons, spermatheca, hypoderm, intestine	Anti-microbial?	14,19
nlp-28 B0213.3 V	QWGYGGYGRGYGGYGGYGRGMYGGYG GMYGGYGRGMYGGWG		Anti-microbial?	14,19, OST
nlp-29 B0213.4 V	QWGYGGYGRGYGGYGGYGRGMYGGYG GMYGGYGRGMYGGYGRGMYGGWG	Hypoderm, intestine	Anti-microbial?	14,19

continued on next page

Table 3. Continued

Gene[#] Cosmid LG	Putative Peptides[@]	Expression Pattern[+]	Function or Phenotype	Ref[##]
nlp-30 B0213.5 V	QWGYGGYGRGYGGYGGYGRGYGGYG GYGGYGRGMWGRPYGGYGWG	Hypoderm	Anti-microbial?	14,19
nlp-31 B0213.6 V	QWGYGGYGRGYGGYGGYGRGYGGYGGYG GYGGYGRGMYGGYGRPYGGYGWG	Hypoderm, embryos	Anti-microbial	14,19
nlp-32 F30H5.2 III	YGGWGGRGGWGRGGGRGYGGRRR GGGWGGRGGGWGRGGGGRGFYGGG		Anti-microbial?	14,19
nlp-33 T19C4.7 V	QWGYGGPYGGYGGGYGGGPWGYGGGW RHWGGYGGGPWGGYGGGPWGGYY	Hypoderm	Anti-microbial?	19, EST, OST
nlp-34 B0213.17 V	PYGYGGYGGW PYGYGWG			EST, AH
nlp-35 C33A12.2 IV	*AVVSGYDNIYQVLAPRF			28,36
nlp-36 B0464.3 III	*SMVARQIPQTVVADH			28

continued on next page

Table 3. Continued

Gene[#] Cosmid LG	Putative Peptides[@]	Expression Pattern[+]	Function or Phenotype	Ref[##]
nlp-37 ***(pdf-2)*** F48B9.4 X	*NNAEVVNHILKNFGALDRLGDVG	BDU, AVG, AIM, RIS, AVD, PVT, PHA, PHB, RID, RIM, AQR, PQR, rectD, rectVL/R, virL/R, post. arcade cells (additional head neurons, PVP)	Overexpression affects locomotion; Receptor: C13B9.4	28,36, 101b,101g
nlp-38 C01A2.7 I	**(ASDDR)VLGWNKAHGLWG *TPQNWNKLNSLWG *SPAQWQRANGLWG			28,36, AH, EST
nlp-39 C54C8.9 I	*EVPNFQADNVPEAGGRV			28
nlp-40 Y74C9A.2 I	**APSAPAGLEEKL(R) *pQPAADTFLGFVPQ			28,36
nlp-41 C04H5.8 II	*APGLFELPSRSV			28,36
nlp-42 Y80D3A.10 V	SALLQPENNPEWNQLGWAWG NPDWQDLGFAWG			AH, EST
nlp-43 C45G9.13 III	X2 KQFYAWAG FYSWEDA	Head neurons, intestine		101h

continued on next page

Table 3. Continued

Gene[#] Cosmid LG	Putative Peptides[@]	Expression Pattern[+]	Function or Phenotype	Ref[#]
nlp-44 Y23B4A.2 X	APHPSSALLVPYPRVG SNILNNNSESQNSVQ LYMARVG AFFYTPRIG		Receptor: K10B4.4	101h, 101i
nlp-45 T01B6.4 X	RNLLVGRYGFRIG			101h
nlp-46 T28H10.4 V	NIAIGRGDGLRPG			101h, 101j
nlp-47 F36H12.1 X	pQMTFTDQWT			AH
pdf-1 T07E3.6 III	*SNAELINGLIGMDLGKLSAVG *SNAELINGLLSMNLNKLSGAG	ADA, PVT, ASI, ASK, PHA, PHB, RMED/V, RID, ADE, PQR, rectD, rectVL/R (PVP)	Involved in locomotion; Receptor: C13B9.4	101b, 101g

[#]Genes for which ESTs, ORFeomes (OST), cDNAs, or encoded peptides have been isolated are in bold. [@]Common sequences among peptides encoded by the same gene are underlined. No. of copies of peptide encoded by gene indicated. A C-terminal glycine donates an amide group during amidation. Some *nlp* peptide predictions have been revised. [+]Based on colocalization with new markers, some expression patterns have been revised from published data. Cells in parentheses are variably expressed and/or tentative identifications. *Peptides have been biochemically isolated; *peptides including residues in parentheses have been isolated. [##]References are as follows: pers comm: AH, Anne Hart; UP, unpublished results; EST or OST in EST or ORFeome databases indicated only if not identified in canonical reference and sequence spans at least one intron. Modified from Li.[10]

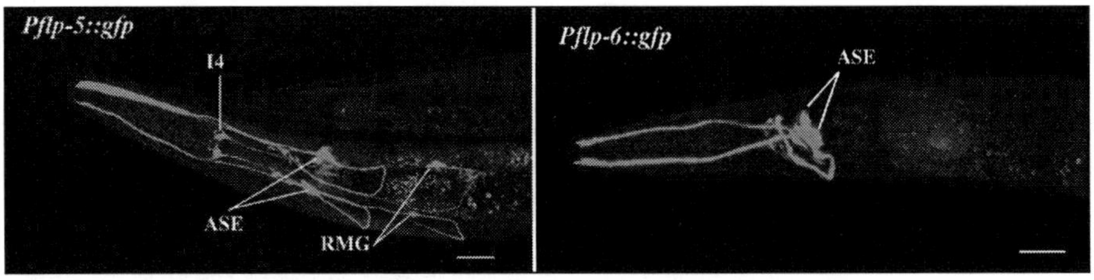

Figure 1. Expression pattern of two *flp* genes. Using a transcriptional reporter for *flp-5* and *flp-6*, the expression pattern of the genes can be seen in the head of the animal. Note that the two genes have different expression patterns, although both genes are expressed in the ASE chemosensory neuron. Scale bar = 50 μm. Modified from reference 16.

mammals, neuropeptides are expressed not only in the nervous system, but also in nonneuronal tissue, such as the intestine, somatic gonad and vulva. These data suggest that the peptides also function as long-range acting hormones.[6,7,14,16]

Cleavage and Processing of Neuropeptides

As in other animals,[22] neuropeptides in *C. elegans* are derived from larger precursor molecules. These precursor molecules are proteolytically cleaved to release the smaller peptides, which may be further modified to attain their final, active state (see Fig. 2). The precursor molecule can contain a single peptide, multiple copies of a single peptide, multiple distinct peptides, or any combination thereof. In other invertebrates and in mammals, the precursor molecule can be differentially cleaved to yield different peptides in different cell types.[22,23] Whether differential processing also occurs in *C. elegans* is unknown.

Cleavage of Neuropeptide Precursors by Proprotein Convertases

After removal of the signal sequence, the precursor molecule must undergo at least two proteolytic cleavage steps to release the peptides. In practice, one scans a precursor molecule for the presence of basic residues; the sequence between a pair of basic residues is a potential neuropeptide. The initial cleavage of the precursor molecule is after a C-terminal basic residue sequence by a Kex-2/subtilisin/proprotein convertase (for reviews see refs. 24, 25) which must itself be activated. This activation is dependent on a chaperonin protein SBT-1 7B2.[26,27] In *C. elegans* predicted and isolated FLP peptides are flanked by at least one basic residue; the most common flanking sequences are lysine -arginine and arginine (see Table 4).[18,28-33] The predicted NLP peptides are also flanked by at least one basic residue[14] and like the FLPs cleavage generally occurs after dibasic residues.[28] However, several NLPs have been isolated where an alanine is the flanking residue.[28] The enzymes responsible for these cleavages are likely related to the subtilisin/kexin-like isozyme 1 (SKI-1/SIP) or apoptosis-regulated convertase-1 (PCSK9/NARC-1) enzymes that cleave at nonbasic residues (for review see ref. 24.)

Table 4. Basic cleavage sites flanking FLPs

Basic Residue(s)	No. of Sites*
KR	86
R	26
K	9
RK	7
KK	7
RR	1
KRR	1

*Total no. of sites examined = 137.
K = lysine, R = arginine.

Four *C. elegans* genes encode proprotein convertases: *kpc-1*, *egl-3/kpc-2*, *aex-5/kpc-3*, *bli-4/kpc-4*. Two general approaches have been used to tease apart the relative contribution of each proprotein convertase in neuropeptide processing. Both of these approaches are dependent on isolating animals in which the respective proprotein convertase gene has been inactivated. In

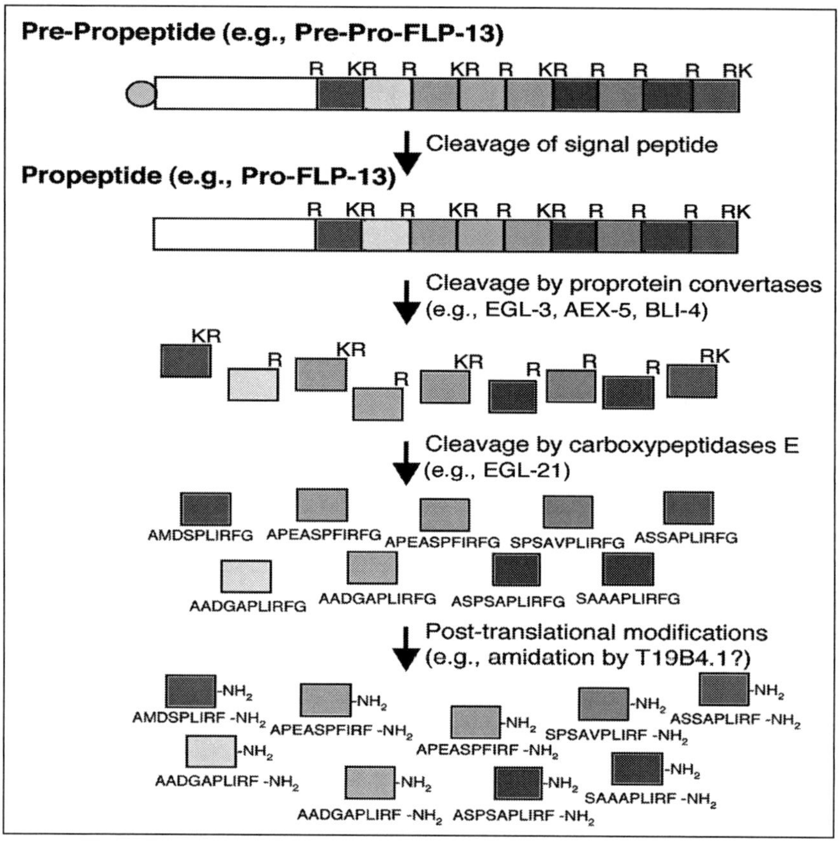

Figure 2. Processing of a neuropeptide precursor molecule: FLP-13 as an example. After translation of the *flp-13* transcript, prepro-FLP-13 is cleaved by signal peptidase to remove the signal sequence. Pro-FLP-13 is further cleaved C-terminal to the mono- and dibasic residues (indicated by K and R) by proprotein convertases. The basic residues are removed by carboxypeptidases to generate the basic neuropeptides. The FLP-13 peptides are further modified by the addition of an amide group, which is donated by the C-terminal glycine, to yield the active peptides. Many of the FLP-13 peptides have been isolated.[10]

one approach, the knockout animals were assessed for behavioral phenotypes and decreases in neuropeptide expression. The second approach was to perform a peptidomic profile of knockout animals and determine whether peptide levels were modified. By these two complementary approaches EGL-3 appears to be the major proprotein convertase for neuropeptide processing. Using an antibody that recognizes only the processed form of a FMRFamide-like peptide,[34] Kaplan and coworkers found that inactivation of *egl-3* led to a significant decrease in the level of FMRFamide-like immunoreactivity, presumably because the FLP precursor molecules could no longer be cleaved.[35] Furthermore, no FLP or NLP peptides were isolated from *egl-3(n729* and *gk238)* in a peptidomic analysis, although these animals had low levels of FMRFamide-like immunoreactivity.[36] Behaviorally, *egl-3* knockout animals show several defects. Wild-type animals move in a sinusoidal waveform and lay between 250-300 eggs. *egl-3* mutants were initially isolated on the basis of an egg-laying defect, whereby the mutants were bloated because they retained eggs rather than releasing them.[37] The mutants also coil as they move rather than moving in a sinusoidal waveform. *egl-3* is expressed in about 100 neurons[35] as well as nonneuronal cells, such as the hypodermis and muscle cells. Because *egl-3* is expressed in a subset of the 302 neurons in *C. elegans* and the *flp* genes are expressed in over 50% of the neurons,[16] other proprotein convertases must be active in other neurons.

Inactivation of *kpc-1* causes mild uncoordination, slow growth and a slightly smaller body size.[38] Based on these phenotypes, Thacker and Rose[38] proposed that KPC-1 cleaves TGF-β family member DAF-7, which is involved in dauer pathway formation (see below) and/or ligands to DAF-2, an insulin-related receptor.[39] The expression pattern of *kpc-1* is unknown. In a peptidomic analysis of *kpc-1 (gk8)* mutants, the profile of *kpc-1* was not significantly different than wild type,[36] suggesting that the relative contribution of KPC-1 in neuropeptide precursor cleavage is minor.

The role of *bli-4/kpc-4* in neuropeptide precursor cleavage is unclear. The *bli-4/kpc-4* locus is extremely complicated because the RNA product undergoes alternative splicing to produce at least nine gene products (for review see ref. 38). Many of these transcripts are differentially expressed in different tissues (for review see ref. 38). Loss of *bli-4/kpc-4* results in lethality because the procollagens within the cuticle are not cleaved and, therefore, the structural integrity of the cuticle cannot be maintained.[40] However, some of the *bli-4/kpc-4* transcripts are expressed in neurons.[38] The peptidomic profile of *bli-4(e937)/kpc-4* animals is similar to that of wild type;[36] however, other alleles that disrupt neural-specific transcripts were not investigated. Hence, the role of *bli-4/kpc-4* in neuropeptide precursor cleavage awaits further clarification.

The proprotein convertase *aex-5/kpc-3* was first identified in a screen for defecation mutants. Because *aex-5/kpc-3* mutants do not undergo anterior body or expulsion contractions during the defecation cycle, the animals show severe constipation.[41] Ablation of the neurons controlling defecation, the GABAergic AVL and DVB neurons, causes a similar constipation defect.[42] The *aex-5/kpc-3* gene is proposed to be a downstream gene within an operon with *unc-54*.[38] Hence, *aex-5/kpc-3* is likely to show the same expression pattern as *unc-54*,[43] which encodes the major myosin heavy chain[44] in body wall, intestinal, defecation and reproductive muscles. AEX-5/KPC-3 is hypothesized to cleave a precursor molecule in muscle to produce a peptide that serves as a retrograde signal for exocytosis.[45] However, contrary to its proposed expression only in muscles, the peptidomic profile of *aex-5/kpc-4* mutants was substantially decreased compared to wild-type animals, providing strong evidence that *aex-5/kpc-3* is also responsible for neuropeptide precursor processing in neurons.

Subsequent Cleavage of Neuropeptide Precursors by Carboxypeptidases

After the proprotein convertases cleave the neuropeptide precursors into smaller fragments, the C-terminal basic residues are removed from the peptide sequences by the activity of carboxypeptidases E. *egl-21* encodes a neural-specific carboxypeptidase E that is expressed in about 60% of the neurons and is considered the primary carboxypeptidase E for neuropeptide processing in *C. elegans*.[46] Inactivation of *egl-21* carboxypeptidase E causes more severe phenotypes and affects more behaviors than those seen in the *egl-3/kpc-2* proprotein convertase mutants. *egl-21* null mutants have defects in egg laying, locomotion, mechanosensation and defecation. Extremely low levels of FMRFamide-like immunoreactivity is detected in *egl-21* mutants, suggesting that the behavioral phenotypes are likely due to low levels of neuropeptide processing of FMRFamide-related and other peptides.[46] Two other carboxypeptidases have been identified in the *C. elegans* genome, but their roles in neuropeptide processing have not been investigated.[46]

Following carboxypeptidase processing, the basic neuropeptide molecule is produced; however, many neuropeptides undergo further processing to attain their biologically active form and/or to confer them protection from degradation. These modifications can occur at the N-terminus, such as pyroglutamylation, at the C-terminus, such as amidation, or within the peptide, such as sulfation or phosphorylation. The most common known modification in *C. elegans* is amidation. A C-terminal glycine donates an amino group in the amidation process, so any peptide sequence with a C-terminal glycine, which includes all FLPs and many of the NLPs, are likely to be amidated. The amidation occurs through the activity of two enzymes in mammals, peptidylglycine-alpha-hydroxylating monooxygenase (PAM) and peptidyl-alpha-hydroxyglycine alpha-amidating lyase (PHM), which act sequentially; the enzymes are synthesized on the same molecule as adjacent domains on a bifunctional protein, peptidyl- α-hydroxyglycine α-amidating lyase (PAM).[47] Han and coworkers[48] identified one PAM-like and one PHM molecule in *C. elegans*. T19B4.1 encodes a monooxygenase that may function as a PAM-like molecule; knockdown of T19B4.1 activity causes resistance to the acetylcholinesterase inhibitor aldicarb, presumably by disrupting peptide processing.[27] Whether T19B4.1 and/or the other identified enzymes are involved in neuropeptide amidation has not been determined.

Neuropeptides Are Released from Dense Core Vesicles

The processing of the neuropeptide precursors begins in the endoplasmic reticulum with the removal of the signal peptide. Subsequent processing occurs as the precursors travel through the Golgi complex and are packaged into vesicles as they are transported down the axon.[22] In contrast to many small classical transmitters that are located in small, clear vesicles, bioactive neuropeptides are located in dense core vesicles, which are derived from the trans-Golgi network. Transport of both small, clear and dense core vesicles are dependent on UNC-104 kinesin, a motor for fast axonal transport in *C. elegans*.[46,49,50] Mutations in *unc-104*, for instance, cause an increase of FMRFamide-like immunoreactivity in neuronal cell bodies.[46,50] However, slow anterograde transport of IDA-1, a tyrosine phosphatase-like receptor that is associated with dense core vesicles, still occurred in *unc-104* mutants, indicating that dense core vesicles also move by slow anterograde transport and that at least two distinct motors can transport dense core vesicles.[51]

Once transported to the terminal, small, clear vesicles are segregated to the synaptic zones, while dense core vesicles are more diffusely situated throughout the terminal at nonsynaptic zones[22] (for review see ref. 52). This segregation of the vesicles allows release of the small clear vesicles with focal increases of calcium at the synaptic zones. By contrast, release of neuropeptides from dense core vesicles appears to be dependent on a general increase of calcium throughout the nerve terminal, which can occur after high levels of stimulation[22] (for review see ref. 52). While many of the molecular components necessary for small clear vesicles have been identified (for review see ref. 53) the exact mechanism for dense core vesicle movement to the cell membrane is unknown. Some components are used for both small clear and dense core vesicles movement. For instance, UNC-13, a high-affinity phorbol ester receptor, is necessary for small, clear vesicle release [54,55] and has a minor role in dense core vesible release; [56,56] complete lack of UNC-13 results in lethality.[57] Three components have been implicated specifically for dense core vesicle release in *C. elegans*: UNC-31 CAPS,[56a,58,58a] IDA-1[59] and PKC-1 protein kinase 1.[56] A cytoplasmic calcium-dependent activator protein (CAPS) specifically bridges dense core vesicles and the plasma membrane in preparation for release in PC12 cells and *Drosophila*.[60,61] Similarly, *C. elegans* UNC-31 CAPS also promotes dense core vesicle release.[56a,58,58a] *unc-31* genetically interacts with *ida-1* and *ida-1* may function to regulate the activity of *unc-31*.[59] IDA-1 is an IA-2 insulinoma-associated protein in the family of tyrosine phosphatase-like receptors. IA-2 proteins are localized to dense core vesicles and, similarly, *ida-1* is expressed in a subset of neurons that also expresses *egl-3/kpc-2*.[51] Loss of *ida-1* enhances weak alleles of genes, such as *daf-2* and *daf-28*, which are involved in insulin-like signaling.[59] Mutations in *pkc-1*, which encodes a protein kinase I, cause an increased accumulation of neuropeptide precursor molecules, but does not appear to affect release of small clear vesicles, suggesting that PKC-1 is specifically necessary for dense core vesicle release.[56] Because *pkc-1* is expressed in the cholinergic, peptidergic ventral cord motor neurons and not the GABAergic, peptidergic motor neurons,[56] other protein kinases are likely to function to promote dense core vesicle release in the GABAergic, peptidergic motor neurons as well as other neuronal cell types.

After release from dense core vesicles, active neuropeptides are removed from the cleft by the action of proteolytic enzymes, one of which may be NEP-1 neprilysin.[27] By contrast to small molecule transmitters that can be recycled at the synaptic cleft, neuropeptides must be synthesized de novo in the cell body and transported down to the axon terminal. Hence, the time to replenish dense core vesicles containing neuropeptides is slower than for small molecule transmitters.

In addition to their use as primary neurotransmitters, neuropeptides also act as hormones, i.e., as long range signaling molecules. In such a capacity, the neuropeptides are released from neuronal or nonneuronal cells to affect target cells that can be some distance away. Historically, some of the first isolated mammalian neuropeptides, such as cholescysteckinin, were initially identified as hormones and later found to be used also at synapses within the nervous system.[22] Similarly, some neuropeptides in *C. elegans*, such as those expressed in the gonad and intestine (see Tables 1-3), are presumed to be hormones. To monitor secretion of neuropeptides into the pseudocoelom, Sieburth et al[56] took advantage of the scavenger activity of the coelomocytes, which continuously endocytose fluid from the pseudocoelom;[62] by monitoring the appearance of GFP-tagged neuropeptide precursors in the coelomocytes, they could deduce the release of the precursors into the pseudocoelom.

Biochemical Isolation of Neuropeptides

Using the above bioinformatics and experimental approaches, researchers predicted that over 250 neuropeptides are produced in *C. elegans*. But the nagging question remained whether all these peptides are actually produced. To address this question the initial approach was to carry out the tedious biochemical isolation of the different peptides. Much of the early work focused on the FLPs and 13 FLPs were painstakingly isolated from tens of grams of worms. Fortunately, the sensitivity and resolution of nanoscale liquid chromatography combined with mass spectrometry has increased over the years, so that the amino acid sequence of peptides could be determined with smaller and smaller amounts of starting material. This later proteomic effort nearly tripled the number of isolated FLPs. To date, 36 FLPs encoded by sixteen *flp* genes (*flp-1,3, 5, 6, 8, 9,11, 13, 14, 16, 18, 19, 22, 24, 26* and *33*) have been isolated[28-33,63,64] (Table 2). The proteomic approach also allowed the identification of a novel *flp* gene, *flp-33*, as well as 29 NLPs encoded by 19 *nlp* genes.[28,64] Two FLPs that are not encoded by any of the identified *flp* genes have also been isolated (N. Marks, A. Maule and A. Stretton, pers comm), underscoring the difficulty of identifying small neuropeptide genes with BLAST searches. These data suggest that many of the FLPs are also likely produced in *C. elegans*. In the future, we may be able to isolate peptides from single cell types by selecting GFP-expressing cells.

The predicted and/or isolated peptides in *C. elegans* have identical or highly similar counterparts in related nematodes, such as *Ascaris suum*[65-68], *Haemonchus contortus*[69,70] and *Panagrellus redivius*.[71-74] In addition, the FLP-12 (AF24) and FLP-21 (AF9) peptides have been isolated from *Ascaris suum*,[66,68] suggesting that these peptides are also produced in *C. elegans*. Similarly, several of the predicted NLPs are similar to peptides isolated from other invertebrates.[14] Using the proteomic approach to isolate peptides from *Ascaris*, Stretton and coworkers found that, like in *C. elegans*, most of the isolated peptides were of the FLP family.[68] These data suggest that many of the predicted neuropeptides are indeed produced and highlight the rich diversity of neuropeptides in *C. elegans* and other nematodes.

Neuropeptide Function

By looking at the expression patterns of the different neuropeptide genes, one can infer possible functions of the neuropeptides. For instance, some of the genes are expressed in sensory neurons responsible for dauer formation and chemosensation. Other neuropeptide genes are expressed in motor neurons responsible for movement. The widespread expression of the different genes suggests that neuropeptides are involved in a multitude of behaviors, including dauer formation, locomotion, egg laying and mechano- and chemosensation. To determine the function of the different neuropeptides, two general approaches have been used. The most common strategy is to decrease or increase the activity of specific neuropeptide genes. Because many of the neuropeptide gene families encode very similar peptides, peptides encoded by different genes may bind to the same receptors. The functional overlap among the peptides, therefore, complicates understanding the roles of different peptides in these knockout animals. The second approach has been to inactivate genes by RNAi, which, unfortunately, does not efficiently downregulate the activity of genes in all neurons.[75] Recently, however, several researchers have identified certain genetic backgrounds that increase the efficiency of RNAi in neurons[27,75,76] and cell-specific RNAi using double-stranded RNA.[76a] Despite these limitations, several neuropeptide genes have been shown to have unique functions.

The Insulin-Like Gene Family

After hatching *C. elegans* passes through four larval stages before emerging as a hermaphrodite or male. If during late first larval and second larval stages environmental conditions are unfavorable, such as in overcrowding or lack of food, a second larval stage animal will enter an alternative life cycle and become a dauer animal rather than a third larval stage animal.[77,78] Whereas wild-type animals have a lifespan of about three weeks, a dauer animal can survive for months. When environmental conditions become favorable again, the dauer animal will re-enter the life cycle as a fourth larval stage animal.[77] Two parallel signaling pathways—the insulin signaling and transforming growth factor β (TGF β) pathways—mediate the decision to enter reproductive growth or the dauer lifecycle (for review see ref. 79). Loss of either pathway results in constitutive dauer formation, indicating that the pathways function independently. However, loss of both pathways causes a stronger dauer effect than loss of only one, indicating that there is crosstalk between the two pathways.[80] Loss of *asna-1*, which encodes an ATPase that acts non-cell autonomously, enhances the effects due to loss of the TGFβ pathway, indicating that ASNA-1 is also involved in reproductive growth as a positive regulator of the insulin signaling pathway.[81] Kenyon and coworkers[82] were the first to recognize that the genes regulating dauer formation overlap with genes that regulate longevity. *daf-2* encodes an insulin-related receptor;[39] decreased *daf-2* signaling promotes dauer formation and extends lifespan.[82] Hence, insulin signaling is necessary for reproductive growth and regulated lifespan.

Before the decision to enter dauer, if a first larval stage animal emerges in the absence of food, the animal arrests growth.[78] The decision to arrest or undergo reproductive growth is also dependent on the DAF-2 insulin-like receptor and ANSA-1 ATPase.[81,83] If food is presented to the starved L1 animals, they resume reproductive growth.[78] Hence, insulin signaling is important at two junctions during the larval lifecycle to determine reproductive growth: L1 arrest and dauer formation.

The 42 insulin-like peptides, encoded by 41 *ins* genes and *daf-28*, represent the largest family of neuropeptides in *C. elegans*.[6,7] It was initially puzzling as to why there would be so many insulin-like peptides and only one insulin-like receptor, DAF-2. Recently, however, a more divergent family of 56 insulin-like receptors has been identified.[84] DAF-2 is the closest homologue to the mammalian insulin-like receptor, but presumably the more divergent receptors are able to bind some of the insulin-like peptides.

Which of the insulin-like peptides activate DAF-2? Monitoring of environmental conditions is mediated by the amphidial chemosensory neurons.[85] Many of the insulin producing genes are expressed in neurons.[6,7] In particular, *ins-1*, *ins-9* and *daf-28* are expressed in ASI and ASJ, which are the critical chemosensory neurons in the decision for dauer formation,[85] as well as other neurons and, in the case of *ins-1* and *daf-28,* nonneuronal tissue, such as intestinal cells.[6,7] Loss of *ins-1*, which encodes the peptide most similar to mammalian insulin, does not affect dauer formation or lifespan.[7] By contrast, the *daf-28(sa191)* mutation causes a transient dauer formation.[86] Because the severity of the phenotypes is greater in *daf-28(sa191)/ozDf2* animals (*ozDf2* carries a deficiency that covers *daf-28*) compared to *ozDf2* heterozygotes, the mutation was proposed to be a dominant negative mutation whereby the *daf-28(sa191)* gene product antagonizes wild-type DAF-28[86] and hence, DAF-2 activity.[6] Determination of the molecular lesion of *daf-28(sa191)* indicated that the mutation disrupts a proprotein convertase cleavage site, such that the insulin B domain is larger than usual;[6] this mutation causes the mature insulin-like peptide to act as a poison product, possibly because ligand binding does not activate the DAF-2 receptor. Overexpression of *daf-28* in a decreased TGFβ signaling background promotes exit from dauer, indicating that increased insulin signaling can bypass the TGFβ pathway[81] and suggesting that DAF-28 normally activates the DAF-2 receptor.[6] As monitored with a $P_{daf\text{-}28}$::GFP transgene, levels of *daf-28* expression are regulated by environmental cues, such that starvation and the dauer pheromone decreases GFP levels and aging increases the number of GFP-expressing cells.[6]

To determine whether the DAF-2 receptor binds insulin-like peptides in addition to DAF-28, the phenotypes caused by overexpression of different *ins* genes using their endogenous promoters were examined in different *daf* backgrounds. Overexpression of *ins-4* or *ins-6* can suppress or partially suppress, respectively, the *daf-28(sa191)* mutation, suggesting that INS-4 or INS-6 can functionally substitute for DAF-28 and activate the DAF-2 receptor when present at high levels.[6] Furthermore, similar to overexpression of *daf-28*,[6] overexpression of *ins-4* can bypass the effects of a mutation in the TGFβ pathway.[81] Several other INS peptides also affect DAF-2 activity. For instance, overexpression of *ins-1* and *ins-18* caused a low level of dauer arrest and enhanced the dauer phenotype of *daf-2* and/or *daf-7* TGFβ mutants, suggesting that INS-1 and INS-18 may downregulate *daf-2* expression or antagonize DAF-2 activity.[7] While overexpression of *ins-9, ins-19, ins-22* and *ins-31* did not affect dauer formation, overexpression of *ins-9* or of *ins-31* and *ins-19* in combination in a wild-type or *daf-2* mutant background caused embryonic or larval arrest, a phenotype similar to one shown by some *daf-2* alleles.[7] INS-9, INS-31 and INS-19, therefore, may signal through DAF-2 to affect other aspects of development.[7] Overexpression of *ins-7, 9, 17, 21, 22* and *23* did not suppress the *daf-28(sa191)* mutation.[6] Collectively, these data indicate that several insulin-like peptides can activate or affect DAF-2 activity to determine developmental growth or dauer formation.

To determine the role of the other insulin-like peptides and identify the receptors through which these peptides act, mutations in the *ins* genes and their corresponding receptors need to be identified. Until such mutants are isolated, the function of this large family of neuropeptides is unknown. Recently, *ins-1* mutants were found to have a defect integrating temperature and food information.[87] Well-fed wild-type animals move to the temperature on which they were cultivated when placed on a thermal gradient, whereas starved animals move away from the temperature on which they were cultivated.[88] This thermotaxis behavior is mediated by the AFD thermosensory neuron, which signals through the AIY interneuron.[89] Wild-type animals move more slowly in the presence of food compared to in the absence of food; this behavior, referred to as the basal movement rate, is dependent on dopamine.[90] Starved animals show an enhanced slowing when encountering food and this slowing behavior is dependent on serotonin.[90] Well-fed *ins-1* mutants show normal thermotaxis behavior and migrate to the temperature on which they were cultivated, showing that the thermotaxis circuit is functional.[87] Starved *ins-1* mutants show enhanced slowing when they encounter food, indicating that they can sense food and recognize their starvation state.[87] Starved *ins-1* mutants, however, move towards, rather than away from their cultivation temperature, suggesting that the mutants are not integrating their starvation state and their thermotaxis responses.[87] Expression of *ins-1* in different neurons could rescue the integration defect.[87] Furthermore, decreased activity of DAF-2 or AGE-1, a phosphoinositide 3-kinase, could suppress the integration defect, indicating that INS-1 normally antagonizes DAF-2 activity. Hence, INS-1 is proposed as a key neuropeptide in the integration of behavior with the functional state of the animal by antagonizing DAF-2 activity.[87]

In other neurons, however, INS-1 may activate the DAF-2 receptor. The ASE chemosensory neurons provide the major sensory response to water soluble attractants;[85] the AIA interneuron provides presynaptic input to ASE.[91] Wild-type animals normally chemotax towards sodium chloride (NaCl)[92] and ablation of ASE significantly decreases this response.[85] However, the state and experience of the animal can modulate this response. For instance, after pre-exposure to NaCl, starved, but not well-fed animals will avoid NaCl in a behavior referred to as salt chemotaxis learning.[93] Animals carrying mutations in several genes involved in DAF-2 signaling, including *daf-2, age-1, pdk-1* phosphoinositide-dependent kinase and *akt-1* AKT kinase, as well as *ins-1* are defective for salt chemotaxis learning,[94] implicating involvement of the DAF-2 pathway in salt chemotaxis learning and suggesting that INS-1 is the ligand for DAF-2 in this pathway. Based on cell-specific transgenic rescues and cell ablations, Tomioka and coworkers[94] proposed a model whereby INS-1 is released from the AIA interneurons to activate DAF-2 receptors in ASER to initiate the DAF-2 signaling cascade. INS-1, therefore, is involved in multiple processes where the state of the animal needs to be integrated with a sensory response. Whether INS-1 activates or antagonizes DAF-2 signaling is dependent on the behavior. This sort of plasticity in response is very common among neuropeptides: the functional output of their action is context dependent.

The *flp* Family

Because the *flp* gene family is expressed in over 50% of the nervous system, including sensory, motor and interneurons,[16,95] the FLPs are likely involved in many behaviors. However, many of the *flp* genes have overlapping expression patterns, indicating that the genes may have overlapping functions. The exact contribution of each gene, as well as the exact contribution of each peptide encoded by a given gene, therefore, are extremely difficult to elucidate. As a first step in this process, deletion mutants are being isolated for each gene. Because of the small size of the *flp* genes, however, deletions usually remove all peptide coding regions. Thus far, deletions in twenty-five *flp* genes have been isolated[96,96a] (*C. elegans* Knockout Consortium, Japanese National Bioresource Project; unpublished obs.). These mutants show a range of phenotypes, a few of which will be discussed. The function of *flp-18* will be disussed below as well as in conjunction with the function of their receptor, NPR-1.

The sinusoidal movement of the animal is controlled by the body wall muscles, which use acetylcholine and GABA as their primary excitatory and inhibitory neurotransmitters, respectively.[97,98] Disruption of either of these transmitter biosynthetic pathways leads to uncoordination.[98,99] However, neuropeptides also influence these motor circuits. Inactivation of *flp-1*, for example, causes several defects, the most visible of which is a hyperactive movement defect when on a solid surface.[96] Similarly, the HSNs regulate egg laying and use serotonin as their major transmitter; however, the HSNs are also cholinergic[100] and express multiple neuropeptide genes.[14,16] Although *flp-1* does not appear to be expressed in the HSNs,[96] *flp-1* mutants show defects in the timing of egg laying,[101] thereby causing a decreased number of eggs laid (unpubl. obs.); FLP-1 peptides are also necessary for down-regulation of egg laying in the absence of food.[101] A gain-of-function mutation in *egl-6*, which encodes a G protein-coupled receptor, causes eggs to be retained and laid at a later developmental stage than normal.[96a] To identify the cognate ligand for EGL-6, peptides encoded by *flp-1* through *flp-23* were overexpressed and screened for egg-laying defects; overexpression of FLP-10 and FLP-17 peptides were found to increase egg retention.[96a] Furthermore, FLP-10 and FLP-17 peptides at nanomolar concentrations activated the EGL-6 receptor co-expressed in oocytes with GIRK channels.[96a] Both *flp-10* and *flp-17* are expressed in the BAG neurons, suggesting that the BAG neurons feed onto the HSN neurons to affect egg laying; in addition, *flp-10* is also expressed in cells of the somatic gonad, suggesting that the somatic gonad plays a hormonal role to regulate egg laying.[96a]

Signaling through the FLP peptides is necessary in certain circuits to integrate multiple sensory inputs. The AIY interneuron receives input from multiple sensory neurons.[91] *flp-18* is expressed in AIY and another interneuron, RIG. Loss of *flp-18* resulted in excess fat accumulation, enhancement of dauer formation in *daf-7* TGFβ mutants, decreased oxygen consumption, decreased response to low levels of chemoattractants, and decreased exploratory behavior after starvation; these defects could be rescued by expression of wild-type *flp-18* in AIY, suggesting that AIY signaling via FLP-18 peptides integrates different sensory inputs to coordinate a behavioral or metabolic response.[101a] Furthermore, FLP-18 peptides activate different G protein coupled receptors to mediate the appropriate responses. Specifically, FLP-18 signaling through NPR-4 mediates the decreased chemosensory and exploratory behavior responses, while signaling through NPR-5 mediates the dauer response; FLP-18 signaling through both NPR-4 and NPR-5 mediates the fat metabolic responses.[101a]

The remaining *flp* mutants are also being characterized on a battery of behavioral assays. When initially examined, most of the *flp* mutants had no obvious movement defect (unpubl obs). However, using a variety of other measures for movement, such as swimming and resistance to serotonin-induced sluggishness, many of the mutants have slight locomotory defects. The swimming assay involves placing animals in physiological buffer and counting the number of thrashes per minute; this assay is more sensitive for detecting locomotory defects than examining an animal's movement on a solid surface. Although wild type while moving on a solid surface, *flp-9* mutants were found to thrash significantly less than wild-type animals in the swimming assay (unpubl obs). Several *flp* mutants have defects on the swimming assay, suggesting that the swimming circuit is modulated by several FLPs. Because many of the FLPs may bind to the same receptor (see below), animals carrying multiple knockouts need to be generated and the receptor to which the peptides bind must be identified to fully understand the role of different FLPs in the swimming circuit.

The *nlp* Family

Like the *flp* genes, the *nlp* genes are widely expressed in neuronal and nonneuronal tissue.[14] Analysis of several *nlp* mutants suggests that the NLP peptides mediate a variety of behaviors. For instance, NLP-12 peptides, which are distantly related members of the cholecystokinin (CCK)/gastrin family, signal through CCK-2, a G protein-coupled receptor, to regulate fat storage.[101b] Signaling through NLP-7 peptides, which are also distantly related CCK-like peptides, affect lifespan extension via dietary restriction.[101c] In microarray analyses to identify genes whose expression levels are changed in response to fungal or bacterial insults, expression of *nlp-29, 31* and *33* was induced, suggesting that NLP-29, 31 and 33 function as anti-microbial peptides.[19] To determine whether the putative anti-microbial peptide NLP-31 could prevent a fungal infection, *C. elegans* were infected with fungal spores and then incubated with NLP-31, which decreased the fungal infection on its outer surface, indicating that NLP-31 has anti-microbial activity.[19] Infection with bacterial strain *Leucobacter chromiireducens* subsp. *solipictus* induced *nlp-29* expression, suggesting that NLP-29 is also involved in immune defense.[101d] In addition to bacterial or fungal infections, physical injury to the cuticle also induced a defense response via induction of *nlp-29* and *nlp-31* expression, suggesting that the anti-microbial peptides are used globally for innate immunity.[101e] Although infection and wound healing initially activate parallel pathways, eventually the two pathways converge onto the p38-MAP kinase pathway.[101e] Loss of NPR-1, a G protein-coupled receptor that binds peptides encoded by *flp-18* and *flp-21*[114,125] (see below), increased susceptibility to bacterial infections, indicating that FLP and NLP peptides are both involved in the immune response.[101f] As would be expected with an anti-microbial role, *nlp-29* and *nlp-31* are expressed in hypodermal cells; however, *nlp-29* and *nlp-31* are also expressed in the intestine and embryo, respectively, indicating that the peptides may also function as neuromodulators.[14] The peptides encoded by *nlp-24, 25, 27, 28* and *30* are similar to those encoded by *nlp-29, 31* and *33*, suggesting that these peptides also have anti-microbial functions.[19] Moreover, *nlp-27* is expressed not only in hypodermal cells but also in neurons, intestinal cells and the spermatheca,[14] again indicating that this class of peptides may function not only as anti-microbial agents but also as neuromodulators.

Behaviors Affected by Multiple Classes of Neuropeptides

Several general assays have been used to perform global screens to identify genes that affect certain behaviors. Using the power of RNAi in a sensitized background to increase the effectiveness of RNAi in neurons, many genes can be screened relatively quickly for behavioral defects. As mentioned above, acetylcholine is the primary excitatory transmitter at the neuromuscular junction and is necessary for the smooth sinusoidal movements of the animals.[99] Aldicarb inhibits acetylcholinesterase, thereby increasing the amount of acetylcholine at the synapse and causing paralysis and lethality.[99,102] As an example of how global screens can be effectively used to identify genes affecting a behavior, Sieburth and coworkers[27] performed a genome-wide RNAi screen on animals in a sensitized background to identify genes whose lowered activity enhanced or suppressed the effects of aldicarb. In addition to the neuropeptide processing enzymes, decreased activity of four neuropeptide genes, two *ins* genes (*ins-22* and *ins-31*), one *flp* gene (*flp-1*) and one *nlp* gene (*nlp-12*), conferred aldicarb-resistance, suggesting that the peptides encoded by these genes modulate acetylcholine signaling.[27] Verification of these effects can be done by examining the mutants, as has been done for *flp-1*. These types of genome-wide screens will allow rapid identification of genes that affect behaviors.

Neuropeptide Receptors

Many of the neuropeptide genes, particularly those of the *flp* and *nlp* families, encode multiple distinct, but similar peptides. Furthermore, not only are multiple neuropeptides expressed in a single cell, but a specific neuropeptide may bind to multiple receptors. As with the insulin-like peptides and their receptors, a complementary strategy to understand the function of specific neuropeptides is to inactivate the receptors to which the peptides bind.

There are approximately 1500 G protein-coupled receptors in *C. elegans*.[2,103-105] Most of these are likely chemoreceptors and only a small subset, 50 plus in number, is predicted to be neuropeptide receptors.[106] The expression pattern of most of these receptors has not been determined. In an RNAi screen to inactivate G protein-coupled receptors that were predicted to bind either a small molecule transmitter or a neuropeptide, Keating and coworkers[107] identified many receptors whose decreased activity caused behavioral defects. For example, decreased activity of six receptors, C16D6.2, C25G6.5, C26F1.6, F35G8.1, F41E7.3 and F59C12.2, resulted in either an increased or decreased brood size.[107] Similarly, decreasing the activity of eight receptors, AC7.1 (tachykinin-like), C15B12.5, C10C6.2, C24A8.4, F15A8.5, F59D12.1, T02E9.1 and T05A1.1, affected locomotion.[107] In cases where mutants were available, the phenotypes were confirmed (e.g., for T05A1.1 and F35G8.1).[107] The ligands for several of these receptors have now been identified (see below).

Many receptors that bind FMRFamide-related peptides have been isolated from both invertebrates and vertebrates[108-112]. Among these, all are G protein-coupled receptors except for a molluscan FMRFamide-gated amiloride-sensitive channel that has homology to the MEC-4 and MEC-10 mechanoreceptors in *C. elegans.*[113] Several strategies have been used to match a FLP ligand to its corresponding receptor. Candidate receptors are transfected into *Xenopus* oocytes or heterologous cells and different readouts are used (see Table 2). Interestingly, a common theme in these studies is that multiple FLPs can activate a single receptor, implying that FLP receptors are fairly promiscuous in their binding partners; these FLPs can be encoded by one gene or by multiple genes. A few examples will be described below.

The Upjohn/Pharmacia group[114-116] selected candidate receptors and chimeric G proteins for transfection into Chinese hamster ovary (CHO) cells; the readout was binding of GTPγS, a nonhydrolyzable form of GTP, to membranes. Upon binding of the appropriate FLP ligand to the receptor, the bound receptor would presumably activate G proteins. GTPγS binding to membranes of transfected cells, therefore, indicates ligand binding. The receptor C10C6.2 bound both FLP-15 peptides, GGPQGPLRFamide and RGPSGPLRFamide, with an EC_{50} (concentration which produces 50% maximal activation) of 250 and 160 nM, respectively.[115] Despite high sequence similarity with other FLPs, particularly at the active C-terminus (such as FLP-21 GLGPRPLRFamide), no other tested FLP activated the receptor.[115]

Mertens and coworkers[117,118] similarly expressed candidate receptors and a G protein ($G_{\alpha16}$) in heterologous cells (HEK or CHO cells), but they screened for an increased calcium response, as monitored by an increase in fluorescence. The concentration of peptides needed to see a signal was 1000-fold higher (usually in the μM rather than nM range) than the Upjohn/Pharmacia system, suggesting that the sensitivity of the fluorescent readout may be lower than that when using GTPγS binding as the readout. Nevertheless, the group identified two ligands that bound with similar EC_{50} values to C26F1.6. FLP-7-1 TPMQRSSMVRFamide activated with an EC_{50} of ~1 μM; surprisingly, FLP-7-2 SPMQRSSMVRFamide, which differs from FLP-7-1 by only one amino acid at the N-terminus, did not active the receptor at concentrations up to 10 μM.[118] By contrast, FLP-11-1 AMRNALVRFamide, which has no sequence similarity to the FLP-7 peptides, activated C26F1.6 with an EC_{50} of ~1.33 μM.[118]

Other FLP receptors can bind ligands with a wide range of EC_{50} values. 15 FLPs encoded by 6 *flp* genes bind to the Y59H11AL.1 receptor with EC_{50} values ranging from 25 nM to 5 μM.[119] The ligands that produced the largest responses, FLP-7-3 SPMERSAMVRFamide (25 nM), FLP-1-8 KPNFMRYamide (100 nM) and FLP-11-1 AMRNALVRFamide (750 nM), have EC_{50} values less than 1 μM and show varying sequence similarity.[119] Note that FLP-7 peptides bind to both C26F1.6 and Y59H11AL.1. *flp-7* is expressed in head and tail neurons[16] (Table 2) and while the expression patterns of these two receptors are currently unknown, these patterns may give clues into the behavioral circuits that they modulate. If receptors generally bind multiple FLPs and a single FLP can also bind multiple receptors, then by using different combinations of FLPs the nervous system can exquisitely fine-tune behavior.

de Bono and Bargmann[120] found that mutations in the NPR-1 gene, which encodes a homologue to the mammalian neuropeptide Y (NPY) receptor, affected aggregation behavior. Wild-type animals tend to be solitary feeders, while *npr-1* mutants aggregate when they feed, a behavior referred to as social feeding, and clump at the borders of the *E. coli* food source, a behavior termed bordering. In addition, *npr-1* mutants had defects in tolerance to alcohol.[121] The social behavior of *npr-1* mutants can be suppressed by mutations in *gcy-35* or *gcy-36*,[122] both of which encode soluble guanylate cyclases.[123] GCY-35 guanylate cyclase binds oxygen, leading to the hypothesis that the aggregation behavior of *npr-1* mutants is related to oxygen levels in the local environment of the animals.[124] Because no NPY appears in the *C. elegans* genome but NPY has some sequence similarity to the FLPs, de Bono and coworkers guessed that one or more of the FLPs would activate the NPR-1 receptor. To identify which FLP, de Bono and colleagues used *Xenopus* oocytes to inject constructs for NPR-1 and an inwardly rectifying potassium channel and screened for receptor activation of the potassium channels by different FLPs.[125] Two groups, Rogers et al.[125] and Kubiak et al,[114] determined that FLP-21 activates NPR-1; in addition, Rogers et al.[125] found that FLP-18 peptides also activated NPR-1, demonstrating again that FLP receptors bind multiple ligands. However, despite all FLP-18 peptides having a C-terminal PGVLRFamide sequence (see Table 2), different FLP-18 peptides activated NPR-1 with different potencies.[126] These differences in activities were attributed to the varying N-terminal sequences, which by NMR analysis showed considerable differences in pH dependence, overall peptide charge, hydrogen bonding, and structural conformation.[126] Animals carrying a mutation in *flp-21* display only mild aggregation compared to *npr-1* mutants,[125] presumably because FLP-18 peptides can still signal through NPR-1. The cells on which FLP-21 and FLP-18 act and how they are regulated remains to be determined.

Pharmacology of FLP Neuropeptides

Another effective approach at examining the role of the different peptides in *C. elegans* is to develop systems in which the peptides can be applied and examined for their effects. For instance, Holden-Dye and coworkers[127,128] have very successfully used the *C. elegans* pharyngeal system to examine the role of FLPs. Using a semi-dissected head preparation, the terminal bulb of the pharynx can be exposed and impaled with a recording electrode or recorded with a suction pipette. The muscles of the pharynx have a basal frequency of firing that can be enhanced or inhibited by bath application of serotonin or octopamine, respectively.[128] Similarly, bath application of different FLPs enhanced or inhibited the frequency of firing. Specifically, peptides encoded by eight *flp* genes, *flp-2, 4, 5, 6, 8, 14, 17* and *22*, have excitatory effects; among these FLP-8 and FLP-17 peptides elicit the most potent effects.[127,128] Peptides encoded by 11 *flp* genes, *flp-1, 3, 9, 11, 13, 14, 15, 17, 18, 19* and *21*, decreased the firing rate; FLP-11 and FLP-13 peptides generated the strongest inhibitory responses.[127,128] Peptides from *flp-7, 10, 12, 20* and *23* were also tested, but had no effect.

To further examine the effects of peptides encoded by the same gene, nine FLPs encoded by *flp-3* and seven FLPs encoded by *flp-13* were examined for their activity on pharyngeal muscle; all were found to have inhibitory effects, suggesting that peptides encoded by the same gene can have the same effects, albeit with different potencies.[127] These data are consistent with the expression of some of these FLPs in the pharynx[16] and suggest that multiple FLPs modulate feeding behavior. Whether the different FLPs signal through one or multiple FLP receptors has yet to be determined.

Parasitic nematodes also have a robust FLP repetoire, which is widely expressed throughout the animals[65-67,69,70,129] (for review see ref. 130). Because the FLPs and their corresponding receptors are possible targets for anti-helminthetic drugs, there is great interest in identifying the function and signaling pathways of these peptides in parasites. Towards this goal the body wall, reproductive and pharyngeal muscles of *Ascaris suum* have been effectively used to examine the pharmacology of multiple FLPs[72,131-134] (for reviews see refs. 130,135-137). For instance, application of KPNFLRFamide (FLP-1-6) and SDPNFLRFamide (FLP-1-4) to somatic muscle cells opens chloride or potassium channels, respectively.[131,138] Application of FLP-6 KSAYMRFamide is more nuanced: on ventral muscles FLP-6 causes contractions while on dorsal muscles FLP-6 causes relaxation.[74] The response to FLPs can also be more complex. For instance, application of KNEFIRFamide (FLP-8) or KHEYLRFamide (FLP-14) to somatic muscle strips elicits biphasic responses consisting of an initial hyperpolarization, followed by an excitatory phase of rhythmic contractions.[65,138] At the cellular level, Stretton and coworkers have begun the physiological characterization of *C. elegans* and *Ascaris* FLPs on the synaptic activity of *Ascaris* motor neurons.[139] Overall, the number of FLPs that can elicit physiological effects on muscles alone is striking and highlights the complex and intricate ways that different FLPs can modulate synaptic and muscle activity.

Conclusion

The sheer number of neuropeptide genes in *C. elegans* is somewhat surprising given that the animal has only 302 neurons and presents the daunting task of determining the function not only of each neuropeptide gene, but of each neuropeptide. Many of the neuropeptides, such as the insulin-like peptides or some of the FMRFamide-related peptides, are similar to mammalian peptides. However, the greatest similarity among the peptides is, as expected, with other nematode peptides. The hope is that complementary approaches in *C. elegans*, which has access to genetic tools, and parasitic nematodes, for which RNAi seems to be effective in many neurons,[140] will allow us to elucidate the roles of many of these complex gene families. In addition, the identification of the peptide receptors, although only slowly progressing, will be a critical step in furthering our understanding of how neuropeptides act and signal in *C. elegans*. The diversity and widespread expression of neuropeptides suggests that neuropeptides are involved in all behaviors in *C. elegans*.

Acknowledgements

We thank K. Ashrafi, A. Hart, A. Maule, P. McVeigh, S. McIntire, L. Schoofs and A. Stretton for sharing unpublished data, O. Hobert with revisions of expression patterns and A. Hart, J. Kaplan and A. Maule for helpful discussions. This work was supported by grants from the NIH to C.L. and the City College of the City University of New York through RCMI.

References

1. Chang MM, Leeman SE. Isolation of a sialogogic peptide from bovine hypothalamic tissue and its characterization as substance P. J Biol Chem 1970; 245:4784-4790.
2. Genome sequence of the nematode C. elegans: a platform for investigating biology. Science 1998; 282:2012-2018.
3. Duret L, Guex N, Peitsch MC et al. New insulin-like proteins with atypical disulfide bond pattern characterized in Caenorhabditis elegans by comparative sequence analysis and homology modeling. Genome Research 1998; 8:348-353.
4. Gregoire FM, Chomiki N, Kachinskas D et al. Cloning and developmental regulation of a novel member of the insulin-like gene family in Caenorhabditis elegans. Biochem Biophys Res Commun 1998; 249:385-390.
5. Kawano T, Ito Y, Ishiguro M et al. Molecular cloning and characterization of a new insulin/IGF-like peptide of the nematode Caenorhabditis elegans. Biochem Biophys Res Commun 2000; 273:431-436.
6. Li W, Kennedy SG, Ruvkun G. daf-28 encodes a C. elegans insulin superfamily member that is regulated by environmental cues and acts in the DAF-2 signaling pathway. Gene Dev 2003; 17:844-858.
7. Pierce SB, Costa M, Wisotzkey R et al. Regulation of DAF-2 receptor signaling by human insulin and ins-1, a member of the unusually large and diverse C. elegans insulin gene family. Genes Dev 2001; 15:672-686.
8. Price DA, Greenberg MJ. Structure of a molluscan cardioexcitatory neuropeptide. Science 1977; 197:670-671.
9. Dockray GJ. The expanding family of -RFamide peptides and their effects on feeding behaviour. Exp Physiol 2004; 89:229-235.
10. Li C. The ever-expanding neuropeptide gene families in the nematode Caenorhabditis elegans. Parasitology 131 Suppl, 2005; S109-127.
11. Li C, Kim K, Nelson LS. FMRFamide-related neuropeptide gene family in Caenorhabditis elegans. Brain Res 1999; 848:26-34.
12. Nelson LS, Kim K, Memmott JE et al. FMRFamide-related gene family in the nematode, Caenorhabditis elegans. Brain Res Mol Brain Res 1998; 58:103-111.
13. Li C, Nelson LS, Kim K et al. Neuropeptide gene families in the nematode Caenorhabditis elegans. Ann N Y Acad Sci 1999; 897:239-252.
14. Nathoo AN, Moeller RA, Westlund BA et al. Identification of neuropeptide-like protein gene families in Caenorhabditis elegans and other species. Proc Natl Acad Sci USA 2001; 98:14000-14005.

15. Blumenthal T. Trans-splicing and polycistronic transcription in Caenorhabditis elegans. Trends Genet 1995; 11:132-136.
16. Kim K, Li C. Expression and regulation of an FMRFamide-related neuropeptide gene family in Caenorhabditis elegans. J Comp Neurol 2004; 475:540-550.
17. McVeigh P, Leech S, Mair GR et al. Analysis of FMRFamide-like peptide (FLP) diversity in phylum Nematoda. Int J Parasitol 2005; 35:1043-1060.
18. Rosoff ML, Burglin TR, Li C. Alternatively spliced transcripts of the flp-1 gene encode distinct FMRFamide-like peptides in Caenorhabditis elegans. J Neurosci 1992; 12:2356-2361.
19. Couillault C, Pujol N, Reboul J et al. TLR-independent control of innate immunity in Caenorhabditis elegans by the TIR domain adaptor protein TIR-1, an ortholog of human SARM. Nature Immunology 2004; 5:488-494.
20. Kim K. Function of a FMRFamide-related neuropeptide gene family in Caenorhabditis elegans. PhD thesis, Boston, MA: Boston University 2003.
21. Sithigorngul P, Stretton AO, Cowden C. A versatile dot-ELISA method with femtomole sensitivity for detecting small peptides. J Immunol Methods 1991; 141:23-32.
22. Strand FL. Neuropeptides 1999. (Cambridge, MA: MIT Press).
23. Sossin WS, Sweet-Cordero A, Scheller RH. Dale's hypothesis revisited: different neuropeptides derived from a common prohormone are targeted to different processes. Proc Natl Acad Sci USA 1990; 87:4845-4848.
24. Scamuffa N, Calvo F, Chretien M et al. Proprotein convertases: lessons from knockouts. FASEB J 2006; 20:1954-1963.
25. Steiner DF. The proprotein convertases. Curr Opin Chem Biol 1998; 2:31-39.
26. Lindberg I, Tu B, Muller L. Cloning and functional analysis of C. elegans 7B2. DNA Cell Biol 1998; 17:727-734.
27. Sieburth D, Ch'ng Q, Dybbs M et al. Systematic analysis of genes required for synapse structure and function. Nature 2005; 436:510-517.
28. Husson SJ, Clynen E, Baggerman G et al. Discovering neuropeptides in Caenorhabditis elegans by two dimensional liquid chromatography and mass spectrometry. Biochem Biophys Res Commun 2005; 335:76-86.
29. Marks NJ, Maule AG, Geary TG et al. APEASPFIRFamide, a novel FMRFamide-related decapeptide from Caenorhabditis elegans: structure and myoactivity. Biochem Biophys Res Commun 1997; 231:591-595.
30. Marks NJ, Maule AG, Geary TG et al. KSAYMRFamide (PF3/AF8) is present in the free-living nematode, Caenorhabditis elegans. Biochem Biophys Res Commun 1998; 248:422-425.
31. Marks NJ, Maule AG, Li C et al. Isolation, pharmacology and gene organization of KPSFVRFamide: a neuropeptide from Caenorhabditis elegans. Biochem Biophys Res Commun 1999; 254:222-230.
32. Marks NJ, Shaw C, Halton DW et al. Isolation and preliminary biological assessment of AADGAPLIRFamide and SVPGVLRFamide from Caenorhabditis elegans. Biochem Biophys Res Commun 2001; 286:1170-1176.
33. Marks NJ, Shaw C, Maule AG et al. Isolation of AF2 (KHEYLRFamide) from Caenorhabditis elegans: evidence for the presence of more than one FMRFamide-related peptide-encoding gene. Biochem Biophys Res Commun 1995; 217:845-851.
34. Marder E, Calabrese RL, Nusbaum MP et al. Distribution and partial characterization of FMRFamide-like peptides in the stomatogastric nervous systems of the rock crab, Cancer borealis and the spiny lobster, Panulirus interruptus. J Comp Neurol 1987; 259:150-163.
35. Kass J, Jacob TC, Kim P et al. The EGL-3 proprotein convertase regulates mechanosensory responses of Caenorhabditis elegans. J Neurosci 2001; 21:9265-9272.
36. Husson SJ, Clynen E, Baggerman G et al. Defective processing of neuropeptide precursors in Caenorhabditis elegans lacking proprotein convertase 2 (KPC-2/EGL-3): mutant analysis by mass spectrometry. J Neurochem 2006; 98:1999-2012.
37. Trent C, Tsuing N, Horvitz HR. Egg-laying defective mutants of the nematode Caenorhabditis elegans. Genetics 1983; 104:619-647.
38. Thacker C, Rose AM. A look at the Caenorhabditis elegans Kex2/Subtilisin-like proprotein convertase family. Bioessays 2000; 22:545-553.
39. Kimura KD, Tissenbaum HA, Liu Y et al. daf-2, an insulin receptor-like gene that regulates longevity and diapause in Caenorhabditis elegans. Science 1997; 277:942-946.
40. Thacker C, Peters K, Srayko M et al. The bli-4 locus of Caenorhabditis elegans encodes structurally distinct kex2/subtilisin-like endoproteases essential for early development and adult morphology. Genes Dev 1995; 9:956-971.
41. Thomas JH. Genetic analysis of defecation in Caenorhabditis elegans. Genetics 1990; 124:855-872.

42. Reiner DJ, Thomas JH. Reversal of a muscle response to GABA during C. elegans male development. J Neurosci 1995; 15:6094-6102.
43. Miller DM 3rd, Ortiz I, Berliner GC et al. Differential localization of two myosins within nematode thick filaments. Cell 1983; 34:477-490.
44. MacLeod AR, Waterston RH, Fishpool RM et al. Identification of the structural gene for a myosin heavy-chain in Caenorhabditis elegans. J Mol Biol 1977; 114:133-140.
45. Doi M, Iwasaki K. Regulation of retrograde signaling at neuromuscular junctions by the novel C2 domain protein AEX-1. Neuron 2002; 33:249-259.
46. Jacob TC, Kaplan JM. The EGL-21 carboxypeptidase E facilitates acetylcholine release at Caenorhabditis elegans neuromuscular junctions. J Neurosci 2003; 23:2122-2130.
47. Eipper BA, Milgram SL, Husten EJ et al. Peptidylglycine alpha-amidating monooxygenase: a multifunctional protein with catalytic, processing and routing domains. Protein Sci 1993; 2:489-497.
48. Han M, Park D, Vanderzalm PJ et al. Drosophila uses two distinct neuropeptide amidating enzymes, dPAL1 and dPAL2. J Neurochem 2004; 90:129-141.
49. Hall DH, Hedgecock EM. Kinesin-related gene unc-104 is required for axonal transport of synaptic vesicles in C. elegans. Cell 1991; 65:837-847.
50. Schinkmann K. FMRFamide-like peptides in the nematodes Caenorhabditis elegans and Caenorhabditis vulgaris. PhD thesis, Boston, MA: Boston University 1994.
51. Zahn TR, Angleson JK, MacMorris MA et al. Dense core vesicle dynamics in Caenorhabditis elegans neurons and the role of kinesin UNC-104. Traffic 2004; 5:544-559.
52. Salio C, Lossi L, Ferrini F et al. Neuropeptides as synaptic transmitters. Cell Tissue Res 2006; 326:583-598.
53. Bonanomi D, Benfenati F, Valtorta F. Protein sorting in the synaptic vesicle life cycle. Progress in Neurobiology 2006; 80:177-217.
54. Ahmed S, Maruyama IN, Kozma R et al. The Caenorhabditis elegans unc-13 gene product is a phospholipid-dependent high-affinity phorbol ester receptor. Biochem J 1992; 287:995-999.
55. Richmond JE, Davis WS, Jorgensen EM. UNC-13 is required for synaptic vesicle fusion in C. elegans. Nat Neurosci 1999; 2:959-964.
56. Sieburth D, Madison JM, Kaplan JM. PKC-1 regulates secretion of neuropeptides. Nat Neurosci 2007; 10:49-57.
56a. Hammerlund M, Watanabe S, Schuske K et al. CAPS and syntaxin dock dense core vesicles to the plasma membrane in neurons. J Cell Biol 2008; 180:483-491.
57. Kohn RE, Duerr JS, McManus JR et al. Expression of multiple UNC-13 proteins in the Caenorhabditis elegans nervous system. Molecular biology of the cell 2000; 11:3441-3452.
58. Ann K, Kowalchyk JA, Loyet KM et al. Novel Ca2+-binding protein (CAPS) related to UNC-31 required for Ca2+-activated exocytosis. J Biol Chem 1997; 272:19637-19640.
58a. Speese S, Petrie M, Schuske K et al. UNC-31 (CAPS) is required for dense-core vesicle but not synaptic vesicle exocytosis in Caenorhabditis elegans. J Neurosci 2007; 27:6150-6162.
59. Cai T, Fukushige T, Notkins AL et al. Insulinoma-Associated Protein IA-2, a vesicle transmembrane protein, genetically interacts with UNC-31/CAPS and affects Neurosecretion in Caenorhabditis elegans. J Neurosci 2004; 24:3115-3124.
60. Grishanin RN, Klenchin VA, Loyet KM et al. Membrane association domains in Ca2+-dependent activator protein for secretion mediate plasma membrane and dense-core vesicle binding required for Ca2+-dependent exocytosis. J Biol Chem 2002; 277:22025-22034.
61. Renden R, Berwin B, Davis W et al. Drosophila CAPS is an essential gene that regulates dense-core vesicle release and synaptic vesicle fusion. Neuron 2001; 31:421-437.
62. Fares H, Grant B. Deciphering endocytosis in Caenorhabditis elegans. Traffic 2002; 3:11-19.
63. Rosoff ML, Doble KE, Price DA et al. The FLP-1 propeptide is processed into multiple, highly similar FMRFamide-like peptides in Caenorhabditis elegans. Peptides 1993; 14:331-338.
64. Husson SJ, Schoofs L. Altered neuropeptide profile of Caenorhabditis elegans lacking the chaperone protein 7B2 as analyzed by mass spectrometry. FEBS Lett 2007; 581:4288-4292.
65. Cowden C, Stretton AO. AF2, an Ascaris neuropeptide: isolation, sequence and bioactivity. Peptides 1993; 14:423-430.
66. Cowden C, Stretton AO. Eight novel FMRFamide-like neuropeptides isolated from the nematode Ascaris suum. Peptides 1995; 16:491-500.
67. Cowden C, Stretton AO, Davis RE. AF1, a sequenced bioactive neuropeptide isolated from the nematode Ascaris suum. Neuron 1989; 2:1465-1473.
68. Yew JY, Kutz KK, Dikler S et al. Mass spectrometric map of neuropeptide expression in Ascaris suum. J Comp Neurol 2005; 488:396-413.
69. Keating CD, Holden-Dye L, Thorndyke MC et al. The FMRFamide-like neuropeptide AF2 is present in the parasitic nematode Haemonchus contortus. Parasitology 1995; 111 (Pt 4), 515-521.

70. Marks NJ, Sangster NC, Maule AG et al. Structural characterisation and pharmacology of KHEYLRFamide (AF2) and KSAYMRFamide (PF3/AF8) from Haemonchus contortus. Mol Biochem Parasitol 1999; 100:185-194.
71. Geary TG, Price DA, Bowman JW et al. Two FMRFamide-like peptides from the free-living nematode Panagrellus redivivus. Peptides 1992; 13:209-214.
72. Maule AG, Geary TG, Bowman JW et al. Inhibitory effects of nematode FMRFamide-related peptides (FaRPs) on muscle strips from Ascaris suum. Invert Neurosci 1995; 1:255-265.
73. Maule AG, Shaw C, Bowman JW et al. The FMRFamide-like neuropeptide AF2 (Ascaris suum) is present in the free-living nematode, Panagrellus redivivus (Nematoda, Rhabditida). Parasitology 1994; 109 (Pt 3):351-356.
74. Maule AG, Shaw C, Bowman JW et al. KSAYMRFamide: a novel FMRFamide-related heptapeptide from the free-living nematode, Panagrellus redivivus, which is myoactive in the parasitic nematode, Ascaris suum. Biochem Biophys Res Commun 1994; 200:973-980.
75. Simmer F, Tijsterman M, Parrish S et al. Loss of the putative RNA-directed RNA polymerase RRF-3 makes C. elegans hypersensitive to RNAi. Curr Biol 2002; 12:1317-1319.
76. Schmitz C, Kinge P, Hutter H. Axon guidance genes identified in a large-scale RNAi screen using the RNAi-hypersensitive Caenorhabditis elegans strain nre-1(hd20) lin-15b(hd126). Proc Natl Acad Sci USA 2007; 104:834-839.
76a. Tavernarakis N, Wang SL, Dorovkov M et al. Heritable and inducible genetic interference by double-stranded RNA encoded by transgenes. Nat Genet 2000; 24:180-183.
77. Cassada RC, Russell RL. The dauerlarva, a post-embryonic developmental variant of the nematode Caenorhabditis elegans. Dev Biol 1975; 46:326-342.
78. Johnson TE, Mitchell DH, Kline S et al. Arresting development arrests aging in the nematode Caenorhabditis elegans. Mechanisms of ageing and development 1984; 28:23-40.
79. Riddle DL, Albert PS. Genetic and environmental regulation of dauer larva development. In: Riddle DL, Blumenthal T, Meyer BJ, Priess JR, eds. C. elegans II. New York: Cold Spring Harbor Laboratory Press, 1997:739-768.
80. Thomas JH, Birnby DA, Vowels JJ. Evidence for parallel processing of sensory information controlling dauer formation in Caenorhabditis elegans. Genetics 1993; 134:1105-1117.
81. Kao G, Nordenson C, Still M et al. ASNA-1 positively regulates insulin secretion in C. elegans and mammalian cells. Cell 2007; 128:577-587.
82. Kenyon C, Chang J, Gensch E et al. A C. elegans mutant that lives twice as long as wild type. Nature 1993; 366:461-464.
83. Gems D, Sutton AJ, Sundermeyer ML et al. Two pleiotropic classes of daf-2 mutation affect larval arrest, adult behavior, reproduction and longevity in Caenorhabditis elegans. Genetics 1998; 150:129-155.
84. Dlakic M. A new family of putative insulin receptor-like proteins in C. elegans. Curr Biol 2002; 12: R155-157.
85. Bargmann CI, Horvitz HR. Control of larval development by chemosensory neurons in Caenorhabditis elegans. Science 1991; 251:1243-1246.
86. Malone EA, Thomas JH. A screen for nonconditional dauer-constitutive mutations in Caenorhabditis elegans. Genetics 1994; 136:879-886.
87. Kodama E, Kuhara A, Mohri-Shiomi A et al. Insulin-like signaling and the neural circuit for integrative behavior in C. elegans. Genes Dev 2006; 20:2955-2960.
88. Hedgecock EM, Russell RL. Normal and mutant thermotaxis in the nematode Caenorhabditis elegans. Proc Natl Acad Sci USA 1975; 72:4061-4065.
89. Mori I, Ohshima Y. Neural regulation of thermotaxis in Caenorhabditis elegans. Nature 1995; 376:344-348.
90. Sawin ER, Ranganathan R, Horvitz HR. C. elegans locomotory rate is modulated by the environment through a dopaminergic pathway and by experience through a serotonergic pathway. Neuron 2000; 26:619-631.
91. White JG, Southgate E, Thomson JN et al. The structure of the nervous system of the nematode Caenorhabditis elegans. Philos Trans R Soc Lond B Biol Sci 1986; 314:1-340.
92. Ward S. Chemotaxis by the nematode Caenorhabditis elegans: identification of attractants and analysis of the response by use of mutants. Proc Natl Acad Sci USA 1973; 70:817-821.
93. Saeki S, Yamamoto M, Iino Y. Plasticity of chemotaxis revealed by paired presentation of a chemoattractant and starvation in the nematode Caenorhabditis elegans. J Exp Biol 2001; 204:1757-1764.
94. Tomioka M, Adachi T, Suzuki H et al. The insulin/PI 3-kinase pathway regulates salt chemotaxis learning in Caenorhabditis elegans. Neuron 2006; 51:613-625.
95. Schinkmann K, Li C. Localization of FMRFamide-like peptides in Caenorhabditis elegans. J Comp Neurol 1992; 316:251-260.
96. Nelson LS, Rosoff ML et al. Disruption of a neuropeptide gene, flp-1, causes multiple behavioral defects in Caenorhabditis elegans. Science 1998; 281:1686-1690.

96a. Ringstad N, Horvitz HR FMRFamide neuropeptides and acetylcholine synergistically inhibit egg-laying by C. elegans. Nature Neuroscience 2008; 11:1168-1176.
97. Alfonso A, Grundahl K, Duerr JS et al. The Caenorhabditis elegans unc-17 gene: a putative vesicular acetylcholine transporter. Science 1993; 261:617-619.
98. McIntire SL, Jorgensen E, Kaplan J et al. The GABAergic nervous system of Caenorhabditis elegans. Nature 1993; 364:337-341.
99. Brenner S. The genetics of Caenorhabditis elegans. Genetics 1974; 77:71-94.
100. Duerr JS, Gaskin J et al. Identified neurons in C. elegans coexpress vesicular transporters for acetylcholine and monoamines. American J Physiology 2001; 280:C1616-1622.
101. Waggoner LE, Hardaker LA, Golik S et al. Effect of a neuropeptide gene on behavioral states in Caenorhabditis elegans egg-laying. Genetics 2000; 154:1181-1192.
101a. Cohen M, Reale V, Olofsson B et al. Coordinated regulation of foraging and metabolism in C. elegans by RFfamide neuropeptide signaling. Cell Metab 2009; 9:375-385.
101b. Janssen T, Husson SJ, Lindemans M et al. Functional characterization of three G protein-coupled receptors for pigment dispersing factors in Caenorhabditis elegans. J Biol Chem 2008; 283:15241-15249.
101c. Park SK, Link CD, Johnson TE Life-span extension by dietary restriction is mediated by NLP-7 signaling and coelomocyte endocytosis in C. elegans. FASEB J 2009. [Epub ahead of print].
101d. Muir RE, Tan M-W. Virulence of Leucobacter chromiireducens subsp. solipictus to Caenorhabditis elegans: Characterization of a novel host-pathogen interaction. Appl Environ Microbiol 2008; 74:4185-4198.
101e. Pujol N, Cypowyu S, Ziegler K et al. Distinct innate immune responses to infection and wounding in the C. elegans epidermis. Curr Biol 2008; 18:481-489.
101f. Styer KL, Singh V, Macosko E et al. Innate immunity in Caenorhabditis elegans is regulated by neurons expressing NPR-1/GPCR. Science 2008; 322:460-464.
101g. Janssen T, Husson SJ, Meelkop E et al. Discovery and characterization of a conserved pigment dispersing factor-like neuropeptide pathway in Caenorhabditis elegans. J Neurochem 2009; 111:228-41.
101h. Husson SJ, Janssen T, Baggerman G et al. Impaired processing of FLP and NLP peptides in carboxypeptidase E (EGL-21)-deficient Caenorhabditis elegans as analyzed by mass spectrometry. J Neurochem 2007; 102:246-60.
101i. Lindemans M, Janssen T, Husson SJ et al. A neuromedin-pyrokinin-like neuropeptide signaling system in Caenorhabditis elegans. Biochem Biophys Res Commun 2009; 379:760-764.
101j. McVeigh P, Alexander-Bowman S, Veal E et al. Neuropeptide-like protein diversity in phylum Nematoda. Int J Parasitol 2008; 38:1493-1503.
102. Nguyen M, Alfonso A, Johnson CD et al. Caenorhabditis elegans mutants resistant to inhibitors of acetylcholinesterase. Genetics 1995; 140:527-535.
103. Robertson HM. Two large families of chemoreceptor genes in the nematodes Caenorhabditis elegans and Caenorhabditis briggsae reveal extensive gene duplication, diversification, movement and intron loss. Genome Research 1998; 8:449-463.
104. Robertson HM. The large srh family of chemoreceptor genes in Caenorhabditis nematodes reveals processes of genome evolution involving large duplications and deletions and intron gains and losses. Genome Research 2000; 10:192-203.
105. Robertson HM. Updating the str and srj (stl) families of chemoreceptors in Caenorhabditis nematodes reveals frequent gene movement within and between chromosomes. Chem Senses 2001; 26:151-159.
106. Bargmann CI. Neurobiology of the Caenorhabditis elegans genome. Science 1998; 282:2028-2033.
107. Keating CD, Kriek N, Daniels M et al. Whole-genome analysis of 60 G protein-coupled receptors in Caenorhabditis elegans by gene knockout with RNAi. Curr Biol 2003; 13:1715-1720.
108. Bonini JA, Jones KA, Adham N et al Identification and characterization of two G protein-coupled receptors for neuropeptide FF. J Biol Chem 2000; 275:39324-39331.
109. Cazzamali G, Grimmelikhuijzen CJ. Molecular cloning and functional expression of the first insect FMRFamide receptor. Proc Natl Acad Sci USA 2002; 99:12073-12078.
110. Duttlinger A, Mispelon M, Nichols R. The structure of the FMRFamide receptor and activity of the cardioexcitatory neuropeptide are conserved in mosquito. Neuropeptides 2003; 37:120-126.
111. Meeusen T, Mertens I, Clynen E et al. Identification in Drosophila melanogaster of the invertebrate G protein-coupled FMRFamide receptor. Proc Natl Acad Sci USA 2002; 99:15363-15368.
112. Tensen CP, Cox KJ, Smit AB et al. The lymnaea cardioexcitatory peptide (LyCEP) receptor: a G-protein-coupled receptor for a novel member of the RFamide neuropeptide family. J Neurosci 1998; 18:9812-9821.
113. Lingueglia E, Champigny G, Lazdunski M et al. Cloning of the amiloride-sensitive FMRFamide peptide-gated sodium channel. Nature 1995; 378:730-733.

114. Kubiak TM, Larsen MJ, Nulf SC et al. Differential activation of "social" and "solitary" variants of the Caenorhabditis elegans G protein-coupled receptor NPR-1 by its cognate ligand AF9. J Biol Chem 2003; 278:33724-33729.
115. Kubiak TM, Larsen MJ, Zantello MR et al. Functional annotation of the putative orphan Caenorhabditis elegans G-protein-coupled receptor C10C6.2 as a FLP15 peptide receptor. J Biol Chem 2003; 278:42115-42120.
116. Lowery DE, Geary TG, Kubiak TM et al. G protein-coupled receptor-like receptors and modulators thereof. (United States: Pharmacia & Upjohn Company) 2003.
117. Mertens I, Meeusen T, Janssen T et al. Molecular characterization of two G protein-coupled receptor splice variants as FLP2 receptors in Caenorhabditis elegans. Biochem Biophys Res Commun 2005; 330:967-974.
118. Mertens I, Vandingenen A, Meeusen T et al. Functional characterization of the putative orphan neuropeptide G-protein coupled receptor C26F1.6 in Caenorhabditis elegans. FEBS Lett 2004; 573:55-60.
119. Mertens I, Clinckspoor I, Janssen T et al. FMRFamide related peptide ligands activate the Caenorhabditis elegans orphan GPCR Y59H11AL.1. Peptides 2006; 27:1291-1296.
120. de Bono M, Bargmann CI. Natural variation in a neuropeptide Y receptor homolog modifies social behavior and food response in C. elegans. Cell 1998; 94:679-689.
121. Davies AG, Bettinger JC, Thiele TR et al. Natural variation in the npr-1 gene modifies ethanol responses of wild strains of C. elegans. Neuron 2004; 42:731-743.
122. Cheung BH, Arellano-Carbajal F, Rybicki I et al. Soluble guanylate cyclases act in neurons exposed to the body fluid to promote C. elegans aggregation behavior. Curr Biol 2004; 14:1105-1111.
123. Morton DB, Hudson ML, Waters E et al. Soluble guanylyl cyclases in Caenorhabditis elegans: NO is not the answer. Curr Biol 1999; 9:R546-547.
124. Gray JM, Karow DS, Lu H et al. Oxygen sensation and social feeding mediated by a C. elegans guanylate cyclase homologue. Nature 2004; 430:317-322.
125. Rogers C, Reale V, Kim K et al. Inhibition of Caenorhabditis elegans social feeding by FMRFamide-related peptide activation of NPR-1. Nat Neurosci 2003; 6:1178-1185.
126. Dossey AT, Reale V, Chatwin H et al. NMR analysis of Caenorhabditis elegans FLP-18 neuropeptides: implications for NPR-1 activation. Biochemistry 2006; 45:7586-7597.
127. Papaioannou S, Marsden D, Franks CJ et al. Role of a FMRFamide-like family of neuropeptides in the pharyngeal nervous system of Caenorhabditis elegans. J Neurobiol 2005; 65:304-319.
128. Rogers CM, Franks CJ, Walker RJ et al. Regulation of the pharynx of Caenorhabditis elegans by 5-HT, octopamine and FMRFamide-like neuropeptides. J Neurobiol 2001; 49:235-244.
129. Cowden C, Sithigorngul P, Brackley P et al. Localization and differential expression of FMRFamide-like immunoreactivity in the nematode Ascaris suum. J Comp Neurol 1993; 333:455-468.
130. McVeigh P, Geary TG, Marks NJ et al. The FLP-side of nematodes. Trends Parasitol 2006; 22:385-396.
131. Bowman JW, Friedman AR, Thompson DP et al. Structure-activity relationships of an inhibitory nematode FMRFamide-related peptide, SDPNFLRFamide (PF1), on Ascaris suum muscle. Int J Parasitol 2002; 32:1765-1771.
132. Fellowes RA, Maule AG, Marks NJ et al. Modulation of the motility of the vagina vera of Ascaris suum in vitro by FMRF amide-related peptides. Parasitology 1998; 116:(Pt 3)277-287.
133. Moffett CL, Beckett AM, Mousley A et al. The ovijector of Ascaris suum: multiple response types revealed by Caenorhabditis elegans FMRFamide-related peptides. Int J Parasitol 2003; 33:859-876.
134. Trailovic SM, Clark CL, Robertson AP et al. Brief application of AF2 produces long lasting potentiation of nAChR responses in Ascaris suum. Mol Biochem Parasitol 2005; 139:51-64.
135. Brownlee D, Holden-Dye L, Walker R. The range and biological activity of FMRFamide-related peptides and classical neurotransmitters in nematodes. Adv Parasitol 2000; 45:109-180.
136. Brownlee DJ, Fairweather I. Exploring the neurotransmitter labyrinth in nematodes. Trends Neurosci 1999; 22:16-24.
137. Day TA, Maule AG. Parasitic peptides! The structure and function of neuropeptides in parasitic worms. Peptides 1999; 20:999-1019.
138. Bowman JW, Friedman AR, Thompson DP et al. Structure-activity relationships of KNEFIRFamide (AF1), a nematode FMRFamide-related peptide, on Ascaris suum muscle. Peptides 1996; 17:381-387.
139. Davis RE, Stretton AO. Structure-activity relationships of 18 endogenous neuropeptides on the motor nervous system of the nematode Ascaris suum. Peptides 2001; 22:7-23.
140. Kimber MJ, McKinney S, McMaster S et al. flp gene disruption in a parasitic nematode reveals motor dysfunction and unusual neuronal sensitivity to RNA interference. FASEB J 2007; 21:1233-1243.

Chapter 7

Control of Nematode Parasites with Agents Acting on Neuro-Musculature Systems:

Lessons for Neuropeptide Ligand Discovery

Richard J. Martin* and Alan P. Robertson

Abstract

There are a number of reasons why the development of novel anthelmintics is very necessary.[1] In domestic animals, parasites cause serious loss of production and are a welfare concern. The control of these parasites requires changes in management practices to reduce the spread of infection and the use of therapeutic agents to treat affected animals. The development of vaccines against parasites is desirable but their development so far has been very limited. One notable exception is the vaccination of calves against infection by *Dictyocaulus viviparous* (lung-worm) which has proved to be very effective.[2] In domestic animals, the total market for anti-parasitic agents (both ecto- and endo-parasites) is in excess of a billion US dollars. In humans there are serious problems of morbidity and mortality associated with parasite infections. 1.6 billion People throughout the world are infected with ascariasis (Fig. 1A) and/or hookworm. Approximately one-third of the world's population is suffering from the effects of intestinal nematode parasites, causing low growth-rates in infants, ill-thrift, diarrhea and in 2% of cases, loss of life. Despite the huge number of affected individuals, the market for anti-parasitic drugs for humans is not big enough to foster the development of anthelmintics because most infestations that occur are in undeveloped countries that lack the ability to pay for the development of these drugs. The major economic motivator then, is for the development of animal anthelmintics.

In both domestic animals and now in humans, there is now a level of resistance to the available anthelmintic compounds.[2] The resistance is either: constitutive, where a given species of parasite has never been sensitive to the compound; or acquired, where the resistance has developed through Darwinian selection fostered by the continued exposure to the anti-parasitic drugs. The continued use of all anthelmintics has and will, continue to increase the level of resistance. Cure rates are now often less than 100% and resistance of parasites to agents acting on the neuromuscular systems is present in a wide range of parasites of animals and humans hosts.[3,4]

In the face of this resistance the development of novel and effective agents is an urgent and imperative need. New drugs which act on the neuromuscular system have an advantage for medication for animals and humans because they have a rapid therapeutic effect within 3 hours of administration. The effects on the neuromuscular system include: spastic paralysis with drugs like levamisole and pyrantel; flaccid paralysis as with piperazine; or disruption of other vital muscular

*Corresponding Author: Richard J. Martin—Department of Biomedical Sciences, Iowa State University, Ames, Iowa. Email: rjmartin@iastate.edu

Neuropeptide Systems as Targets for Parasite and Pest Control, edited by Timothy G. Geary and Aaron G. Maule.

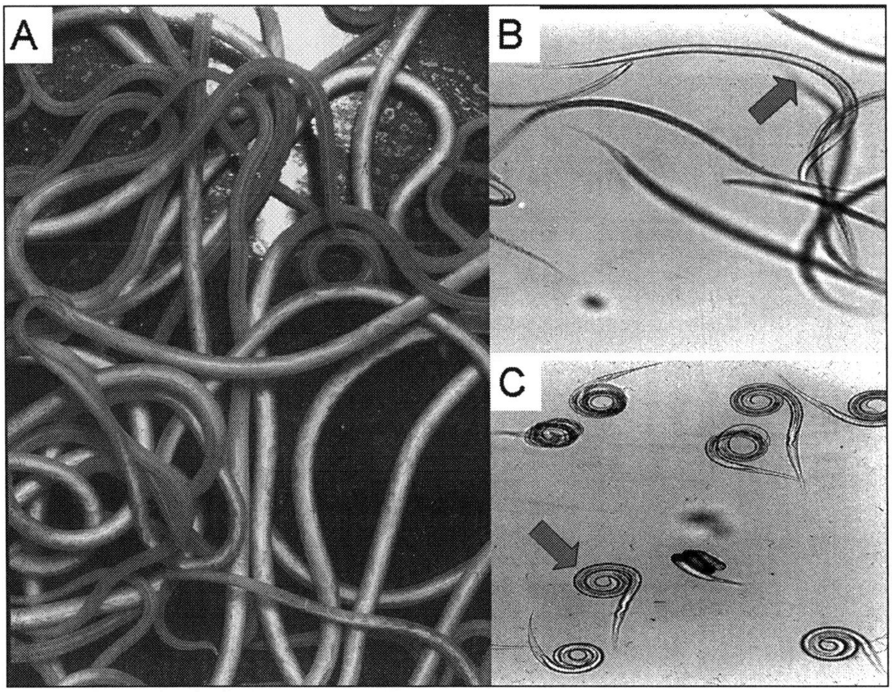

Figure 1. A) Adult *Ascaris suum*. These large intestinal nematode parasites of the pig are nearly identical to *Ascaris lumbricoides* found in the human intestine. B) L3 larvae of *Ostertagia ostertagiae* swimming freely in tap water. C) L3 larvae of *Ostertagia ostertagiae* showing tight coiling or spastic paralysis following treatment with 10 μM levamisole.

activity as with ivermectin. Figure 1 B and C, illustrates an example of a spastic effect of levamisole on infectious L3 larvae of *Ostertagia ostertagiae*, a parasite of pigs. The effect was produced within minutes of the in vitro application of levamisole.

In this chapter we comment on the properties of existing agents that have been used to control nematode parasites and that have an action on neuromuscular systems. We then draw attention to resistance that has developed to these compounds and comment on their toxicity and spectra of actions. We hope that some of the lessons that the use of these compounds has taught us may to be applied to any novel neuropeptide ligand that may be introduced. Our aim is then is to provide some warning signs for recognized but dangerous obstacles.

The Existing Anti-Nematodal Drugs with Effects on Neuromuscular Systems

There are a limited number of classes of anti-nematodal drugs that are available.[1,5,6] It is usually assumed that compounds of the same class have the same mode of action, have similar spectra of actions, toxicities and can show cross-resistance. This assumption arises from the view that they have the same molecular sites of action. This is a commonly used assumption because it has the advantage of predicting spectra of actions and toxicities but is a limiting simplification because compounds that belong to the same class can have different receptor selectivities,[7,8] can show differences in the details of their pharmacokinetic properties[9-11] toxicities and spectra of action. The different classes of anthelmintic drugs that have actions on nematode neuromuscular systems are now listed:

GABA Agonists

Piperazine is a GABA agonist that activates and gates GABA receptor channels on nematode muscle to produce an inhibitory effect and flaccid paralysis of the parasite.[12,13] The drug is effective against larger nematodes found in the gastro-intestinal tract. The GABA agonists have not been developed beyond piperazine which has a limited spectrum of action. Although diethylcarbamazine is a piperazine derivative it is not a GABA agonist. Diethylcarbamazine appears to act as a lipooxygenase inhibitor and inhibits the production of leukotrienes so that it modifies the innate immune response.[14] The modification of the innate immune response changes the host parasite balance and may be responsible for the elimination of microfilaria in the blood of hosts. Diethylcarbamazine is metabolized to a number of products in the mammalian host;[15,16] one of these products is piperazine so that its mode of action against some nematodes, like hookworm, in the GI tract may be mediated by the generation of piperazine. Given the limited number of GABA agonists and the limited use of this class of compounds, it is possible this class of compounds could be developed in the future as novel anthelmintics. The use of piperazine has demonstrated that agents that cause muscle relaxation may be used as effective anthelmintics. Novel neuropeptide ligands that have similar effects (e.g., PF4[17,18] but whose effects are mediated by other receptors[19] might be developed.

Nicotinic Agonists

Levamisole and pyrantel are examples of nicotinic agonists (Fig. 2) that selectively gate acetylcholine ion-channels on nematode muscle and on nematode neurons to produce spastic paralysis and inhibition of egg-laying. There are a number of other nicotinic agonists including: butamisole, morantel, bephenium, oxantel and thenium that have been used or are still in current use. Their

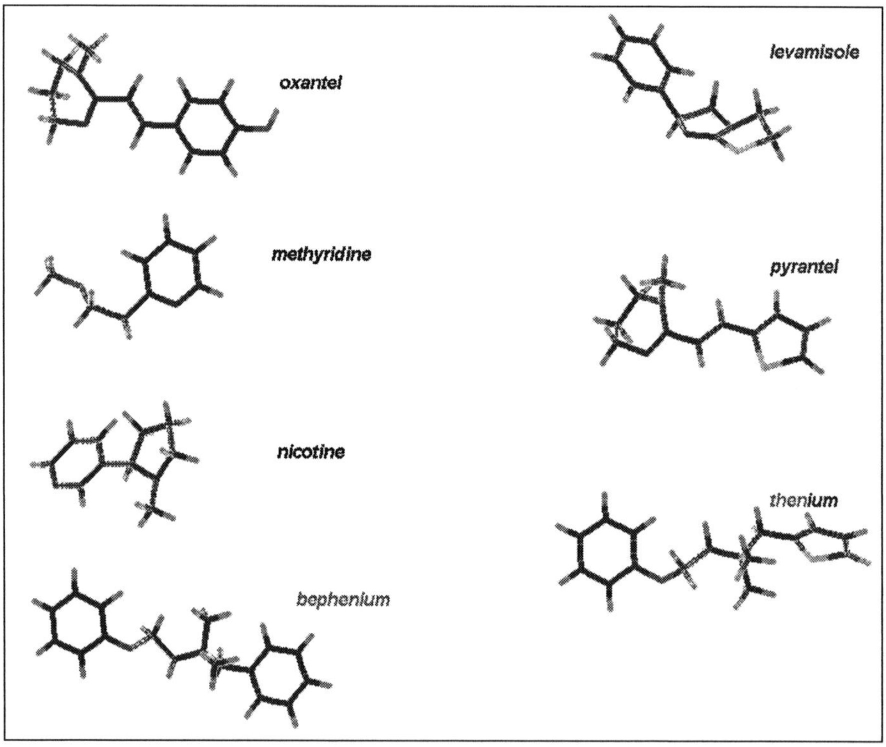

Figure 2. Chemical structure of nicotinic anthelmintics.

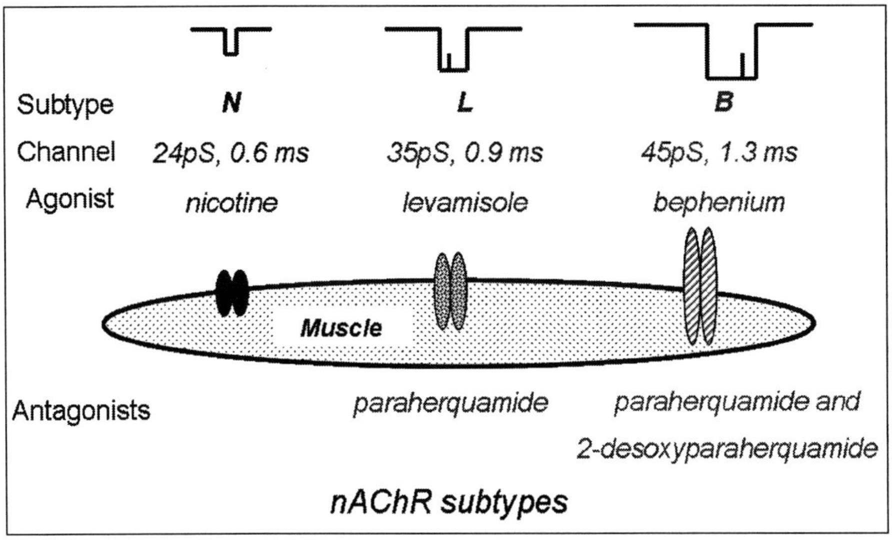

Figure 3. Subtypes of nicotinic acetylcholine receptor found on *Ascaris* muscle. There is the N-subtype, preferentially sensitive to nicotine, the L-subtype, preferentially sensitive to levamisole and antagonized by paraherquamide and the B-subtype, preferentially sensitive to bephenium and antagonized by paraherquamide and desoxyparaherquamide.

spectra of actions are broader than piperazine but depend on their pharmacokinetic distribution and receptor selectivity. Preparations of pyrantel (pamoate or embonate) are limited to the G.I tract oral administration and so are limited in action to effects on gastro-intestinal nematodes with no effect on lungworms. Levamisole, however, is distributed more widely and can be administered by injection or by pour-on preparation. Levamisole reaches therapeutic concentrations in the lungs. It can be used to treat lungworm in some species like cattle. There are different subtypes of nicotinic receptor, in parasitic nematodes which have differences in pharmacological selectivities. We have observed 3 subtypes on *Ascaris* muscle (Fig. 3). There are the N-subtypes (nicotine and oxantel preferring), the L-subtype (levamisole and pyrantel preferring) and the B-subtypes (bephenium-preferring) in *Ascaris suum*.[20] It is interesting that levamisole resistant *Haemonchus contortus* remain sensitive to bephenium[21] and that *Trichuris trichuris* is not sensitive to pyrantel but is sensitive to oxantel.[22] These observations illustrate the fact that even if the mode of action of a class of compounds appears to be similar, it is possible that the different members of the class of compounds have different selectivities for different subtypes of the receptor target site. The same phenomena could occur for novel neuropeptide ligands.

Nicotinic Antagonists

Paraherquamide, desoxyparaherquamide and phenothiazine can act as nicotinic acetylcholine channel antagonists[7] phenothiazine.[23] Phenothiazine is the oldest of this group of compounds and its use is perhaps of only historical interest. It was used more extensively to treat gastro-intestinal nematodes of sheep, cattle and horses[24] but resistance and toxicity were limiting factors. It has now fallen into disuse. Paraherquamide and desoxyparaherquamide have some useful anthelmintic effects but currently, have not been developed commercially for therapeutic purposes. All of these nicotinic antagonists inhibit muscle motility and lead to paralysis and elimination of the nematodes.

Organophosphorous Compounds

Metriphonate and dichlorvos are examples of organophosphorous cholinesterase antagonists that have selective effects against both insect pests and nematodes but fortunately have only a limited effect on host cholinesterases. The organophosphorous compounds selectively and irreversibly block the breakdown of acetylcholine by cholinesterases. There is then an accumulation of acetylcholine at nerve terminal sites and this over stimulates nicotinic and muscarinic receptors on postsynaptic nerves, muscle and glandular tissue. It fatally disrupts the normal operation of nerve and muscle systems of pests and parasites by gating nicotinic receptor channels open and stimulating G-protein activated acetylcholine receptors. Organophosphorous compounds produce uncoordinated muscle and enteric activity in the parasite or pest and destroy its normal physiological control. These compounds still have small but limited use for the treatment of internal insect parasites like stomach bots of equidae. Although their spectrum of action includes both insect and nematode parasites, the effect of the organophosphorous compounds on host cholinesterases at higher drug doses, means that host toxicity is a problem. More recently, organophosphorous drugs have also been found to be associated with demyelinating neurotoxicities in humans exposed to these compounds. These toxicities can appear seen several weeks after exposure. Any novel neuropeptide agent which acts on very similar target sites in the parasite and the host will obviously carry with it the potential for toxic effect. The selectivity for the parasite receptor must be much higher than for the host receptors.

Avermectins

This class of compounds includes: ivermectin, milbemycin, moxidectin, doramectin, abamectin and selamectin. The mode of action of this group of compounds is considered to be due to the selective activation of glutamate-gated chloride channels that are only found in invertebrates including nematodes and insects.[1,6,25-27] Unfortunately these glutamate-gated chloride channels are also found in fresh water invertebrates including crustaceans. There is a possibility of toxicity if the avermectins get into pond and river water. Avermectins are also found in animal feces and affects on the biology of the dung beetles have been of concern.[28]

The avermectins can also activate some subtypes of $GABA_a$ ion-channel receptors at higher doses as well as inhibit N-subtype (ACR-16) α7-like nicotinic acetylcholine receptor channels[29] of *C. elegans*. Depending on the species of nematode parasite, avermectins inhibit pharyngeal pumping and feeding,[30-32] or will inhibit egg-laying, or will inhibit muscle motility.[33] The effects on the different tissues of the nematode parasite might be explained by the presence of glutamate-gated chloride channels in these tissues (Fig. 4). The distribution of the glutamate-gated chloride channels appears to vary between nematode parasites and the sensitivity of the different tissues in the different nematodes also varies. The result is that the different species of nematode will be affected in different ways by a particular avermectin. For example pharyngeal pumping in *Haemonchus contortus* is very sensitive to ivermectin.[30] The hookworm is less sensitive to avermectins[34] and *Ancylostoma ceylanicum* is 40-50 times more sensitive that *Necator americanus*. The avermectins have a very broad spectrum of action and that includes most nematode parasites, biting and sucking insect parasites but the avermectins do not have an action against trematodes like *Fasciola* hepatica or *Schistosoma mansoni.* The avermectins are much more potent than the other classes of anthelmintic. The very broad-spectrum of action and potency of the avermectin class of anthelmintics, has not so far not been bettered. The potency and very broad spectrum has given rise to their extensive use, resistance against these compounds and the development of several therapeutic compounds from this class of drugs.[35] Even though the avermectins are very potent, the development of resistance to them means that novel agents are still required, even if they are less potent and have a narrower spectrum of action than the avermectins.

Benzimidazoles

Thiabendazole, albendazole and triclabendazole and others belonging to this class of anthelmintics, act by binding to β-tubulins and inhibiting the formation of microtubules.[36-39] The functions of microtubules are many and include intracellular transport, vesicular transport, cell division, cell shape and synapse formation. Benzimidazoles are slower acting than other anthelmintics and will upset the general biochemistry and homeostasis of nematodes and trematodes including their neuromuscular systems. In nematodes, they will inhibit egg production and muscle movement

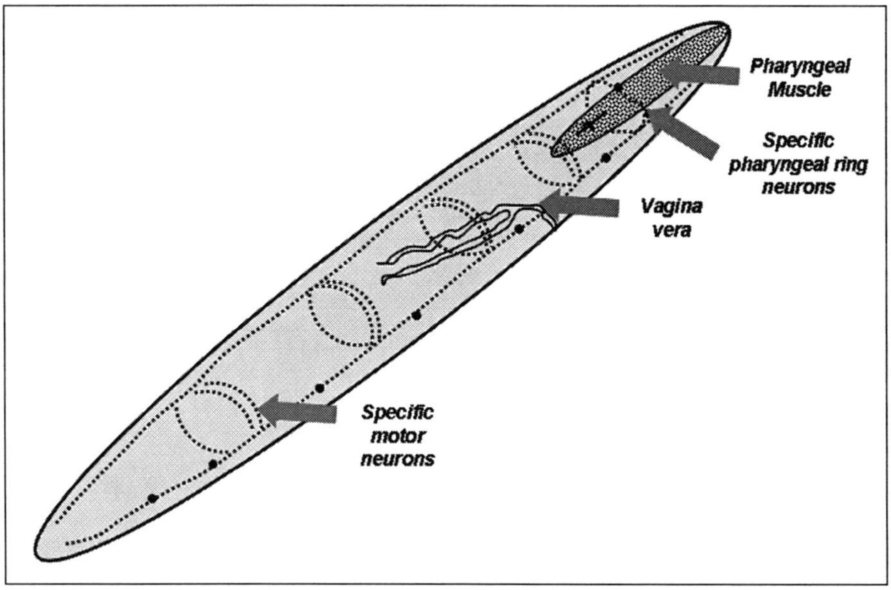

Figure 4. Diagram showing the predicted locations of avermectin receptors in a generalized parasitic nematode. The main locations include the pharynx, motor neurons and the vagina vera. The diagram shows the two nerve cords and connecting commissures. Regions of receptor localization are marked with an arrow.

after a delay of 12 hours following administration of the drug to the patient or host animal. The spectrum of action of the more recent benzimidazoles like albendazole is broad.[40] The benzimidazoles are effective against nematodes but hypobiotic nematodes and trematodes require higher doses. Some benzimidazoles show evidence of teratogenic effects if they are administered in the first trimester of pregnancy. This toxic effect includes swollen chosto-chondral junctions (ribs) and an inability to extend limbs fully.

Parasites that are resistant to benzimidazoles are now fairly common and widespread in most countries and most host species. Resistance has been associated with changes in the structure of the β-tubulin genes with the β-tubulins becoming similar to host β-tubulin.[41,42]

Latrotoxin Receptor Agonist

Emodepside,[43] developed by Bayer under the influence of Achim Harder,[44] inhibits neuromuscular transmission[45] and appears to act by activating a latrotoxin-like G-protein receptor that causes release of transmitter vesicles from synaptic terminals giving rise to paralysis. It may also activate a potassium current in muscle of nematodes. Emodepside has been introduced recently to the market. Because of its more limited spectrum of action it is combined with praziquantel to produce a broader spectrum product against both roundworms and tapeworms. This is an exciting novel agent and the further development of its use and perhaps other analogues might follow if this product is successful in the market.

Glutamate Gated Cation Channels Agonists and Antagonists

Kainate and quisqualate are found in significant therapeutic concentrations in seaweeds found in Asia.[46] The extracts of the seaweeds were used successfully as anthelmintics before the active ingredients, kainate and quisqualate, were discovered to be agonists on excitatory glutamate receptors in the vertebrate CNS Little development of these compounds as anthelmintics has been carried out. The presence of excitatory glutamate gated channels in *C. elegans* and transporters I

Ascaris suggests that these ion-channel receptors or transporters might be considered as suitable target sites for potential drugs acting on the neuromuscular system of parasitic nematodes. If these compounds were to be developed for anthelmintic use, then selective compounds that are only selective for nematode glutamate receptor channels without an effect on host receptors will be necessary. MK-801 is a glutamate antagonist[47,48] and that has a potent anti-nematodal action seen in *C. elegans*. The glutamate antagonists have not yet been developed as anthelmintics.

Of the list of the 8 classes of drugs used for the treatment of parasites, 3 classes, (nicotinic agonists, avermectins and benzimidazoles) are used extensively to eliminate nematode parasites because of their efficacy, potency and wider spectrum of action. The remainder, GABA agonists, nicotinic antagonists, latrotoxin receptor agonists and glutamate agonists, currently, are not so widely used. Their use in the future will depend on the development of suitable agents. The organophosphorous compounds are less likely to be developed because of concerns over toxicity in animal hosts and humans.

Resistance Is Predicted

General Comments

Resistance to most chemotherapeutic agents including any novel neuropeptide ligands that may be discovered and developed for the control of anti-parasitic agents should be anticipated. Right from the days of Paul Erich, it was known that resistance to therapeutic agents could develop in parasitic species. Use of *Trypan red* by,[49] to cure mice infected with *mal de Caderas* (trypanosomiasis), marked the beginning of modern 'chemotherapy', the process by which diseases, including those produced by parasites, are cured by treatment with chemical agents. Very soon after the discovery of the action of Trypan red, acquired resistance was observed. In 1905, Franke and Roehl, while working work with Ehrlïch, discovered that mice with trypanosomiasis, initially treated with Trypan red at low doses, became unresponsive to the original curative dose.[50]

'Resistance occurs when a greater frequency of individuals in a population of parasites, usually affected by a dose or concentration of compound are no longer affected' and resistance is inherited.[51] We have just mentioned how Franke and Roehl,[50] while working work with Ehrlïch, discovered that mice with trypanosomiasis, treated with Trypan red at low doses, became unresponsive to the original curative dose. This is probably the first account of the development of resistance to chemotherapeutic agents, not only for nematode parasites but more generally for all therapeutic agents including bacteria viruses and malignant cancers.

The development of resistance to one agent within a chemical class, say albendazole, is expected to result in resistance to another member of that same class, say thiabendazole: we may refer to this as *Cross-Resistance* other individuals refer to this a side-resistance. Similar cross-resistance may be expected within the different classes of compound, say for example between the avermectins, ivermectin and doramectin. The cross-resistance may arise if the different compounds within the same class act on the protein target and its affinity for the drug class changes; or if uptake, metabolism or excretion of the drug class changes in the parasite to limit toxicity. If parasites become resistant to several different classes of anthelmintic then we refer to this as *Multiple-resistance*. Examples include *Haemonchus contortus* isolates, some of which show resistance to different classes including, avermectins, benzimidazoles and nicotinic anthelmintics.[33,52-54]

Polygenic Resistance

Because resistance is inherited and carried genetically, it should be detectable by appropriate molecular markers or assays. However, to date, the development of molecular markers has met with limited success, perhaps because the resistance is polygenic.[55-57] It appears that many genes are involved in the therapeutic response and these have still be characterized. The detection of resistance to benzimidazoles has been possible in some isolates by looking for single nucleotide polymorphisms (SNPs) of the β-tubulin genes[58] in some species of parasitic nematodes.[38,59-61] A phenylalanine to tyrosine change at position 200 (P200Y) in isotype 1 β-tubulin (equivalent to TTC to TAC in the nucleotide of isotype 1—there being 4 isotypes present in some parasitic nematodes but 6 in

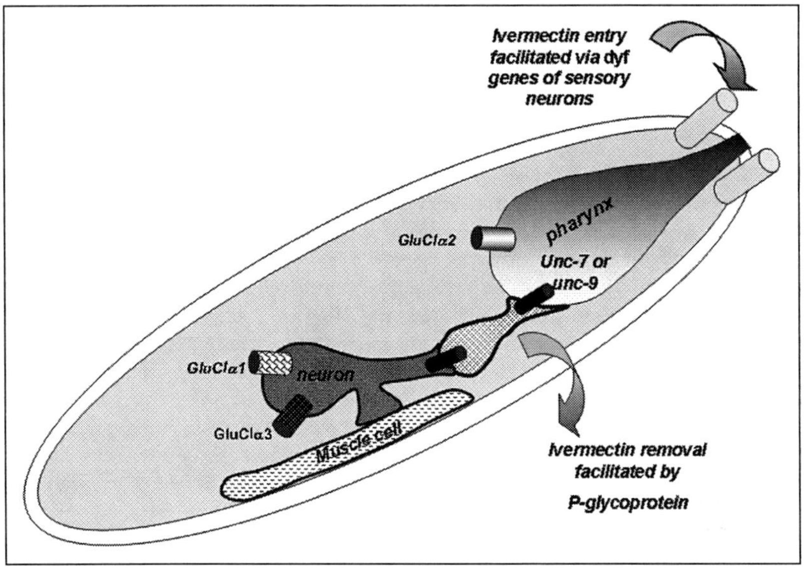

Figure 5. A diagram of the protein products of genes involved in ivermectin resistance in *C. elegans*. Derived from reference 62.

C. elegans) has been associated with resistance in some isolates of *Haemonchus* contortus. We know however, that all benzimidazole resistance is not associated with this change. Resistance to avermectins has been suggested to be involved with a number polymorphisms of genes including GABA subunit genes P-glycoproteins[62] glutamate-gated chloride channel subunits[63] polymorphism on *GluClα3β* (L256F) in some instances and innexins and dye filling genes.[64] It seems likely that there will be many alleles and many SNPs that are associated with resistance, in part because the target sites of the anthelmintics are not single receptors because there is considerable redundancy with different isotypes of the receptors being present in the parasite. There are 4 β-tubulin genes producing 4 target sites for benzimidazoles in parasitic nematodes and more than 4 genes (equivalent to *C. elegans glc-1, glc-2, avr-14 and avr-15*) that form the ion-channel target sites of the avermectins.[65] The distribution of some of the proteins involved in avermectin resistance in *C. elegans* is illustrated in (Fig. 5). If we consider the number of genes that are responsible for the expression and formation of the target site of an anthelmintic, along with genes that affect responses to the target site of the anthelmintic, like the down-stream effectors, then it is not surprising that anthelmintic resistance has been found to be polygenic in parasitic nematodes. Since resistance has followed the introduction of all anthelmintics, we should also anticipate that resistance to any novel anthelmintic that is a neuropeptide ligand will also occur and will involve changes in several genes.

Fitness

It has usually been assumed that resistance is associated with a loss of fitness due to the loss of a drug target site. This loss of fitness is expected to allow reversion, a return to sensitivity when the anthelmintic is withdrawn. This logic ignores the possibility and perhaps the necessity, of the parasites to retain its fitness in order to survive in vivo. So, in addition, the parasites are expected to accumulate other alleles (adaptation alleles) not directly needed for the resistance but that are required to recover the fitness of the parasite. Suppose that if a nematode parasite becomes resistant to levamisole by reducing the number of nicotinic acetylcholine receptors most sensitive to levamisole (*L-subtype*)[20] to retain fitness, it might recover a level of fitness by increasing the number of muscarinic acetylcholine receptors on the muscle and/or reducing the number of

GABA receptors on the muscle. So we predict that resistance is associated with two types of genetic modifications: the alleles associated directly with resistance and the alleles associated with adaptation or recovery of fitness. Both are required to allow a successful resistant nematode to survive. We would expect to see the same process to be associated with the development of resistance to any novel neuropeptide receptor anthelmintic.

The Development of Resistance

The studies on the genome of parasitic nematodes including *Haemonchus contortus* have revealed the presence of: 'Extreme polymorphism' at the nucleotide level. The single nucleotide variation in nematodes is large[66-69] and varies from 0.05 to 2% in shotgun sequences. This implies that many alleles that could contribute to resistance are already present in the population of parasitic nematodes. These resistance alleles are present at a low level before the selection pressure of anthelmintics is applied but increase with the maintained use of the anthelmintic.

It is assumed that resistance in the parasite population occurs when a selective agent (the anthelmintic) kills susceptible worms but allows the resistant worms that carry resistance alleles to survive and to reproduce. The use of anthelmintics will provide a powerful selection pressure for the increase in the frequency of resistance alleles and encourage the dispersal of those alleles throughout the population. Figure 6 illustrates how this may occur with the anthelmintic selecting for the resistant genes. If each resistance allele only contributes a small component to the resistance—one example might be an allele that increases the expression of a P-glycoprotein transporter which could reduce the concentration of the avermectins in the parasite by 50%—then a low dose of the anthelmintic will encourage the accumulation of this allele and other low level resistance alleles. If the dose of anthelmintic is then gradually increased with each passage of the life-cycle of the nematode, then the low level resistance genes will accumulate and concentrate in the surviving population. The simultaneous presence of lower-level resistance genes, particularly if they are recessive, will give rise

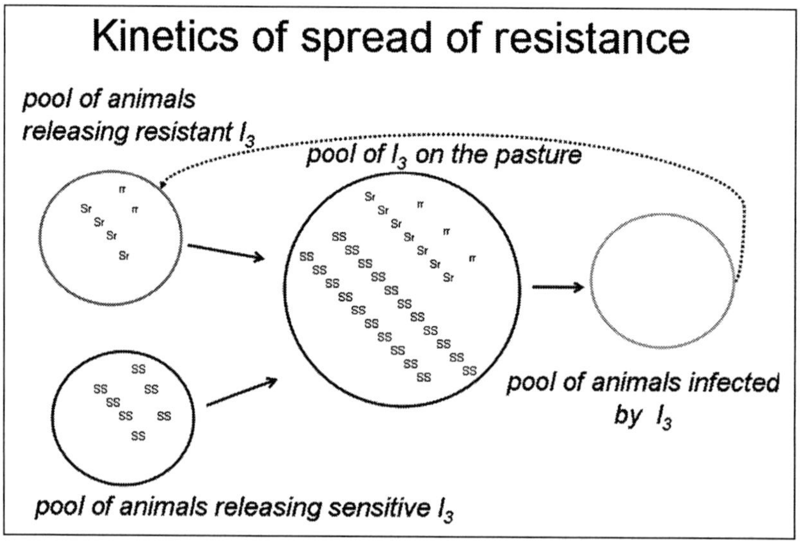

Figure 6. Diagram showing the pool of resistance genes in L3 larvae on pasture and how they may be affected by contamination from animal grazing the pasture. The resistant alleles [r] are assumed to be recessive to the sensitive [S] alleles. The genetic composition of the pool of infective L3 larvae on the pasture (center pool) will depend on the source of contamination. The % resistant alleles on the pasture will be affected by the production of eggs from infected animals (left circle) and whether these animals produce sensitive or resistant alleles. Anthelmintic treatment of these contaminating animals will select for resistant alleles.

to a much higher level of resistance.[64] So we can start to see that there will be different stages of the development of resistance. We can start with *emergence*, the first stage, followed by *development*, the second stage with accumulation of resistance alleles, followed by *full resistance*, where clinical resistance is seen with little response to therapeutic doses of the anthelmintic. The resistance will develop more quickly in a population if low or under-dosing is practiced. Sub-therapeutic doses occur if the drug preparation has a very long half-life ($t_{1/2}$) so that the drug concentrations tails off slowly over a long time so that there is a long period of sub-therapeutic concentrations present. The same sorts of problems and the development of resistance should be anticipated for a new product developed as a neuropeptide ligand.

The rate of accumulation of resistance alleles for a novel neuropeptide anthelmintic ligand will depend on a number of factors.[65] These include:

- The use of low or sub-therapeutic anthelmintic doses.
- The speed of the life cycle of the parasite—resistance will appear more quickly if there is rapid life cycle of the nematode parasite. It will also be faster if there is a direct life cycle. An indirect-life cycle will slow the accumulation of resistance alleles.
- The presence of isolated pockets of sensitive parasites (refugia) that are not subject to the selection pressure. If these parasites have a fitness advantage they will continue to dilute the resistance alleles and limit the rate of appearance of resistance. Refugia will also dilute out the resistance alleles. Refugia will slow the accumulations of resistance alleles. One approach to limit the speed of development of resistance is to treat selected, clinically affected animals and leave those that are not clinically affected to provide refugia for sensitive alleles. Only a small proportion of the population, (Fig. 7), shows a high level of infection, the remainder have a low level of infection. The distribution of infected animals is described by a negative binomial (few animals with high levels of infections others with decreasing levels of infection). It turns out that a number of factors, like resistance of individual host animals to parasite infection, the dose level of the infection and the status of the host immune system means that few animals harbor high infection levels but most do not. To limit the accumulation of resistance it is considered better to treat those animals with high infection rates and to leave those with low infection rates to provide the refugia of sensitive parasites. However it will be necessary to treat animals or patients that are showing clinical signs. A particular scheme, FAMACHA[70] where this is practiced is in the control of *Haemonchus contortus* were many of the anthelmintics may become ineffective because of rapid development of resistance. In order to maintain the refugia and to control more pathogenic nematode parasites, narrow spectrum anthelmintics effective against the pathogenic species may be considered useful to introduce if the development costs can be reduced.
- To reduce the probability of obtaining full clinical resistance, it is more desirable if the anthelmintic has several sites of action that require several genes changes to be present simultaneously before the resistance appears clinically. It would be useful if the novel neuropeptide ligand had an effect on multiple isotypes of receptor rather that a single receptor coded for by one gene. Targeting multiple protein receptors from genes that are genetically separate will require that any resistance is polygenic. The greater the number of separate genes then the slower will be the development of resistance. We have referred to an anthelmintic that requires the development of many genes to allow resistance as the MISA (multiple independent site of action) anthelmintics.[71] A MISA neuropeptide ligand as an anthelmintic is more desirable and will limit the rate of appearance of resistance.
- Combinations of two or more anthelmintic that have actions at target sites that are genetically separate will reduce the rate of accumulation of resistance. If possible a combination of novel neuropeptide receptor ligands would be desirable to reduce the rate of accumulation of resistance alleles. Pharmaceutical companies may be reluctant to do this because of licensing costs but they might be able to combine two active molecules into a single molecule to overcome the need to license two separate drugs.

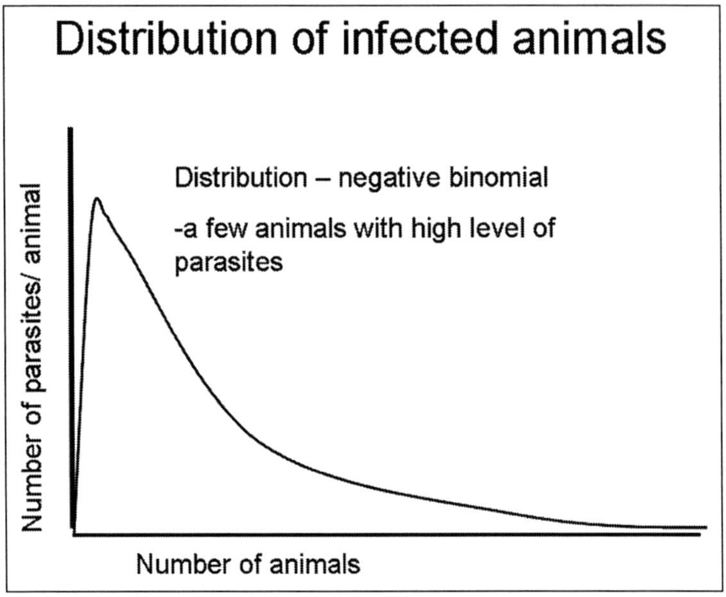

Figure 7. Diagram of the negative binomial distribution that describes the distribution of the numbers of parasites in the animals of a flock or herd. Only a few animals have a high number of parasites; the remainder have lower levels of infection. It is suggested that these very highly parasitized animals should be treated selectively as per the FAMACHA system.

Cross Resistance of Novel Neuropeptide Anthelmintic with Existing Anthelmintics

We have mentioned already that cross-resistance can occur with the same class of anthelmintic agents. We may also expect to see cross-resistance if there is an overlap in the mechanism of action of the novel neuropeptide receptor ligand with existing agents. For example we know that levamisole and pyrantel activate subtypes on nicotinic acetylcholine channels on nematode muscle cells to produce contraction. Suppose a novel anthelmintic that is an agonist of the neuropeptide AF2 receptor were developed successfully as an anthelmintic. AF2 increases contractions in some model nematodes, presumably by activating a G-protein coupled receptor.[72] Muscle contraction involves an increase in the cytosolic calcium which is coupled to the proteins that produce contraction. Suppose also that resistance to levamisole and pyrantel is produced by a null mutant or allele of a homologue of the *unc-69* gene[73] that produces calcium induced calcium release to amplify the increase in cytosolic calcium and facilitates contraction. This resistant allele would be less sensitive to levamisole and pyrantel as well as our novel anthelmintics like AF2 receptor ligand and would show cross-resistance. So the lesson here would be to minimize the selection target sites for drug development that have the potential for overlapping mechanisms with existing anthelmintics.

Broad Spectrum or Narrow Spectrum

A small molecule neuropeptide receptor ligand may be easier to be developed that has a narrow spectrum perhaps with an activity against a few species of parasite. This is based on the knowledge that the distribution of the neuropeptides within the same neurons in different species of nematode appears to be unique.[74] Many identical neurons in the parasitic nematode *A. suum* and in the model nematode *C. elegans* do not have the same neuropeptide transmitter present. This implies that the receptors of the neuropeptide may be different between nematode species and that the effect of a small molecule ligand agonist will vary between nematode species, depending on the distribution of the receptors. It will be necessary to overcome this potential limitation when developing an anthelmintic compound that is active against a wide range of species (broad spectrum).

Given the inevitable costs associated with discovering and developing a small molecule ligand it will be important to find compounds that have a broad spectrum that kill most parasites and all the significant pathogenic parasites. Drugs that can be used as a preventative as well as curative agent have a marketing advantage especially if they can act as an endo-parasiticide and an ecto-parasiticide like the avermectins (moxidectin and ivermectin). It will be necessary to focus on the neuropeptides and their receptor ligands that act across a range of parasites and phyla.[75] Some peptides have effects in arthropods as well as in *A. suum*[76] but we do not know if they are mediated by the same receptor. In addition it will be important for the active agents to have effects in a range of domestic species, including cattle, which on a worldwide basis is the most significant economic species (in terms of potential revenue to the pharmaceutical company).

Potential Problems Associated with Neuropeptide Receptors as Target Sites for Anthelmintics

We have described the mode of action of existing anthelmintics and we can see that many have an action on ligand-gated ion-channels. Only emodepside may activate a G-protein coupled receptor to exert its effect. Most of the neuropeptides activate receptors that are expected to be GPCRs.[72] There are a number of potential problems that will need to be navigated to allow a successful anthelmintic to be developed.

- RNAi knock down has been used to reduce the expression of putative neuropeptide receptors in the model nematode, *C. elegans*.[77] This large ranging RNA interference survey did not find many observable changes (phenotypes) that were produced by the RNAi. Another study aimed at 60 putative neuropeptide receptor GPCRs and neurotransmitters[78] found only 13 phenotypes produced by RNAi that were characterized by uncoordinated movement and changes in egg lying. The main point that can be drawn form these two studies is that RNAi experiments have failed to identify behavioral changes associated with knock down of most of the neuropeptides or putative GPCR receptors. We can predict on a basis of these observations that antagonists of neuropeptides or their receptors is unlikely to produce an effect. Agonists however are very different. We know that the neuropeptides including the FaRPs produce dramatic effects when applied to neuromuscular preparation of nematodes.[79] Thus we expect that antagonists of the neuropeptide receptors will show very little response and be less likely to be successful anthelmintics. In contrast agonists of the neuropeptide receptors are expected to be more successful as anthelmintics.
- GPCRs are noted for their desensitization. It will be necessary to check receptor tachyphylaxis does not limit the novel neuropeptide agonist action as an anthelmintic.
- Neuropeptides may activate several receptors either in the same species or across phyla.[70,75] A synthetic drug selected for activating one receptor may activate other receptors in the same species or across species and phyla. This will be an advantage if there is a synergistic effect between the activated receptors in the same species. There is also a potential for antagonism between activated receptors in the same species.

High-Throughput Screens for Neuropeptide Ligands

The selection of the anthelmintic neuropeptide agonist will probably be achieved through high-throughput primary screens rather than using slower primary screens based on animal studies.[1,80,81] The consequence is that there is no-longer a minimal screen for toxicity and pharmacokinetics. It is missed and has to be done later. This means that methods for screening for active compounds can lead to more false starts. The development of the present successful anthelmintics (nicotinic anthelmintics, benzimidazoles and avermectins) were based on whole animal studies. The development of the avermectins for example used *Nippostrongylus* in mice as a screen.

The new high throughput screens rely on modern molecular and genetic techniques with cloning and expression of G Protein-coupled receptors and even some ion-channels into yeast, bacteria and insect or mammalian cell systems. When these cell systems are grown in multi-well plates and the receptors coupled to growth or a fluorescent signal[82] they can be screened by thousands of compounds very rapidly.

It is pointed out that many of the potential GPCRs of *C. elegans* have been identified by bioinformatics, their natural ligand and functions in the nematodes remain to characterize in most instances. Expression of the nematode G Protein-coupled receptors in cell systems have not been easy. There have been difficulties with culture temperatures, perhaps affecting protein folding; there have been difficulties with protein trafficking with different signal sequence being present in nematodes and mammalian preparations.[83] Despite this, let us take it that these systems can be developed for the screening of G protein receptors and ion-channels of nematode and insect neuropeptide receptors. Let us also take it that the high-throughput screening systems will allow an active ligand to be recognized and a significant 'hit' to be identified. It turns out there is still an enormous amount of work to convert the hit into an effective pharmaceutical product. That work will include minimizing any toxicity and getting the optimizing pharmacokinetics of the compounds for the best delivery of the drug.

Toxicity and Safety

The three cardinal features required for successful therapeutic drug are *safety* (lack of significant toxicity to host and environment); *efficacy* (the drug must work in a high proportion of treatments or better than other agents available); and appropriate *cost* (the drug must be affordable by the users and allow the pharmaceutical companies to recover development, production and future investment costs). We will now comment on safety.

The anthelmintic agent will need to have minimal (ideally be free from) toxic effects to the host. The toxicities to be considered are both the acute toxicities and chronic toxicities so that there is a low risk of acute overdose and low risk associated with longer term administration since the drug may be administered over a protracted period. The freedom of toxicity is necessary to establish for all the species to which the anthelmintic is to be administered to as well as establishing safety guidelines for the use and handlers of the drug.

Drugs administered to animals that are used for meat or milk will require additional review for safety because of their consumption by humans. Residues and *no-detectable effect* levels will need to be established to determine how long the therapeutic drug requires to be eliminated from the meat and milk.[84] The aim is to determine how long a safety period between administration of the drug and the killing of the production animal needs to be for safe use (establish the *preslaughter* period). The requirement for the measurement of residues in food animals makes the introduction and licensing more expensive for food animals, so the development of drugs for companion animals that are not consumed is less expensive and may be introduced initially (c.f. emodepside for use in cats). There are some interesting examples of toxicities associated with specific anthelmintics. Some of the toxicities are associated with the mode of action of the drug and might be predicted for any novel neuropeptide receptor ligand that is developed.

Levamisole is a nicotinic anthelmintic that is widely distributed in the host animal so that it has effects on gastro-intestinal and lungworms present. It has a selective effect on the nematode nicotinic receptors but the therapeutic index in not high and depends on the species of the host. Horses are sensitive to levamisole[85] and show pronounced gastro-intestinal effects, presumably due to stimulation of neuronal nicotinic receptors of the parasympathetic and sympathetic nervous system. The effect includes sweating and colic. Other species are less sensitive but levamisole will produce symptoms if the therapeutic dose is increased by more that threefold. The other nicotinic anthelmintics are less likely to produce these symptoms because their pharmaceutical preparations require oral administration and limit the distribution of the drug to the gastro-intestinal tract. This means that they are safer but their action is confined to gastro-intestinal parasites.

The avermectin class of anthelmintics (e.g., ivermectin) is excluded from the central nervous system by the blood brain barrier which possesses an exclusion pump. If the avermectins cross the blood brain barrier they can have effects on ion-channels including the $GABA_a$ receptors and product CNS depression. There are breed specificities, in Collie dogs[86,87] and Murray cattle[88] where it appears that a decreased blood brain barrier of the choroid plexus because there is a reduced P-glycoprotein like transporter present. This means that these species may allow high brain concentrations of the avermectins during treatment and thus show significant toxicity. There are several reports of neurological effects in humans following treatment with ivermectin for onchocerciasis. If we transfer this lesson to any novel neuropeptide receptor agonists, then we will have to be careful to keep in mind that in addition to species specific toxicities, we will have to consider the possibility of breed specific toxicities.

The benzimidazoles, including thiabendazole and albendazole, may produce teratogenic effects if they are administered in early of gestation.[89,90] This toxic effect presumably relates to the disruption of vital microtubules required for the proper formation of the ribs and limbs. Again the use of any novel anthelmintic will need to tested for any teratogenic effects and avoided in necessary during pregnancy.

The environment is becoming a more important constituent that we must protect. The appearance of any drug residues, ether the metabolite or parent compound should not persist in the environment like the chlorinated hydrocarbons (dieldrin, aldrin and DDT) and should be quickly degraded. Even if they are degraded within a few days or a week, it will be important that toxic effects on environmentally sensitive species are not produced.[84] Here the avermectins serve as an example. Ivermectin has pronounced effects on insects and crustacean as well as the parasitic nematode that is under treatment. Contamination of water in rivers will have a deleterious effect on some crustaceans. Residues in fecal remains or 'dung pats' have the potential to damage and upset the live-cycles of insects, like the dung beetle. Again we can learn from the experience with the avermectins and anticipate and avoid any negative impact that may occur with a novel neuropeptide receptor agonists.

Conclusion

We have considered the lessons that established anthelmintic drugs which act on neuromuscular systems can provide for us. These lessons predict that for any novel neuropeptide receptor ligand that is introduced will give rise to resistance but that we can delay the onset of that resistance by careful use. They also predict that species, breed and environmental toxicities will need to be considered along with the testing and review of residues in meat and milk to determine withdrawal periods and preslaughter times. The use and development of established anthelmintics also show us that the cost of drug development favors the marketing of drugs with a wide-spectrum of action. The possibility of differences in neuropeptide receptor functions in different species of nematode means that any neuropeptide receptor target or small molecule neuropeptide could limit the spectrum of action of the drug.

Acknowledgement

Supported by NIH R01 AI047194-06A1-06.

References

1. Geary TG, Sangster NC and Thompson DP. Frontiers in anthelmintic pharmacology. Vet Parasitol 1999; 84:275-295.
2. Bain RK. Irradiated vaccines for helminth control in livestock. Int J Parasitol 1999; 29:185-191.
3. Albonico M, Bickle Q, Haji HJ et al. Evaluation of the efficacy of pyrantel-oxantel for the treatment of soil-transmitted nematode infections. Trans: Roy Soc Trop Med 2002; 685-690.
4. Reynoldson JA, Behnke JM, Pallant LJ et al. Failure of pyrantel in treatment of human hookworm infections (Ancylostoma duodenale) in the Kimberley region of north west Australia. Acta Trop 1997; 68:301-312.
5. Martin RJ. Modes of action of anthelmintic drugs. Vet J 1997; 154:11-34.

6. Kohler P. The biochemical basis of anthelmintic action and resistance. Int J Parasitol 2001; 31:336-345.
7. Robertson AP, Clark CL, Burns TA et al. Paraherquamide and 2-deoxy-paraherquamide distinguish cholinergic receptor subtypes in Ascaris muscle. J Pharmacol Exp Ther 20021; 302:853-860.
8. Martin RJ, Bai G, Clark CL et al. Methyridine (2-[2-methoxyethyl]-pyridine]) and levamisole activate different ACh receptor subtypes in nematode parasites: a new lead for levamisole-resistance. Br J Pharmacol 2003; 140:1068-1076.
9. Edwards G, Breckenridge AM. Clinical pharmacokinetics of anthelmintic drugs. Clin Pharmacokinet 1988; 15:67-93.
10. Baggot JD, Mckellar QA. The absorption, distribution and elimination of anthelmintic drugs: the role of pharmacokinetics. J Vet Pharmacol Ther 1994; 17:409-419.
11. Hennessy DR. Physiology, pharmacology and parasitology. Int J Parasitol 1997; 27:145-152.
12. Martin RJ. Electrophysiological effects of piperazine and diethylcarbamazine on Ascaris suum somatic muscle. Br J Pharmacol 1982; 77:255-265.
13. Martin RJ. gamma-Aminobutyric acid- and piperazine-activated single-channel currents from Ascaris suum body muscle. Br J Pharmacol 1985; 84:445-461.
14. Piper PJ, Temple DM. The effect of lipoxygenase inhibitors and diethylcarbamazine on the immunological release of slow reacting substance of anaphylaxis (SRS-A) from guinea-pig chopped lung. J Pharm Pharmacol 1981; 33:384-386.
15. Roy TK, Sharma S, Srivastava VM. Comparative tissue distribution and urinary excretion of diethylcarbamazine and centperazine. Indian J Med Res 1981; 74:565-571.
16. Lee S, Casteel DA, Fleckenstein L. Specific gas chromatographic analysis of diethylcarbamazine in human plasma using solid-phase extraction. J Chromatogr B Biomed Sci Appl 1997; 704:181-185.
17. Holden-Dye L, Brownlee DJ, Walker RJ. The effects of the peptide KPNFIRFamide (PF4) on the somatic muscle cells of the parasitic nematode Ascaris suum. Br J Pharmacol 1997; 120:379-386.
18. Fellowes RA, Maule AG, Marks NJ et al. Nematode neuropeptide modulation of the vagina vera of Ascaris suum: in vitro effects of PF1, PF2, PF4, AF3 and AF4. Parasitology 2000; 120 (Pt 1):79-89.
19. Purcell J, Robertson AP, Thompson DP et al. PF4, a FMRFamide-related peptide, gates low-conductance Cl(-) channels in Ascaris suum. Eur J Pharmacol 2002; 456:11-17.
20. Qian H, Martin RJ, Robertson AP. Pharmacology of N-, L- and B- subtypes of nematode nAChR resolved and the single channel level. FASEB Journal 2006; 20:E1-E9.
21. Sangster NC, Davis CW, Collins GH et al. Effects of cholinergic drugs on longitudinal contraction in levamisole-susceptible and -resistant Haemonchus contortus Int J Parasitol 1991; 21:689-695.
22. Lee EL, Iyngkaran N, Grieve AW et al. Therapeutic evaluation of oxantel pamoate (1, 4, 5, 6-tetrahydro-1-methyl-2-[trans-3-hydroxystyryl] pyrimidine pamoate) in severe Trichuris trichiura infection. Am J Trop Med Hyg 1976; 25:563-567.
23. Mitchell SC. The toxicity of phenothiazine. Drug Metabol Drug Interact 1994; 11:201-235.
24. Boddie GF. Veterinary Therapeutics Oliver and Boyd 1952; London.
25. Arena JP, Liu KK, Paress PS et al. The mechanism of action of avermectins in Caenorhabditis elegans: correlation between activation of glutamate-sensitive chloride current, membrane binding and biological activity. J Parasitol 1995; 81:286-294.
26. Cully DF, Paress PS, Liu KK et al. Identification of a Drosophila melanogaster glutamate-gated chloride channel sensitive to the antiparasitic agent avermectin. J Biol Chem 1996; 271:20187-20191.
27. Martin RJ, Robertson AP, Wolstenholme AJ. Mode of action of the macrocyclic lactones. ed. Vercruysse J, Rew RS 2002; 125-140. Wallingford: CABI.
28. Steel JW, Waring RH. Ecological Impact of Macrocyclic lactones on Dung Fauna. In Macrocyclic lactones in antiparasitic therapy ed. Vercruysse J, Rew RS 2002; 141-162. Wallingford: CABI.
29. Raymond V, Mongan NP, Sattelle DB. Anthelmintic actions on homomer-forming nicotinic acetylcholine receptor subunits: chicken alpha7 and ACR-16 from the nematode Caenorhabditis elegans. Neuroscience 2000; 101:785-791.
30. Geary TG, Sims SM, Thomas EM et al. Haemonchus contortus: ivermectin-induced paralysis of the pharynx. Exp Parasitol 1993; 77:88-96.
31. Martin RJ. An electrophysiological preparation of Ascaris suum pharyngeal muscle reveals a glutamate-gated chloride channel sensitive to the avermectin analogue, milbemycin D. Parasitology 1996; 112 (Pt 2):247-252.
32. Sheriff JC, Kotze AC, Sangster NC et al. Effects of macrocyclic lactone anthelmintics on feeding and pharyngeal pumping in Trichostrongylus colubriformis in vitro. Parasitology 2002; 125:477-484.
33. Gill JH, Kerr CA, Shoop WL et al. Evidence of multiple mechanisms of avermectin resistance in haemonchus contortus—comparison of selection protocols. Int J Parasitol 1998; 28:783-789.
34. Richards JC, Behnke JM, Duce IR. In vitro studies on the relative sensitivity to ivermectin of Necator americanus and Ancylostoma ceylanicum. Int J Parasitol 1995; 25:1185-1191.

35. Geary TG. Ivermectin 20 years on: maturation of a wonder drug. Trends Parasitol 2005; 21:530-532.
36. Mckellar QA and Scott EW. The benzimidazole anthelmintic agents—a review. J Vet Pharmacol Ther 1990; 13:223-247.
37. Martin RJ, Robertson AP, Bjorn H. Target sites of anthelmintics. Parasitology 1997; 114 Suppl: S111-S124.
38. Lubega GW, Klein RD, Geary TG et al. Haemonchus contortus: the role of two beta-tubulin gene subfamilies in the resistance to benzimidazole anthelmintics. Biochem Pharmacol 1994; 47:1705-1715.
39. Geary TG, Nulf SC, Exander-Bowman SJ et al. Cloning and characterization of cDNAs encoding beta-tubulin from Dirofilaria immitis and Onchocerca volvulus. J Parasitol 1998; 84:356-360.
40. Gunawan M, Sangster NC, Kelly JD et al. The efficacy of fenbendazole and albendazole against immature and adult stages of benzimidazole-resistant sheep trichostrongylids. Res Vet Sci 1979; 27:111-115.
41. Sangster NC, Prichard RK, Lacey E. Tubulin and benzimidazole-resistance in Trichostrongylus colubriformis (Nematoda). J Parasitol 1985; 71:645-651.
42. Sangster NC, Gill J. Pharmacology of anthelmintic resistance. Parasitol Today 1999; 15:141-146.
43. Von Samson-Himmelstjerna G, Harder A, Sangster NC et al. Efficacy of two cyclooctadepsipeptides, PF1022A and emodepside, against anthelmintic-resistant nematodes in sheep and cattle. Parasitology 2005; 130:343-347.
44. Harder A, Von Samson-Himmelstjerna G. Cyclooctadepsipeptides—a new class of anthelmintically active compounds. Parasitol Res 2002; 88:481-488.
45. Willson J, Amliwala K, Harder A et al. The effect of the anthelmintic emodepside at the neuromuscular junction of the parasitic nematode Ascaris suum. Parasitology 2003; 126:79-86.
46. Davis RE. Action of excitatory amino acids on hypodermis and the motornervous system of Ascaris suum: pharmacological evidence for a glutamate transporter. Parasitology 1998; 116 (Pt 5):487-500.
47. Schaeffer JM, Bergstrom AR, Turner MJ. MK-801 is a potent nematocidal agent. Characterization of MK-801 binding sites in Caenorhabditis elegans. Biochem J 1989; 260:923-926.
48. Schaeffer JM, Bergstrom AR, Frazier EG et al. Nematocidal activity of MK-801 analogs and related drugs. Structure-activity relationships. Biochem Pharmacol 1994; 48:411-418.
49. Ehrlich P, Shiga K. Farbentherapeutische Versuche bei Trypanosomenerkrankung. 1904; 329-392.
50. Browning C. Experimental chemotherapy in Trypanosome Infections. 1907; 1405-1409.
51. Prichard RK. Anthelmintic resistance in nematodes: extent, recent understanding and future directions for control and research. Int J Parasitol 1990; 20:515-523.
52. Waruiru RM, Ngotho JW, Mukiri JG. Multiple anthelmintic resistance in Haemonchus contortus on a sheep farm in Kenya. Vet Res Commun 1997; 21:483-491.
53. Sangster NC, Bannan SC, Weiss AS et al. Haemonchus contortus: sequence heterogeneity of internucleotide binding domains from P-glycoproteins. Exp Parasitol 1999; 91:250-257.
54. Zajac AM, Gipson TA. Multiple anthelmintic resistance in a goat herd. Vet Parasitol 2000; 87:163-172.
55. Martin PJ, Mckenzie JA. Levamisole resistance in Trichostrongylus colubriformis: a sex-linked recessive character. Int J Parasitol 1990; 20:867-872.
56. Jackson F. Anthelmintic resistance—the state of play. Br Vet J 1993; 149:123-138.
57. Coles GC, Jackson F, Pomroy WE et al. The detection of anthelmintic resistance in nematodes of veterinary importance. Vet Parasitol 2006; 136:167-185.
58. Lacey E, Prichard RK. Interactions of benzimidazoles (BZ) with tubulin from BZ-sensitive and BZ-resistant isolates of Haemonchus contortus. Mol Biochem Parasitol 1986; 19:171-181.
59. Lubega GW, Prichard RK. Beta-tubulin and benzimidazole resistance in the sheep nematode Haemonchus contortus. Mol Biochem Parasitol 1991; 47:129-137.
60. Geary TG, Nulf SC, Favreau MA et al. Three beta-tubulin cDNAs from the parasitic nematode Haemonchus contortus. Mol Biochem Parasitol 1992; 50:295-306.
61. Roos MH, Kwa MS, Veenstra JG et al. Molecular aspects of drug resistance in parasitic helminths. Pharmacol Ther 1993; 60:331-336.
62. Blackhall WJ, Liu HY, Xu M et al. Selection at a P-glycoprotein gene in ivermectin- and moxidectin-selected strains of Haemonchus contortus. Mol Biochem Parasitol 1998a; 95:193-201.
63. Blackhall WJ, Pouliot JF, Prichard RK et al. Haemonchus contortus: selection at a glutamate-gated chloride channel gene in ivermectin- and moxidectin-selected strains. Exp Parasitol 1998b; 90:42-48.
64. Dent JA, Smith MM, Vassilatis DK et al. The genetics of ivermectin resistance in Caenorhabditis elegans. Proc Natl Acad Sci. USA 2000; 97:2674-2679.
65. Wolstenholme AJ, Fairweather I, Prichard R. Drug resistance in veterinary helminths. Trends Parasitol 2004; 20:469-476.
66. Van DV, De VE. A single nucleotide polymorphism map of the mitochondrial genome of the parasitic nematode Cooperia oncophora. Parasitology 2004; 128:421-431.

67. Hoglund J, Engstrom A, Morrison DA et al. Genetic diversity assessed by amplified fragment length polymorphism analysis of the parasitic nematode Dictyocaulus viviparus the lungworm of cattle. Int J Parasitol 2004; 34:475-484.
68. Eng JK, Prichard RK. A comparison of genetic polymorphism in populations of Onchocerca volvulus from untreated- and ivermectin-treated patients. Mol Biochem Parasitol 2005; 142:193-202.
69. Hoglund J, Morrison DA, Mattsson JG et al. Population genetics of the bovine/cattle lungworm (Dictyocaulus viviparus) based on mtDNA and AFLP marker techniques. Parasitology 2006; 133:89-99.
70. Kaplan RM, Burke JM, Terrill TH et al. Validation of the FAMACHA eye color chart for detecting clinical anemia in sheep and goats on farms in the southern United States. Vet Parasitol 2004; 123:105-120.
71. Martin RJ, Murray I, Robertson AP et al. Anthelmintics and ion-channels: after a puncture, use a patch. Int J Parasitol 1998; 28:849-862.
72. Greenwood K, Williams T, Geary T. Nematode neuropeptide receptors and their development as anthelmintic screens. Parasitology 2005; 131 Suppl: S169-S177.
73. Jones AK, Sattelle DB. Functional genomics of the nicotinic acetylcholine receptor gene family of the nematode. Caenorhabditis elegans Bioessays 2004; 26:39-49.
74. Yew JY, Kutz KK, Dikler S. Mass spectrometric map of neuropeptide expression in Ascaris suum. J Comp Neurol 2005; 488:396-413.
75. Mousley A, Maule AG, Halton DW et al. Inter-phyla studies on neuropeptides: the potential for broad-spectrum anthelmintic and/or endectocide discovery. Parasitology 2005; 131 Suppl: S143-S167.
76. Mousley A, Marks NJ, Halton DW et al. Arthropod FMRFamide-related peptides modulate muscle activity in helminths. Int J Parasitol 2004; 34:755-768.
77. Kamath RS, Fraser AG, Dong Y et al. Systematic functional analysis of the Caenorhabditis elegans genome using RNAi. Nature 2003; 421:231-237.
78. Keating CD, Kriek N, Daniels M et al. Whole-genome analysis of 60 G protein-coupled receptors in Caenorhabditis elegans by gene knockout with RNAi Curr Biol 2003; 13:1715-1720.
79. Maule AG, Bowman JW, Thompson DP et al. FMRFamide-related peptides (FaRPs) in nematodes: occurrence and neuromuscular physiology. Parasitology 1996; 113 Suppl: S119-S135.
80. Maule AG, Mousley A, Marks NJ et al. Neuropeptide signaling systems—potential drug targets for parasite and pest control. Curr Top Med Chem 2002; 2:733-758.
81. Geary TG, Conder GA, Bishop B. The changing landscape of antiparasitic drug discovery for veterinary medicine. Trends Parasitol 2004; 20:449-455.
82. Stables J, Green A, Marshall F et al. A bioluminescent assay for agonist activity at potentially any G-protein-coupled receptor. Anal Biochem 1997; 252:115-126.
83. Geary TG, Kubiak TM. Neuropeptide G-protein-coupled receptors, their cognate ligands and behavior in Caenorhabditis elegans. Trends Pharmacol Sci 2005; 26:56-58.
84. Mckellar QA. Ecotoxicology and residues of anthelmintic compounds. Vet Parasitol 1997; 72:413-426.
85. Dipietro JA, Todd KS, JR. Anthelmintics used in treatment of parasitic infections of horses. Vet Clin North Am. Equine Pract 1987; 3:1-14.
86. Hopper K, Aldrich J, Haskins S.C. Ivermectin toxicity in 17 collies. J Vet Intern Med 2002; 16:89-94.
87. Nelson OL, Carsten E, Bentjen SA et al. Ivermectin toxicity in an Australian Shepherd dog with the MDR1 mutation associated with ivermectin sensitivity in Collies. J Vet Intern Med 2003; 17:354-356.
88. Seaman JT, Eagleson JS, Carrigan MJ et al. Avermectin B1 toxicity in a herd of Murray Grey cattle. Aust Vet J 1989; 64:284-285.
89. Whittaker SG, Faustman EM. Effects of albendazole and albendazole sulfoxide on cultures of differentiating rodent embryonic cells. Toxicol Appl Pharmacol 1991; 109:73-84.
90. Capece BP, Navarro M, Arcalis T et al. Albendazole sulphoxide enantiomers in pregnant rats' embryo concentrations and developmental toxicity. Vet J 2003; 165:266-275.

Chapter 8

Neuropeptide Signaling in Insects

Miriam Altstein* and Dick R. Nässel

Abstract

Neuropeptides represent the largest single class of signal compounds and are involved in regulation of development, growth, reproduction, metabolism and behavior of insects. Over the last few years there has been a tremendous increase in our knowledge of neuropeptide signaling due to genome sequencing, peptidomics, gene micro arrays, receptor characterization and targeted gene interference combined with physiological and behavior analysis. In this chapter we review the current knowledge of structure and distribution of insect neuropeptides and their receptors, as well as their diverse functions. We also discuss peptide biosynthesis, processing and expression, as well as classification of insect neuropeptides. Special attention is paid to the role insect neuropeptides play as potential targets for pest management and as a basis for development of insect control agents employing the rational/structural design approaches.

Introduction

Neuropeptides and peptide hormones play critical roles in regulation of almost every aspect of insect life.[1-4] Thus, secreted peptides orchestrate important events during development and are vital regulators of adult physiology and behavior. Of special interest here is that many aspects of growth, reproduction and homeostasis rely on peptide hormones. Therefore, insects do not only provide good models for analysis of basic endocrine mechanisms of general interest, but we can also utilize our knowledge for the generation of insect control agents based on antagonists that cause interference with peptide signaling pathways.

Over the years intense research has targeted peptide function and endocrine regulation in a large variety of insects, many of which are severe medical and agricultural pests. One rational for this has been that peptides and their receptors are more species specific than classical neurotransmitters and monoamines and thus targeted interference will be less wide and unspecific. By means of traditional biochemical and molecular techniques a large number of neuropeptides and hormonal peptides have been identified from a variety of insects and their putative functions tested in different bioassays. More recently, several complete insect genomes have been sequenced and provided information about genes encoding both peptides and G-protein coupled receptors (GPCRs) likely to have peptide ligands.[1,5-7] So far, most of the sequenced insect genomes are derived from insects that are not pests; exceptions are the mosquitos *Anopheles gambiae* and *Aedes aegypti*. Even so, the available genomic information provides us with a great resource for identifying components in peptide signaling in a range of insects and is of great use also for research on pest insects.

Based on information on annotated genomes of several species of *Drosophila* and from *Anopheles gambiae*, *Aedes aegypti*, *Bombyx mori* and *Apis mellifera* we know that in each species there are about 30-40 genes encoding neuropeptide precursors and a slightly larger number of genes encoding peptide GPCRs.[1,5,6,8] Since several of these peptide precursors encode more than one predicted neuropeptide it is possible that there are more than 40 functional neuropeptides

*Corresponding Author: Miriam Altstein—Department of Entomology, The Volcani Center, Bet Dagan, 50250 Israel. Email: vinnie2@agri.gov.il

Neuropeptide Systems as Targets for Parasite and Pest Control, edited by Timothy G. Geary and Aaron G. Maule.

or peptide hormones in a species. As discussed in more detail below, it is important to confirm the expression of processed peptides from nervous or other tissues by mass spectrometry, since it turns out that not all predicted peptide cleavage sites on the precursors are utilized.[4,9] The genomic information combined with available biochemical expression data provides us with a starting point for analysis of peptide functions in insects. As a first step of such analysis the novel peptides need to be tested in bioassays or in activation of orphan GPCRs in cellular expression systems.[1,8]

Neuropeptides: Their Biosynthesis, Processing and Expression

Insect neuropeptides consist of 5 to about 80 amino acid residues linked by peptide bonds. Some larger proteins are also known to act as hormones. Examples of these are prothoracicotropic hormone (PTTH) and the heterodimeric cystin knot protein bursicon, both known to have important roles during development.[2] The molecular structures of insect neuropeptides are immensely varied and they also display a great diversity in their distribution patterns, modes of action and their functional roles.[2,3]

It is not clear how many functional neuropeptides there are in a given species. We can make an estimate by looking into genomic information for different insects. In *Drosophila* there are about 35 neuropeptide encoding genes and 48 encoding peptide and protein GPCRs and these numbers are about 36 and 37, respectively, in the honey bee *Apis mellifera*.[1,5,6] However, as we shall see in the following, the number of functional peptides processed from the genes is not totally predictable and there is some variation between insect species. Furthermore, the pairing between all known peptides and specific peptide GPCRs has not been completed for any insect species yet; some orphan peptide GPCRs need to be matched up with their naturally occurring neuropeptide ligands and for some peptides receptors are still unidentified. In addition, some neuropeptides, such as insulin-like peptides, exert their activity through tyrosin kinase receptors,[4,10] and yet others may activate guanylate cyclase-type receptors. There are no genes in insects encoding orthologs of the ion channel-type of FMRFamide receptors[5] known from mollusks.[11]

Insect neuropeptides, like those of other animals, are processed from larger precursor proteins encoded by genes. Some neuropeptides are present in single copies on the precursors, but often several copies of identical or slightly diversified peptides (isopeptides) can be seen. So far the largest number of insect peptides that was shown to be processed from a single precursor was 23 peptides from a cockroach FMRFamide precursor.[12] Since it has been shown that predicted peptide cleavage sites on precursor proteins are not always utilized, it is critical that genomic data is followed up by biochemical determination of peptide complement in the same species. This information about the "peptidome" can now be obtained by very sensitive mass spectrometry techniques. Thus, for *Drosophila* and *Apis* the genome predictions have been tested against peptidome analysis.[6,9,13,14] There is some ambiguous information for some of the larger peptides and some peptides may have escaped detection, but estimates of processed peptides expressed in tissues can now be made more accurately. So, for instance in the honey bee about 100 peptides derived from 36 genes have been identified by mass spectrometry.[6] In the cockroach *Periplaneta americana* about 80 neuropeptides have been identified biochemically, but genomic information is more scarce for this species.[12,15] A further problem that has been only partly addressed is to what extent all expressed neuropeptides, including closely related isoforms, play functional roles in the organism. Most of the neuropeptides originally identified by biochemical means display activities in bioassay systems, but not all peptides predicted from genomic data and then confirmed by mass spectrometry, have been tested. One estimate of the complexity of peptide signaling in a species might be derived from the number of neuropeptide GPCRs. If so, the number of distinct peptide signaling systems would be in the order of about 40-50 in a species. In this number we have taken into account that a few neuropeptides are known to activate more than one GPCR.[1,2] The complexity may be further increased given the fact that GPCRs can diversify the functions by coupling to different G-proteins thus stimulating different downstream secondary messenger pathways. In the next section the peptide genes and GPCRs will be presented.

Many of the neuropeptides are C-terminally alpha-amidated and display other posttranslational modifications of structures that generate varying degrees of stability to peptidases and which are also important for their biological activity.[4] Some peptides contain cysteins that form disulfide bridges that provide structural constraints. For several of the insect neuropeptides the structure-activity relationship (SAR) has been extensively analyzed, either in bioassays or by testing of recombinant GPCRs expressed in cells.[16-25] Examples of peptides analyzed in this respect are: peptides related to pyrokinins (PKs) and pheromone biosynthesis activating neuropeptide (PBAN), dark color-inducing neurohormone (DCIN), tachykinin-related peptides, corticotropin releasing factor (CRF)-like diuretic hormones (DH) and adipokinetic hormones (AKH). For these peptides we thus have information about active cores, as well as residues critical for activity and metabolic stability. These structural data have provided important information on functional similarities within neuropeptide families in different taxa and also about possible sites of peptide modifications to provide stable agonists and antagonists.

In the insect central nervous system (CNS) neuropeptides are produced by neurosecretory cells and interneurons and more rarely by motoneurons or sensory neurons.[3,26] Peptides can also be detected in endocrine cells of the intestinal tract or at other peripheral sites.[3,27] The major release site for peptide hormones produced in neurosecretory cells are: the corpora cardiaca and allata, segmental neurohemal organs associated with the ventral nerve cord (perivisceral organs; PVOs), as well as axon terminations on the anterior aorta, peripheral nerves (including abdominal heart nerves), the intestine and body wall muscles.[14] Thus, peptides can act as circulating hormones, as local neurohormones that are released nonsynaptically within the nervous system, or at muscles or glands. Neuropeptides are also known to be released nearby synapses and act as synaptic modulators or even as co-transmitters, thereby modifying the action of fast-acting "classical" neurotransmitters. These modes of actions in the CNS have been studied more extensively in mammals, mollusks and crustaceans[28-31] and less data is available for insects.

Each neuropeptide or set of peptides derived from a single precursor is distributed in a stereotypic pattern in specific neurons, neurosecretory cells or endocrine cells.[3,26,32] Commonly there are very few neurons or neurosecretory cells expressing each neuropeptide precursor. For example, only two neurons express eclosion hormone in *Drosophila*[33] and four neurons produce SIFamide in the same insect.[34] Commonly there are in the order of 20-50 peptidergic neurons, but in some cases a few hundred can be seen (e.g., tachykinin-related peptides or allatotropin in locusts).[26] Based on the pattern of distribution in the CNS and periphery (if applicable) one can propose that some peptides may play multiple functional roles and others may be only circulating hormones or signal in restricted circuits within the CNS.

Neuropeptides Families, GPCRs and Peptide Functions

Traditionally insect neuropeptides were grouped into "neuropeptide families" which were based either upon homologies in sequences or functional similarities shared by peptides of different taxa. This followed the same principles that were used for vertebrate peptide families. Now that genomic data are available for several organisms it seems more relevant to compare neuropeptide precursor genes and peptides derived from these genes in different species. In Table 1 the genes encoding neuropeptides in *Drosophila* are listed as an example. Orthologs of most of the same genes can be identified in the other insect species analyzed. Some genes could not be found in *Drosophila* although they are known in other insects: genes encoding allatotropin, orcokinin and PBAN. Conversely some of the *Drosophila* genes were not detected in honey bees: genes encoding proctolin, leucokinin, myoinhibitory peptides (MIP) and allatostatin C.[6] Thus, the total number of insect peptide genes appears to be larger than that seen in a single species. It is also likely that all the neuropeptides have not yet been identified in any insect species (including *Drosophila*).

Over the last few years quite a number of the *Drosophila* GPCRs classified as peptide and protein receptors, based on sequence homologies,[5] have been deorphaned, i.e.,their ligands identified. Thus, out of 48 *Drosophila* peptide GPCRs, more than 25 have their ligands identified.[1,2] The GPCRs in *Anopheles* and *Apis* and others can be classified and tentatively assigned ligands by sequence

Table 1. Neuropeptide precursor genes identified from D. melanogaster[1]

Neuropeptide Precursor (Peptides)	Gene	CG number[2]	Sequence[3]
Adipokinetic hormone	*akh*	CG1171	pQLTFSPDWa
Amnesiac product(s)	*amn*	CG11937	*
Allatostatin-A (Drostatin-A-1-4)	*ast*	CG13633	VERYAFGLa
Allatostatin-B[4] (Drostatin-B-1-5)	*mip*	CG6456	AWQSLQSSWa
Allatostatin-C (Drostatin-C)	*ast2*	CG14919	pEVRYRQCYFNPISCF
Bursicon	*burs*	CG13419	*
Partner of Bursicon (bursicon beta)	*pburs*	CG15284	*
Capability (CAP-1-2, PK-1)	*capa*	CG15520	GANMGLYAFPRVa TGPSASSGLWFGPRLa
Crustacean cardioactive peptide	*ccap*	CG4910	PFCNAFTGCa
Corazonin	*crz*	CG3302	pQTFQYSRGWTNa
Diuretic hormone (CRF-like; DH_{44})	*Dh*	CG8348	*
Diuretic hormone (Calcitonin-like)	*Dh31*	CG13094	TVDFGLARGYSGTQEAKH- RMGLAAANFAGGPa
dFMRFamides (dFMRFa-1-8)	*fmrfa*	CG2346	DPKQDFMRFa
Drosulfakinins (DSK-1-2)	*dsk*	CG18090	FDDYGHMRFa
Dromyosuppressin (DMS)	*dms*	CG6440	TDVDHVFLRFa
Ecdysis triggering horm. (ETH-1-2)	*eth*	CG18105	DDSSPGFFLKITKNVPRLa
Eclosion hormone	*eh*	CG6400	*
Hugin/pyrokinin-2 (hug-γ, PK-2)	*hug,*	CG6371	LRQLQSNGEPAYRVRTPRLa SVPFKPRLa
Insulin-like peptide 1	*Dilp-1*	CG14173	*
Insulin-like peptide 2	*Dilp-2*	CG8167	*
Insulin-like peptide 3	*Dilp-3*	CG14167	*
Insulin-like peptide 4	*Dilp-4*	CG6737	*
Insulin-like peptide 5	*Dilp-5*	AE003550[5]	*
Insulin-like peptide 6	*Dilp-6*	CG14049	*
Insulin-like peptide 7	*Dilp-7*	CG13317	*
Ion transport peptide (CHH-like)	*itp*	CG13586	*
IPNamide (of NPLP-1 precursor)	*nplp1*	CG3441	NVGTLARDFQLPIPNa
Leucokinin-like	*lk*	CG13480	NSVVLGKKQRFHSWGa
Neuropeptide F (long)	*npf*	CG10342	SNSRPPRKNDVNTMADA- YKFLQDLDTYYGDRARVRFa
Neuropeptide F (short) (sNPF-1-4)	*snpf*	CG13968	AQRSPSLRLRFa PQRLRWa
Pigment-dispersing factor	*pdf*	CG6496	NSELINSLLSLPKNMNDAa
Proctolin	*Proct*	CG7105	RYLPT
Prothoracicotropic hormone (PTTH)	*ptth*	CG13687	*
SIFamide	*Ifamide*	CG4681	AYRKPPFNGSIFa
Tachykinin-related (DTK-1-6)	*dtk*	CG14734	APTSSFIGMRa

[1]Compiled from refs 3-5. [2]Celera Genomics accession numbers. [3]Sequences given for representative peptides of each precursor. [4]These peptides are also designated myoinhibitory peptides (MIPs). [5]GenBank accession number for gene cluster. *Sequences too long to be given here.

homologies.[1,35] GPCRs from other insects have been identified by more traditional homology cloning and tests of ligands.[8] The identification of peptide GPCRs is of immense value since they can now be expressed in cell systems for assays of activation or blocking by receptor-selective and/or non-selective-agonists and antagonists.

Table 2. Functions of neuropeptides and hormonal peptides in insects[1]

Functions	Peptides
Development	PTTH, allatotropin, allatostatins
Molting	PTTH, allatotropin, allatostatins, eclosion hormone, ecdysis triggering hormone, pre-ecdysis triggering hormone, CCAP, corazonin, FMRFa, myoinhibitory peptide, bursicon
Feeding	NPF, sNPF, Hugin (pyrokinin)
Growth	Insulin-like peptides
Reproduction[2]	Neuroparsins, insulin-like peptides, PBAN, sNPF, SIFamide
Metabolism[3]	AKH, insulin-like peptides
Water and ion regul.	Diuretic hormones (DH_{44}, DH_{31}), CAPA[4], leucokinin, ion transport peptide
Specific behaviors	IPNamide, SIFamide, PDF, NPF
Myotropic	Proctolin, FMRFamides, myosuppressins, PKs, and many others
Multifunctional	Allatotropin, CCAP, tachykinin-related, sNPF, proctolin
Pigmentation	Melanization and reddish coloration hormone (PK), DCIN/corazonin

Note that these peptides may have several additional functions. References and acronyms are given in the text. Both reproduction physiology and reproductive behavior. Carbohydrate and lipid metabolism. Capability (CAPA) gene encodes perviscerokinins (CAP_{2B}) and pyrokinin.

Can neuropeptides and their genes be loosely organized into functional groups? Are there peptides that are primarily involved in regulation of development and others that regulate reproduction and so on? In an attempt to organize peptides somewhat into functional categories we will list peptides after certain functions that have been assigned to them (Table 2). This may also provide an idea of the complex regulation of various aspects of insect physiology and behavior. It is likely that there are some peptide hormones that may sub-serve a single hormonal function (or be part of a single hormonal cascade). This is underscored by their very restricted distribution in a small number of neurosecretory cells (and no presence in interneurons). Such peptides may have distinct hormonal roles orchestrating single aspects of insect life. For example, eclosion hormone and ecdysis triggering hormone in moths and flies are present in small populations of neurosecretory cells and display distinct functions in ecdysis behavior.[27] It is, however, likely that many of the neuropeptides (and peptide hormones) sub-serve multiple functions since they are produced by multiple and diverse interneuron types. For these peptides the distribution of release sites and their receptors in circuits within the CNS determine their functions.[26] Additionally we know that several insect neuropeptides can act both within the CNS and at peripheral targets further expanding possible regulatory roles.[3] The functions assigned to peptides in Table 2 are based on studies of various insect species, including *Drosophila*. Many of these peptides are involved in several functions and are thus listed under more than one category.[3,4,28,36,37]

So what are neuropeptides doing in insects? Many of the peptides have only been investigated in vitro and about half of the known peptides display myostimulatory or myoinhibitory activities. Here, we will only discuss peptides where in vivo functions can be suggested. Insect peptides have been shown to play major roles in regulation of molting,[27] feeding and growth,[38-41] reproduction,[4,36] pheromone production,[22,42] pigmentation,[17,43,44] metabolism of lipids and carbohydrates,[45,46] water and ion transport.[18]

Furthermore it has been shown by targeted gene interference that specific behaviors in *Drosophila* are regulated by neuropeptides. Pigment-dispersing factor (PDF) is an output peptide from the lateral clock neurons of the brain regulating circadian locomotor activity under constant light conditions.[47] Two further peptides have been implicated in the *Drosophila* clock: IPNamide and Neuropeptide F (NPF). SIFamide is present in four brain neurons with extensive arborizations that are especially important for male reproductive behavior.[34] NPF is critical for regulation of feeding, foraging and social feeding behavior and there is a convergence of NPF and *Drosophila* insulin-like peptide (DILP) signaling in regulation of motivated food ingestion.[40,41] Another peptidergic system has been implicated in feeding in *Drosophila*: about 20 neurons in the subesophageal ganglion that express the neuropeptide precursor gene *hugin* (*hug*) expressing a pyrokinin.[39] The *hug*-derived peptide is important for initiation of feeding, dependent on food quality and it acts in circuitry that modulate feeding behavior based on chemosensory and nutrient signals. Furthermore, peptides of the sNPF (short neuropeptide F) gene appear to be important regulators of larval and adult feeding.[48] A peptide similar to the sNPFs has been identified in the mosquito *Aedes* and plays a role in female host seeking behavior.[49] The pyrokinins, NPFs and sNPFs are likely to play similar roles in other insect species and are thus relevant lead peptides for disruption of insect viability.

Insect Neuropeptides as Potential Targets for Pest Management

As evident from the recent studies described above, insect neuropeptides regulate many physiological and behavioral processes during development, reproduction and senescence and maintain growth, homeostasis, osmoregulation, water balance, metabolism and visceral activities. Peptides involved in regulation of vital functions are prime targets for advancement of the understanding of the physiology of insects and also targets for the development of novel insect-control strategies based on interference with their activity.

Although insect neuropeptides have been studied intensely in the past few decades, the mechanisms by which they exert their action are far from being fully characterized or understood. The possibilities of gaining a better insight into the mode of action and of exploiting insect neuropeptides for pest management rely primarily on our understanding the cellular and molecular basis of their actions. One way of obtaining a better insight into the mode of activity and functional diversity of peptides is by use of receptor-selective agonists and antagonists. Despite the vast scientific and insecticidal/insect control potential of antagonists (and to some extent agonists), their application has not been widely implemented so far in insects. This is mainly because of lack of defined methods for obtaining antagonists on the basis of a known neuropeptide agonist and because of the inability to predict which conformation will lead to a highly potent inhibitory or stimulatory receptor-selective activity. In addition, peptides are highly susceptible to proteolytic degradation and have a poor bioavailability. Therefore their conversion into an insecticide prototype requires rendering them resistant to peptidase degradation and to design them with a high bioavailability. Similar problems are also common in the pharmaceutical industry where immense efforts are being made in attempts to convert mammalian neuropeptides into therapeutic drugs. For many years, the most common approach used by the pharmaceutical industry for drug discovery was based on random screening of large chemical libraries of non-peptide compounds and further optimization of a lead molecule with respect to selectivity and pharmacokinetic properties. This approach has produced receptor-subtype-specific bioavailable ligands with nanomolar affinity (similar to that of the endogenous ligand) for peptide receptors some of which have been approved for clinical application and some which are in clinical trials.[50] An example of this is aprepitant (MK 869), a neurokinin-1 antagonist used for the treatment of chemotherapy-induced emesis and treatment of major depressions.[51] In the past decade, a parallel approach, based on rational drug design (or structure-based design) has evolved which integrates and implements the vast amount of information on the genes encoding GPCRs as well as the SAR of neuropeptides and their receptors. The approach has been applied to somatostatin, bradykinin, neurokinin and luteinizing hormone releasing hormone (LHRH) and resulted in the discovery of a few highly potent agonists and antagonists.[52]

In the past few years a novel integrated approach termed backbone cyclic neuropeptide-based antagonist (BBC-NBA) has been developed in which rationally designed BBC conformational libraries were synthesized, based on a detailed SAR study[53] of the insect neuropeptide PK/PBAN family and screened for occurrence of antagonists.[54,55] The backbone cyclization approach resulted in the discovery of conformationally constrained, highly potent, selective and nonselective, metabolically stable and highly bioavailable BBC PK/PBAN antagonists active in the nmole range.[42,42b,54,56,57] Recently, a few PK/PBAN GPCRs have been cloned from various moth species[58-61] revealing structural differences within receptors that mediate different functions (e.g., sex pheromone production, melanization and pupal diapause).[61] The information gained on the selective and nonselective conformationally constrained BBC antagonists as well as the structural information of the PK/PBAN receptors can further serve for rational design of nonpeptidergic small molecule libraries (NPSML). In these, the bioactive biophores (deduced from structural analysis of the BBC antagonists and from their interaction with the receptors) will be applied on simple inexpensive scaffolds for the development of highly potent, metabolically stable, bioavailable and inexpensive insect specific and environment friendly insect control agents. An alternative approach to address the difficulties associated with the development of improved insect neuropeptide active compounds (agonists) was introduced with the same neuropeptide family by Nachman, Altstein and coworkers. The approach was based on design of pseudopeptides in which various amino acids have been substituted in a manner that rendered the molecule more stable to peptidase attack and more bioavailable.[62-65]

Conclusion

Due to the vast amount of information currently available on insect neuropeptides and the restricted space in this review we have highlighted only some of the major issues related to this important group of signaling molecules. Many topics have been omitted and they can be found in the reviews that are cited in this chapter. Much of the recent progress in revealing specific functions of neuropeptide signaling in vivo has been made in *Drosophila* by means of targeted interference with genes of peptide precursors or GPCRs or by cell-specific expression of apoptosis genes.[2,33,40,41,47] In parallel with these studies the in vitro characterization of peptide GPCRs and analysis of peptide and GPCR distributions in various insects has advanced our understanding of neuropeptide signaling tremendously. Comparative experimental studies, combined with information from annotated genomes from multiple insect species, will also improve our insight into the evolution of neuropeptide signaling.

The use of neuropeptides as a basis for drug design made a leap forward in the past decade due to the vast amount of novel information on GPCR and neuropeptide genes and their sequences. This information, together with the rapid developments in bioinformatics, molecular engineering, proteomics and chemical analysis (mainly liquid chromatography coupled to mass spectrometry) generated large amounts of data. This provides a basis for a better understanding of signaling mechanisms of mammalian and insect neuropeptides as well as for development of drugs and insect control agents based on rational/structural design. Both the pharmaceutical and agrochemical fields are still in their infancy and although this strategic approach has been used to develop a few vertebrate neuropeptide antagonist and antagonists the technology has not yet been optimized. Thus, new approaches to the generation of neuropeptide agonists/antagonists and to their further conversion into NPSM compounds with desired features need to be developed. It is anticipated that once these strategies have been worked out and the approaches expanded, it will be possible to implement them to a large variety of neuropeptides for tailoring highly potent drugs or insect control agents.

Addendum

In the last few years, a number of important advances have been made in insect neuropeptide research. A few relevant examples are given here. Sequencing of a few more insect genomes have been completed and thus new information about insect neuropeptide and GPCR genes is available (see refs. 66-68). From the new data we can conclude that some neuropeptide genes may have been lost over evolution in certain species, others seem to have diversified. A promising novel approach that employs quantitative mass spectrometry has been utilized to analyze peptide expression in honey bees under different foraging conditions.[69] This study indicates that peptide expression is dynamic in adult insects. An elegant technique has been developed that makes it possible to determine the peptidome of selected neuron types.[70] The technique is based on marking genetically defined neuron populations with GFP (by Gal4-UAS technique), followed by dissociation of neurons and fluorescent cell sorting, and then analysis by mass spectrometry. In principle it is now possible to determine the pantheon of neuropeptides in major types of neurons that can be defined by promoter Gal4 lines or other means of fluorescent marking in vivo. A novel type of peptide receptor has been identified in endocrine cells of the Oriental fruit-fly, a membrane bound receptor guanylate cyclase activated by eclosion hormone.[71] This finding may be of significance for specific chemical interference with development in pest insects. Finally, major advances have been made in understanding insulin signaling in insects, especially in regulation of growth, metabolism and life span (see refs. 72-74). Another novel avenue of research relates to the bioavailability of neuropeptides. Recent studies have shown that linear and cyclic peptides of different length and polarities are highly bioavailable and can penetrate through the cuticle when applied in aqueous or organic solutions, and reach and activate the target organ.[75-77] These results contradict the common notion that peptides have low bioavailability, and may lead to a dramatic simplification of the strategies needed to be employed for design of neuropeptide based agonist and antagonists insect control agents.

References

1. Hauser F, Cazzamali G, Williamson M et al. A review of neurohormone GPCRs present in the fruitfly Drosophila melanogaster and the honey bee Apis mellifera. Prog Neurobiol 2006; 80:1-19.
2. Johnson EC. Postgenomic approaches to resolve neuropeptide signaling in Drosophila. In: Satake H, ed. Invertebrate Neuropeptides and Hormones: Basic Knowledge and Recent Advances. Trivandrum: Transworld Research Network 2006:179-224.
3. Nässel DR. Neuropeptides in the nervous system of Drosophila and other insects: multiple roles as neuromodulators and neurohormones. Prog Neurobiol 2002; 68:1-84.
4. Taghert PH, Veenstra JA. Drosophila neuropeptide signaling. Adv Genet 2003; 49:1-65.
5. Hewes RS, Taghert PH. Neuropeptides and neuropeptide receptors in the Drosophila melanogaster genome. Genome Res 2001; 11:1126-1142.
6. Hummon AB, Richmond TA, Verleyen P et al. From the genome to the proteome: Uncovering peptides in the Apis brain. Science 2006; 314:647-649.
7. Riehle MA, Garczynski SF, Crim JW et al. Neuropeptides and peptide hormones in Anopheles gambiae. Science 2002; 298:172-175.
8. Meeusen T, Mertens I, De Loof A et al. G protein-coupled receptors in invertebrates: A state of the art. Int Rev Cytol 2003; 230:189-+.
9. Baggerman G, Boonen K, Verleyen P et al. Peptidomic analysis of the larval Drosophila melanogaster central nervous system by two-dimensional capillary liquid chromatography quadrupole time-of-flight mass spectrometry. J Mass Spectrom 2005; 40:250-260.
10. Garofalo RS. Genetic analysis of insulin signalling in Drosophila. Trends Endocrinol Metablo 2002; 13:156-162.
11. Cottrell GA. The first peptide-gated ion channel. J Exp Biol 1997; 200:2377-2386.
12. Predel R, Neupert S, Wicher D et al. Unique accumulation of neuropeptides in an insect: FMRFamide-related peptides in the cockroach, Periplaneta americana. Eur J Neurosci 2004; 20:1499-1513.
13. Baggerman G, Cerstiaens A, De Loof A et al. Peptidomics of the larval Drosophila melanogaster central nervous system. J Biol Chem 2002; 277:40368-40374.
14. Predel R, Wegener C, Russell WK et al. Peptidomics of CNS-associated neurohemal systems of adult Drosophila melanogaster: A mass spectrometric survey of peptides from individual flies. J Comp Neurol 2004; 474:379-392.
15. Predel R, Neupert S, Roth S et al. Tachykinin-related peptide precursors in two cockroach species. FEBS J 2005; 272:3365-3375.
16. Altstein M, Dunkelblum E, Gabay T et al. PBAN-Induced sex-pheromone biosynthesis in Heliothis-Peltigera: Structure, dose and time-dependent analysis. Arch Insect Biochem Physiol 1995; 30:307-319.
17. Altstein M, Gazit Y, Ben Aziz O et al. Induction of cuticular melanization in Spodoptera littoralis larvae by PBAN/MRCH: Development of a quantitative bioassay and structure function analysis. Arch Insect Biochem Physiol 1996; 31:355-370.
18. Coast GM, Orchard I, Phillips JE et al. Insect diuretic and antidiuretic hormones. Advan In Insect Phys 2002; 29:279-409.
19. Gäde G. The explosion of structural information on insect neuropeptides. Fortschr Chem Org Naturst 1997; 71:1-128.

20. Gäde G, Goldsworthy GJ. Insect peptide hormones: a selective review of their physiology and potential application for pest control. Pest Manag Sci 2003; 59:1063-1075.
21. Nachman RJ, Roberts VA, Lange AB et al. Active conformation and mimetic analog development for the pyrokinin-PBAN-diapause-pupariation and myosuppressin insect neuropeptide families. Phytochemicals for Pest Control 1997; 658:277-291.
22. Rafaeli A, Jurenka R. PBAN regulation of pheromone biosynthesis in female moths. Insect Pheromone Biochemistry and Molecular Biology. NY: Academic Press, 2003; 107-36.
23. Schoofs L, Vandenbroeck J, Deloof A. The myotropic peptides of Locusta migratoria: Structures, distribution, functions and receptors. Insect Biochem Mol Biol 1993; 23:859-881.
24. Vanden Broeck J. Neuropeptides and their precursors in the fruitfly, Drosophila melanogaster. Peptides 2001; 22:241-254.
25. Yerushalmi Y, Bhargava K, Gilon C et al. Structure-activity relations of the dark-colour-inducing neurohormone of locusts. Insect Biochem Mol Biol 2002; 32:909-917.
26. Nässel DR, Homberg U. Neuropeptides in interneurons of the insect brain. Cell Tissue Res 2006; 326:1-24.
27. Ewer J, Reynolds S. Neuropeptide control of molting in insects. In: Pfaff DW, Arnold AP, Fahrbach SE et al, eds. Hormones, Brain and Behavior. San Diego: Academic Press, 2002:1-92.
28. Kastin A. Handbook of biologically active peptides. San Diego: Elsevier, 2006.
29. Kiss T, Pirger Z. Neuropeptides as modulators and hormones in terrestrial snails: their occurrence, distribution and physiological significance. In: Satake H, ed. Invertebrate Neuropeptides And Hormones: Basic Knowledge and Recent Advances. Transworld Research Network, Trivandrum, 2006:75-110.
30. Nusbaum MP, Blitz DM, Swensen AM et al. The roles of cotransmission in neural network modulation. Trends Neurosci 2001; 24:146-154.
31. Strand FL. Neuropeptides: Regulators of Physiological Processes. Cambridge, MA: The MIT Press, 1999.
32. Davis NT, Homberg U, Teal PEA et al. Neuroanatomy and immunocytochemistry of the neurosecretory system of the subesophageal ganglion of the tobacco hawkmoth, Manduca sexta: Immunoreactivity to PBAN and other neuropeptides. Microscopy Res Tech 1996; 35:201-229.
33. McNabb SL, Baker JD, Agapite J et al. Disruption of a behavioral sequence by targeted death of peptidergic neurons in Drosophila. Neuron 1997; 19:813-823.
34. Terhzaz S, Rosay P, Goodwin SF et al. The neuropeptide SIFamide modulates sexual behavior in Drosophila. Biochem Biophys Res Commun 2007; 352:305-310.
35. Hill CA, Fox AN, Pitts RJ et al. G protein-coupled receptors in Anopheles gambiae. Science 2002; 298:176-178.
36. De Loof A, Baggerman G, Breuer M et al. Gonadotropins in insects: An overview. Arch Insect Biochem Physiol 2001; 47:129-138.
37. Schoofs L, Clynen E, Cerstiaens A et al. Newly discovered functions for some myotropic neuropeptides in locusts. Peptides 2001; 22:219-227.
38. Broughton SL, Piper MD, Ikeya T et al. Longer lifespan, altered metabolism and stress resistance in Drosophila from ablation of cells making insulin-like ligands. Proc Natl Acad Sci USA 2005; 102:3105-3110.
39. Melcher C, Pankratz MJ. Candidate gustatory Interneurons modulating feeding behavior in the Drosophila brain. Plos Biology 2005; 3:1618-1629.
40. Wu O, Wen T, Lee G et al. Developmental control of foraging and social behavior by the Drosophila neuropeptide Y-like system. Neuron 2003; 39:147-161.
41. Wu O, Zhao Z, Shen P. Regulation of aversion to noxious food by Drosophila neuropeptide Y- and insulin-like systems. Nature Neurosci 2005; 8:1350-1355.
42. Altstein M. Role of neuropeptides in sex pheromone production in moths. Peptides 2004; 25:1491-1501.
42b. Altstein M, Hariton A. Rational design of insect control agents: The PK/PBAN family as a study case. In: Ishaaya I, Horowitz R, eds. Biorational Control of Arthropod Pests: Application and Resistance Management. New York: Springer, 2009:49-81.
43. Matsumoto S, Isogai A, Suzuki A et al. Purification and properties of the melanization and reddish colouration hormone (MRCH) in the armyworm, Leucania Separata (lepidoptera). Insect Biochem 1981; 11:725-733.
44. Tanaka S. Endocrine Mechanisms Controlling Body-Color Polymorphism in Locust. In: Loof AD, Hoffman KH, eds. Arch Insect Biochem Physiol 2001; 139-49.
45. Lee G, Park JH. Hemolymph sugar homeostasis and starvation-induced hyperactivity affected by genetic manipulations of the adipokinetic hormone-encoding gene in Drosophila melanogaster. Genetics 2004; 167:311-323.

46. Rulifson EJ, Kin SK, Nusse R. Ablation of insulin-producing neurons in flies: growth and diabetic phenotypes. Science 2002; 296:1118-1120.
47. Renn SC, Park JH, Rosbash M et al. A pdf neuropeptide gene mutation and ablation of PDF neurons each cause severe abnormalities of behavioral circadian rhythms in Drosophila. Cell 1999; 99:791-802.
48. Lee KS, You KH, Choo JK et al. Drosophila short neuropeptide F regulates food intake and body size. J Biol Chem 2004; 279:50781-50789.
49. Brown MR, Klowden MJ, Crim JW et al. Endogenous regulation of mosquito host-seeking behavior by a neuropeptide. J Insect Physiol 1994; 40:399-406.
50. Hökfelt T, Bartfai T, Bloom F. Neuropeptides: opportunities for drug discovery. Lancet Neurol 2003; 2:463-472.
51. Quartara L, Altamura M. Tachykinin receptors antagonists: from research to clinic. Curr Drug Targets 2006; 7:975-992.
52. Gurrath M. Peptide-binding G protein-coupled receptors: new opportunities for drug design. Curr Med Chem 2001; 8:1605-1648.
53. Zeltser I, Gilon C, Ben-Aziz O et al. Discovery of a linear lead antagonist to the insect pheromone biosynthesis activating neuropeptide (PBAN). Peptides 2000; 21:1457-1465.
54. Altstein M, Ben-Aziz O, Daniel S et al. Backbone cyclic peptide antagonists, derived from the insect pheromone biosynthesis activating neuropeptide, inhibit sex pheromone biosynthesis in moths. J Biol Chem 1999; 274:17573-17579.
55. Altstein M. Novel insect control agents based on neuropeptide antagonists—The PK/PBAN family as a case study. Journal of Molecular Neuroscience 2003; 22:147-157.
56. Altstein M, Ben-Aziz O, Zeltser I et al. Inhibition of PK/PBAN-mediated functions in insects: Discovery of selective and nonselective inhibitors. Peptides 2007; 28:574-584.
57. Ben-Aziz O, Zeltser I, Bhargava K et al. Backbone cyclic pheromone biosynthesis activating neuropeptide (PBAN) antagonists: Inhibition of melanization in the moth Spodoptera Littoralis (Insecta, Lepidoptera). Peptides 2006; 27:2147-2156.
58. Cazzamali G, Torp M, Hauser F et al. The Drosophila gene CG9918 codes for a pyrokinin-1 receptor. Biochem Biophys Res Commun 2005; 335:14-19.
59. Choi MY, Fuerst EJ, Rafaeli A et al. Identification of a G protein-coupled receptor for pheromone biosynthesis activating neuropeptide from pheromone glands of the moth Helicoverpa zea. Proc Natl Acad Sci U S A 2003; 100:9721-9726.
60. Hull JJ, Ohnishi A, Moto K et al. Cloning and characterization of the pheromone biosynthesis activating neuropeptide receptor from the silkmoth, Bombyx mori—Significance of the carboxyl terminus in receptor internalization. J Biol Chem 2004; 279:51500-51507.
61. Zheng L, Lytle C, Njauw C-N et al. Cloning and characterization of the pheromone biosynthesis activating neuropeptide receptor gene in Spodoptera littoralis larvae. Gene 2007; 393:20-30.
62. Nachman RJ, Teal PEA, Ujvary I. Comparative topical pheromonotropic activity of insect pyrokinin/PBAN amphiphilic analogs incorporating different fatty and/or cholic acid components. Peptides 2001; 22:279-285.
63. Nachman RJ, Teal PEA, Strey A. Enhanced oral availability/pheromonotropic activity of peptidase-resistant topical amphiphilic analogs of pyrokinin/PBAN insect neuropeptides. Peptides 2002; 23:2035-2043.
64. Nachman RJ, Ben-Aziz O, Davidovitch M et al. Biostable β-amino acid PK/PBAN analogs: Agonist and antagonist properties. Peptides 2008; 30:608-615.
65. Nachman RJ, Teal PEA, Ben Aziz O et al. An amphiphilic, PK/PBAN analog is a selective pheromonotropic antagonist that penetrates the cuticle of a heliothine insect. Peptides 2009; 30:616-621.
66. Hauser F, Cazzamali G, Williamson M et al. A genome-wide inventory of neurohormone GPCRs in the red flour beetle Tribolium castaneum. Front Neuroendocrinol 2008; 29:142-165.
67. Roller L, Yamanaka N, Watanabe K et al. The unique evolution of neuropeptide genes in the silkworm Bombyx mori. Insect Biochem Mol Biol 2008; 38:1147-1157.
68. Yamanaka N, Yamamoto S, Zitnan D et al. Neuropeptide receptor transcriptome reveals unidentified neuroendocrine pathways. PLoS ONE 2008; 3:e3048.
69. Brockmann A, Annangudi SP, Richmond TA et al. Quantitative peptidomics reveal brain peptide signatures of behavior. Proc Natl Acad Sci USA 2009; 17;106:2383-2388.
70. Yew JY, Wang Y, Barteneva N et al. Analysis of neuropeptide expression and localization in adult Drosophila melanogaster central nervous system by affinity cell-capture mass spectrometry. J Proteome Res 2009; 8:1271-1284.
71. Chang JC, Yang RB, Adams ME, Lu KH. Receptor guanylyl cyclases in Inka cells targeted by eclosion hormone. Proc Natl Acad Sci USA 2009; 106:13371-13376.

72. Barnes AI, Wigby S, Boone JM et al. Feeding, fecundity and lifespan in female Drosophila melanogaster. Proc Biol Sci 2008; 275:1675-1683.
73. Broughton S, Partridge L. Insulin/IGF-like signalling, the central nervous system and aging. Biochem J 2009; 418:1-12.
74. Giannakou ME, Partridge L. Role of insulin-like signalling in Drosophila lifespan. Trends Biochem Sci 2007; 32:180-188.
75. Hariton A, Ben-Aziz O, Davidovitch M et al. Bioavailability of insect neuropeptides: The PK/PBAN family as a case study. Peptides 2009a; 30:1034-1041.
76. Hariton A, Ben-Aziz O, Davidovitch M et al. Bioavailability of β-amino acid and C-terminally derived PK/PBAN analogs. Peptides 2009b; 30:2174-2181
77. Hariton A, Ben Aziz O, Davidovitch M, Altstein M. Bioavailability of backbone cyclic PK/PBAN antagonists: inhibition of sex pheromone biosynthesis elicited by the natural mechanism in Heliothis peltigera females. FEBS J 2010; 277:1035-1044.

Chapter 9

Neuropeptide Physiology in Insects

William G. Bendena*

Abstract

In a search for more environmentally benign alternatives to chemical pesticides, insect neuropeptides have been suggested as ideal candidates. Neuropeptides are neuromodulators and/or neurohormones that regulate most major physiological and behavioral processes in insects. The major neuropeptide structures have been identified through peptide purification in insects (peptidomics) and insect genome projects. Neuropeptide receptors have been identified and characterized in *Drosophila* and similar receptors are being targeted in other insects considered to be economically detrimental pests in agriculture and forestry. Defining neuropeptide action in different insect systems has been more challenging and as a consequence, identifying unique targets for potential pest control is also a challenge. In this chapter, neuropeptide biosynthesis as well as select physiological processes are examined with a view to pest control targets. The application of molecular techniques to transform insects with neuropeptide or neuropeptide receptor genes, or knockout genes to identify potential pest control targets, is a relatively new area that offers promise to insect control. Insect immune systems may also be manipulated through neuropeptides which may aid in compromising the insects ability to defend against foreign invasion.

Introduction

To protect agricultural crops, forests and forest carbon sinks, large-scale use of pesticides are required worldwide. The global pest control industry, including insecticides, herbicides and fungicides, is worth over 30 billion (US) dollars, with over $10 billion spent worldwide annually on pest control products.[1] Insect pests have a particularly significant impact on forests, causing major losses of wood production each year by reducing tree growth, killing trees, or causing wood degradation. In Canada, for example, forest area lost to insect destruction is far more extensive and economically devastating than area lost to wild fires. Climate change is exacerbating this impact as rising temperatures and higher concentrations of greenhouse gases have led to reproductive increases in native mountain pine beetles, spruce budworms, jack pine budworms, tent caterpillars and other pests in climate-changed forests, taking advantage of milder winters, hot dry summers and ozone/carbon dioxide-induced changes to pest resistance in trees.

With the globalization of markets, there also has been a sharp increase in the number of invasive, nonnative insect species being introduced across foreign lands, further threatening resource production as well as ecosystem health and biodiversity. At the same time, there is increasing consumer resistance to the perceived negative health and environmental costs of using traditional chemical pesticides to control these various threats.

Insecticides are vital in protection of human and animal/livestock health against insect vectors carrying disease. Most notable diseases are Typhus (bacteria transmitted by lice, fleas or larval mites/chiggers), Lyme disease (spirochete bacteria transmitted by ticks), Chagas' disease

*William G. Bendena—Department of Biology, Queen's University, Kingston, ON, Canada, K7L 3N6. Email: william.bendena@queensu.ca

Neuropeptide Systems as Targets for Parasite and Pest Control, edited by Timothy G. Geary and Aaron G. Maule.

(Trypanosome parasite transmitted by assassin or kissing bugs) Malaria (Plasmodium parasite transmitted by mosquitoes), Leishmaniasis (Trypanosome parasite transmitted by tsetse flies) and Encephalitis, Yellow fever and Dengue fever (viruses transmitted by mosquitoes). As a measure of the enormity of the problem; at the end of 2004, there were 350-500 million clinical malarial cases in 107 countries and greater that 1 million deaths annually. The vast majority of these deaths are in young children.[2]

Traditional insecticides are either synthetic or naturally-derived chemical compounds that were developed to target either components of the insect nervous or endocrine system, or select insect enzymes. Despite widespread use and effectiveness, chemical insecticides have been criticized as being harmful to humans, animals and ecosystems, as well as to beneficial insects and organisms. A further drawback is insecticide persistence in the environment that can lead to agricultural disturbance. Many insecticides or their breakdown products also resemble natural hormones and thus can function as endocrine disrupters that disrupt fertility in select species.[3]

Acquisition through selection of insecticide resistance is another major concern. As one example, white flies exposed to an insecticide over a 12 year period have developed 2000-fold resistance.[4] Understanding of the molecular basis of resistance and how insect genomes respond to insecticide assault should help overcome this drawback.[5,6]

Ion channels and receptors of the nervous system are major insecticide targets. Organochlorine (DDT: p,p'-dichlorodiphenyl-trichloroethane) and pyrethroid insecticides target the voltage-gated sodium channel. Binding to components of the channel prolongs sodium entry into nerve axons and dendrites thus inducing repetitive nerve firing and hyperexcitability that leads to paralysis and death.[7] Agricultural use of DDT has been banned in many countries since 1972. However, due to uncontrolled spread of mosquitoes carrying malarial vectors, the World Health Organization has approved DDT use for lack of effective alternatives.[8,9]

Other nervous system-directed insecticides target receptors that mediate the action of neurotransmitters; acetylcholine, γ-aminobutyric acid (GABA) and glutamate. Nervous system targets are not unique to insects; however, bio-rational chemical insecticides are considered those that through structure-activity studies demonstrate a greater affinity for insect receptors/channels over the mammalian counterpart.[4]

In widespread use are insecticides that affect insect development. Development in insects is dependent on two hormones, the steroid hormone 20-hydroxyecdysone and the sesquiterpenoid juvenile hormone. Ecdysone agonists (tebufenozide, methoxyfenozide and halofenozide) are effective in controlling lepidopteran pests and juvenile hormone mimics (pyriproxyfen, fenoxycarb and methoprene) affect a wide spectrum of insect pests as they suppress embryogenesis and metamorphosis.[10] As the insect exoskeleton is composed of chitin, chitin synthesis inhibitors (benzoylphenyl ureas and buprofezin) are also effective.[11]

As neuropeptides control critical physiological processes, including metabolism, homeostasis, development, reproduction and behaviour, they would appear to be ideal candidates for pest management strategies. Insect specific neuropeptides offer insect target specificity and environmental compatibility. Direct use of neuropeptides for insect control is impractical because the insect cuticle contains an apolar lipid matrix that inhibits penetration of polar compounds like peptides. The insect gut, hemolymph and membranes of a number of tissues contain peptidases that rapidly degrade peptides into constituent amino acids should they enter the insect via a topical route or through ingestion. This was demonstrated in *Manduca sexta* larvae where only 1% of applied proctolin penetrated the cuticle and larvae fed proctolin had only 5% remaining within the gut after 2-5 hours.[12]

Despite these limitations, an integrative approach merging an understanding of neuropeptide structure-activity, biosynthesis, release, receptor identity and activation and mechanism(s) of physiological action, may uncover unique targets that can be exploited for pest control. This chapter will examine a select group of neuropeptides and illustrate their structural or physiological properties that could serve as prototype insect control agents[13] with a view to recognizing future areas for development.

Neuropeptide Biosynthesis and Action

Most insect neuropeptides appear to exhibit multiple functions (Table 1). Functions have been largely defined in individual species and therefore, it is unclear at present if such actions can be inferred across all insects. Neuropeptide nomenclature can be confusing as the name may not necessarily reflect the main biological action of the neuropeptide. The name assigned to a neuropeptide is generally based on the functional assay used in the peptides initial discovery. Several recent reviews cover the physiology and molecular biology of the most highly characterized insect neuropeptides.[13-17] Other recent reviews focus on individual peptides such as the allatostatins,[18,19] CAPA peptides,[20] proctolin,[15,21] neuropeptide Y (NPY),[22] allatotropin,[23] insulin-like peptides[24] or a specific physiological function such as water balance[25-28], ecdysis[29,30] and reproduction.[31]

Neuropeptides are secretory peptides that are initially translated as large biologically inactive prepropolypeptides (preprohormones) that undergo post-translational processing (cleavage) in the endoplasmic reticulum-golgi network and electron-dense secretory granules. The nature of this biosynthetic process provides opportunities for peptide regulation and diversity at several levels, including RNA alternative splicing, alternative or developmentally regulated prepropolypeptide processing, peptide modification and degradation. Each preprohormone begins with a secretory signal sequence. In insects, this sequence appears unique for specific neuropeptide precursor types and the amino acids that form the signature for signal sequence cleavage are not necessarily conserved. In general signal peptides recognize GlyArg peptide bonds. The secretory neuropeptides within the prepropolypeptide precursor may be flanked by either mono-(Arg), di-(LysArg, ArgLys, ArgArg)[32] or tribasic amino acids(ArgArgArg, LysArgArg).

Based on vertebrate processing models, mono and dibasic motifs are recognized by subtilisin-like proprotein convertases (PCs) that are packaged with the prohormone in electron-dense secretory granules. Tribasic motifs may be cleaved by similar convertase enzymes known as furins within the trans-golgi network. *Drosophila melanogaster* has three PC genes, two furin-genes and *amontillado*, encoding the *Drososphila* ortholog of mammalian PC2. The action of PCs/furins is regulated by a class of serine protease inhibitors known as serpins. In *Drosophila,* one endoplasmic reticulum localized serpin Spn42A when overexpressed in selected neurons inhibits furin/PC1 and/or PC2 with resultant accumulation of abnormally processed neuropeptides. Serpin action in neuropeptide processing may provide a target site for control if select serpins regulate individual neuropeptide pathways. The presence of a cleavage site does not necessarily mean that it is used in all cells or cell types and thus must be verified by peptide identification.[33] Bioinformatic programs have been developed to help predict genome identified vertebrate prohormone processing site cleavages.[34]

Approximately 90% of all known or predicted processed insect neuropeptides are carboxy-terminal α-amidated. This modification, when present, is functionally required for neuropeptide activity and/or receptor binding. Within the prepropolypeptide, neuropeptides destined for C-terminal amidation have the C-terminal signature sequence Gly followed by a di-or tri-basic endoproteolytic cleavage site. After endoproteolytic cleavage the C-terminal glycine is acted upon by two enzymes; peptidylglycine α-hydroxylating monooxygenase (PHM) that forms a hydroxylated intermediate followed by peptidyl α-hydroxyglycine α-amidating lyase (PAL) that cleaves the intermediate producing amidated peptides and glyoxylate. In *Drosophila,* there is one gene for PHM and two genes for PAL.[35] In the mammalian enzyme equivalent both enzymatic activities occupy different domains of the same protein. A mutational study of PHM in *Drosophila* indicated that the modifying activity was important at all stages of development and that the expression of properly modified secretory peptides is critical for developmental transitions.[36]

In select neuropeptides, further modification may occur including sulfation, conversion of the N-terminal amino acid to a cyclic pyroglutamate residue and disulfide bridge formation.[37] Removal of such modifications, in general, will affect function either through alteration of folding, altering the susceptibility to degradation or interaction with their cognate receptor. Processing of a prepropolypeptide may release a single known peptide or several copies of a single peptide or may release several different peptides of varying sequence. Multiple copies of similar peptides within a precursor (paralogues) likely resulted from duplication and mutation through evolutionary time.

Table 1. Major classes of neuropeptides identified in insects

Neuropeptide	Physiological Function	Prototype or Consensus Sequence*
Adipokinetic hormone AKH	Mobilize energy substrates Locmi AKHI stimulates immune system-*Locusta* Hyperactive behavior in response to starvation-*Drosophila*	Locmi AKHI: pQLNFTPNWGT-NH2 Manse AKH: pQLTFTSSWG-NH2 Locmi AKHII: pQLNFSAGW-NH2 Locmi AKHIII: pQLNFTPWW-NH2 Locmi AKHIV: pQVTFSRDWSP-NH2 Drome AKH: pQLTFSPDW-NH2 Anoga AKHI: pQLTFSPDW-NH2 Anoga AKHII: pQVTFSRDWNA-NH2
Allatostatin-A	Inhibits Juvenile hormone (JH) biosynthesis-cockroach and crickets Inhibits muscle (hindgut) contraction-cockroach and locust Inhibits vitellogenin production-cockroach Regulates gut enzyme release	C-Terminal consensus sequence Y/FXFGL-NH2
Allatostatin-B	Inhibits JH biosynthesis-crickets Inhibits spontaneous contractions of hindgut and oviduct-Locusta and foregut-cockroach Prothoracicostatic activity-Lepidoptera	C-Terminal consensus equence W(X6)W-NH2
Allatostatin-C	Inhibits JH biosynthesis-lepidoptera, mosquitoes Inhibits larval heartbeat-Drosophila Inhibits spontaneous foregut contraction and feeding-Lepidoptera Regulation of migratory flight-lepidoptera	pQ(V/I)R(F/Y)RQCYFNPISCF-OH moths V2, F4 Drome V2, Y4 Anoga I2, Y4

continued on next page

Table 1. Continued

Neuropeptide	Physiological Function	Prototype or Consensus Sequence*
Allatotropin	Stimulates JH biosynthesis-lepidoptera, mosquito Cardioexcitatory activity-Manduca Stimulates oscillations of the ventral diaphragm-Pseudeletia Stimulates myotropic contractions of the foregut in larval H. armigera Inhibits ion transport across the larval midgut epithelium-Manduca	*Manse* GFKNVEMMTARGF-NH_2
Baratin	Modulator of motor patterns in abdominal ganglia-cockroach Leucophaea maderae	DNSQWGGFA
Bombyxin Heterodimer	Energy Metabolism-lowers hemolymph carbohydrate levels-moths Stimulates ecdysteroid biosynthesis-*Bombyx*	Bommo A1B QQPQRVHTYCGRHLARTLADLCWEAGVD Bommo A1A GIVDECCLRPCSVDVLLSYC
Bursicon	A heterodimer required for sclerotization, cuticle tanning and wing expansion after eclosion	Bomme Bursicon FPVTGHEVQLPPGTKFFCQECQMTAVIHVLKHRGCKPKAI PSFACIGKCTSYVQVSGSKIWQMERTCNCCQESGEREATVVLFCPDAQ- NEEKRFRKVSTKAPLQCMCRPCGSIEESSIIPQEVAGYSEEGPLYNHFRKSL Bomme Partner of Bursicon EENCETVASEVHVTKEEYDEMGRLLRSCSGEVSVNKCEGMCNSQVHPSISSP- TGFQKECFCCREKFLRERLVTLTHCYDPDGIRFEDEENALMEVRLREPDEC- ECYKCGDFSR
Corazonin	Cardioacceleration-Periplaneta Melanization-Locusta His7 Ecdysis initiation Circadian rhythm and diapause-Manduca	pQTFQYS(R/H)GWTN-NH_2 Locmi H7

continued on next page

Table 1. Continued

Neuropeptide	Physiological Function	Prototype or Consensus Sequence*
Crustacean cardioactive peptide (CCAP) = Cap2a	Stimulates AKH release-Locusta Ecdysis and Eclosion	PFCNAFTGC-NH_2
Diuretic Hormone (DH)	Control of Malpighian tubule fluid secretion Regulates water and ion balance	CRF-related: ManseDH RMPSLSIDLPMSVLRQKLSLEKERKVHALRAAANRNFLNDI-NH2 Manse-DP2 SFSVNPAVDILQHRYMEKVAQNNRNNLNRV-NH2 Calcitonin-like Dippu-DH31 GLDLGLSRGFSGSQAAKHLMGLAAANYAGGP-NH2 Kinins: Leuma-K-1 DPAFNSWG-NH2 Cap2b Manse-CAP2b pQLYAFPRV-NH2
Anti-Diuretic Hormones (ADH)	Control of water and ion reabsorption. Regulate Malpighian tubule secretion and fluid and ion uptake by the intestine (ileum).	Tenmo-ADFa VVNTPGHAVSYHVY-COOH Tenmo-ADFb YDDGSYKPHIYGF-COOH Schgr-ITP SFFDIQCKGVYDKSIFARLDRICEDCYNLFREPQLHSLCRSDCFKSPYFKGCL-QALLLIDEEEKFNQMVEIL-NH2 Schgr-ITP-L SFFDIQCKGVYDKSIFARLDRICEDCYNLFREPQLHSLCRKDCFTSDYFKGCI-DVLLLQDDMDKIQSWIKQIHGAEPGV-COOH
Ecdysis Triggering Hormone (ETH) and Pre-ecdysis triggering hormone (PETH) Eclosion Hormone (EH)	EH is an initiator of ecdysis which is an innate behaviour leading to shedding of the cuticle. In moths,two pre-ecdysis (contractions) controlled by pre-ecdysis triggering hormone (PETH) and ecdysis triggering hormone (ETH)	Manse PETH SFIKPNNVPRV-NH2 Bommo ETH NEAFDEDVMGYVIKSNKNIPRM-NH2 Manse ETH SNEAISPFDQGMMGYVIKTNKNIPRM-NH2 Drome ETH1 DDSSPGFFLKITKNVPRL-NH2 Drome ETH2 GENFAIKNLKTIPRI-NH2

continued on next page

Table 1. Continued

Neuropeptide	Physiological Function	Prototype or Consensus Sequence*
Myosuppressins	May act as myoinhibitors or myostimulators dependent on muscle and species examined.	Leucomyosuppressin pQ/P/T)DV(D/V)H(V/S)FLRF-NH2
	Inhibit AKH release-Schistocerca	P1 SchistoFLRFamide
		T1 Dromyosuppressin
		V4, S6 ManducaFLRFamide
Extended FMRFamides	Differing effects on muscle contracton including heart, gut and skeletal-*Drosophila*	SVKQDFMHF-NH2
		DPKQDFMRF -NH2
		TPAEDFMRF-NH2
		SDNFMRF-NH2
		SPKQDFMRF-NH2
		PDNFMRF-NH2
		SAPQDFVRS-NH2
		MDSNFIRF -NH2
Extended FL/IRFamides	Oviduct contraction-*Locusta*	Manse FLRFamide II GNSFLRF-NH2
	Antennal heart contraction-*Periplaneta*	Manse FLRFamide III DPSFLRF-NH2
	Ecdysis-*Manduca*	Peram FMRF4 GGRSNDNFIRF-NH2
		Peram FMRF 7 DRSDNFIRF-NH2
		Peram FMRF11 ARPDNFIRF-NH2
		Locmi AFIRF-NH2
		Locmi GQERNFLRF-NH2
Long Neuropeptide F	Regulates feeding	Drome-NPF SNSRPPRKNDVNTMADAY-KFLQDLDTYYGDRARVRF-NH2
		Anoga-NPF LVAARPQDSDAASVAAAI-RYLQELETKHAQHARPRF-NH2
		Aedae-NPF SFTDARPQDDPTSVAEAI-NH2

continued on next page

Table 1. Continued

Neuropeptide	Physiological Function	Prototype or Consensus Sequence*
Short Neuropeptide F	Regulates appetite and feeding	Drome sNPF1 AQRSPSLRLRF-NH2 Drome sNPF2 SPSLRLRF-NH2 Lepde- NPF1 ARGPQLRLRF-NH2 Lepde- NPF2 APSLRLRF-NH2 Aedae head peptide 1 pQRPHypSLKTRF-NH2 Aedae head peptide 2 TRF-NH2
Orkokinins	Modulates circadian clock controlling locomoter activity-Leucophaea	Blage NFDEIDRSGFNS Schgr NFDEIDRSGFSGFV
Periviscerokinins CAP = CAP2b	Increases the frequency of contraction of the heart and stimulates amplitude and tonus of the foregut. Cardioacceleratory-*Periplaneta, Manduca* Diuretic-Diptera	C-terminal consensus PRV-NH2 or PVMRN-NH2 or PFGRT-NH2
Proctolin	Stimulates muscle contraction Stimulates AKH release-Locusta	RYLPT
Prothoracicotropic hormone (PTTH) (Homo-dimer)	Stimulate ecdysteroid synthesis and secretion by prothoracic gland	GNIQVENQAIPDPPCTCKYKKEIEDLGENSVPRFIETRNCNKTQQPTCRPPYI-CKESLYSITILKRRETKSQESLEIPNELKYRWVAESHPVSVACLCTRDYQL-RYNNN
Pyrokinin = Cardioacceleratory (CAPA)	Mediates visceral muscle contraction *Locusta* Puparation acceleration activity-*Sarcophaga*	C-terminal consensus WFGPRL/V-NH2

continued on next page

Table 1. Continued

Neuropeptide	Physiological Function	Prototype or Consensus Sequence*
Sulfakinins	Increase hindgut contractions-cockroach, *Locusta* Increase heartbeat frequency-cockroach Reduce food intake-cockroach, Locusta Stimulate gut α-amylase release-lepidoptera	C-terminal consensus sequence Y(SO3H)GHMRF-NH_2
Tachykinin	Stimulates gut muscle contraction-cockroach, *Drosophila* Stimulates oviduct contraction-*Locusta*, cockroach Olfactory processing	C-terminal consensus FXGXR-NH_2

*Insect abbreviations: Locmi *Locusta migratoria*, *Schgr Schistocerca gregaria*, Manse *Manduca sexta*, Bommo *Bombyx mori*, Drome *Drosophila melanogaster*, Anoga *Anopholes gambiae*, Aedae *Aedes aegypti*, Blage *Blattella germanica*, Leuma *Leucophaea maderae*, Peram *Periplaneta americana*, Lepde *Leptinotarsa decemlineata*.

Similar precursors in other insect species (orthologues) may be structurally but not necessarily functionally similar.

Neuropeptides are stored in large electron-dense granules at release sites in axon terminals until depolarization results in exocytosis. Exocytosis may occur at synaptic specializations in axonal terminals or at nonsynaptic release sites throughout the neuron. There is very little known about the components involved in synapse function in insects. A genetic screen in *Caenorhabditis elegans* of genes required for function or development of a neuromuscular junction identified 185 genes that decreased acetylcholine secretion.[38] As several unique genes were found it may be that insects will have unique regulatory targets at the synapse level.

Neuropeptides function through interacting with a receptor that is typically a G-protein-coupled receptor (GPCR). Bioinformatic mining of the *Drosophila* genome[39,40] identified and catagorized 44 neuropeptide GPCRs of the 160 GPCRs found in the genome. Thirty five receptors for neuropeptides have been bioinformatically identified in the honeybee *Apis mellifera*.[41] De-orphaning of receptors has followed based on expression of the receptor in a vertebrate, mammalian or insect cell lines, coupled to functional assays to match the neuropeptide with a specific receptor.[41,42]

Evidence suggests that insect some neuropeptide receptors may be promiscuous interacting with several different neuropeptide structures.[43] Pharmacological evidence has suggested different peptides interacting with the same receptor can activate different signal transduction pathways.[44] Outside of insects with sequenced genomes (*Drosophila melanogaster*, *Anopheles gambiae*, *Apis mellifera*) very few nondipteran insect receptors have been isolated and functionally assayed. Sequence relatedness to *Drosophila* receptors should aid in their identification. Receptor activation usually stimulates a cascade of second messengers which can include cAMP, diacylglycerol and inositol triphosphate. Second messengers can be associated with release of intracellular Ca^{2+} stores. Our information on the cascade of events that follow is a result of molecular/genetic studies in *Drosophila*.

Potential Physiological Targets for Pest Control

Interruption in biosynthesis or physiological action for any given neuropeptide should have an impact on insect health or behaviour. Selected physiological processes that could serve as targets for pest control have been chosen to illustrate the complexity of neuropeptide action as well as the diversity of action within different insects. Processes chosen include energy metabolism, muscle contraction, water balance, growth and reproduction, as well as, growth and development.

Energy Metabolism

Homeostatic regulation of blood energy substrates is a fundamental physiological process in both vertebrates and invertebrates. For example, in mammals unregulated sugar levels can lead to diseases such as diabetes. In mammals, glucagon and insulin are key endocrine effectors that regulate blood glucose levels. Insects produce functional homologues such as the insulin-related peptide, bombyxin, that when injected into the silkmoth, *Bombyx mori*, lowers hemolymph carbohydrate levels in a dose-dependent manner.[45] Carbohydrate hemolymph levels are raised in response to hypertrehalosaemic hormone similar to the action of vertebrate glucagon . Based on peptide sequence similarity hypertrehalosaemic hormone is part of a large family of peptides that collectively are referred to as the adipokinetic hormones (AKHs).

AKHs are produced in the neurosecretory cells of the glandular lobe of the corpora cardiaci (CC). The CC functions as a neurohaemal organ, storing and releasing material from the brain[46] and products of its intrinsic secretory activity.[47] Neuropeptides originating from neurosecretory cells of the brain send axons to the glandular lobe that may regulate the release of AKHs. In locusts, AKH release appears to be stimulated by neuropeptides proctolin,[48,49] crustacean cardioactive peptide[50] and the desert locust *Locusta migratoria* tackykinin I and II.[51,52] SchistoFLRFamide and FMRFamide, however, inhibit release into the hemolymph.[53]

Artificial intervention in the synthesis of these peptides would thus influence the functioning of AKHs. AKHs are a potential target for chemical intervention for control as their primary function is to regulate metabolism of energy stores. Intervention could serve to reduce/eliminate flight, escape or foraging activities. In response to flight, or other energy expensive activities, AKHs mobilize energy reserves from storage organs such as the fat body to the hemolymph.[54,55] The number of AKH peptides and the energy substrate mobilized appear to differ between insect species. In *Locusta*, four AKH peptides have been purified but three are involved in energy substrate mobilization. During locust flight, the carbohydrate, trehalose, is released to the hemolymph through the activation of fat body glycogen phosphorylase during initial flight. During sustained flight, lipid (diacylglycerol) is released through the activation of fat body triacylglycerol lipase.[56] *Locusta* AKH I appears to be most active in lipid mobilization, AKH II most active in carbohydrate mobilization and AKH III appears to be a modulator that provides energy during periods of rest. The fruit fly, *Drosophila* has one AKH and the malarial mosquito two AKH peptides. In *Drosophila*[57,58] and *Anopheles*,[59] trehalose and lipid are mobilized during flight. In the American cockroach, *Periplaneta americana* two AKH peptides function to mobilize only carbohydrate (hypertrehalosaemic) during flight.[60] Certain beetles may mobilize only the amino acid proline (hyperprolinaemic) as an energy substrate.[55]

AKHs have additional roles that may be complementary to energy metabolism through direct interaction with the nervous system.[61,62] Through genetic manipulations in *Drosophila* AKHs appear to carry out a central function involving hyperactive behavior in response to starvation. This is turn may mobilize the insect to find food.[58] In *Periplaneta*, this locomotor behavioral effect appears to be regulated only by one of the two AKHs which can cross the blood-brain barrier to act on the abdominal dorsal unpaired medial (DUM) neurons. DUM neurons release the biogenic amine octopamine which act on muscles of the ventilatory system and on the heart.[63,64]

The AKH G-protein coupled receptor has been cloned[65,66] from *Periplaneta* and its localization in fat body and DUM neurons verified.[65] In *Periplaneta* both peptides activate G protein Gs but AKH1 is also a potent activator of Gq which suggests that different signal transduction pathways can be activated.[65] A similar AKH receptor has been cloned from the moth *Bombyx*.[67] Genome projects have also uncovered similar receptors in *Drosophila*,[67,68] *Anopheles*[69] and *Apis* (Genbank GeneID 551388). Information regarding the signal transduction pathways activated by these receptors is still limited.

Muscle Contraction

Neuropeptides that influence muscle contraction have been suggested as potential targets for insect control as a variety of physiological processes could be altered. An ideal result in terms of pest control would be to achieve paralysis, reduced feeding, aberrant heart beat and/or death. As examples, contraction of foregut, midgut and/or hindgut influences feeding and/or digestion, contraction of skeletal and flight muscle influences locomotion, contraction of heart muscle influences hemolymph circulation, contraction of ovarian muscle affects reproduction and regulate behavioral cascades such as ecdysis. There are several peptide families that exert their influence on muscle contraction in insects. These include the FMRFamide-related peptides (FaRPs), pyrokinins, tachykinins, proctolin, CAPA peptides, corazonin, orkokinins and allatostatins. Unfortunately, there is little in the literature that relates the expression of all of these peptides within a given insect.

The FaRPs are a large superfamily of peptides found throughout the metazoa that share a carboxy-terminal-RFamide.[70] The variation in FaRP family members and their range of biological activities suggest that FaRPs may be ideal candidates for pest control. In insect this family includes N-terminally extended FMRFamides (thus far identified only in Dipterans) and FL/IRFamides, the myosuppressins, sulfakinins and neuropeptide Fs (NPFs). A single *Drosophila* FMRFamide gene specifies a precursor that upon processing would release 11 extended RFamide peptides (there are 5 copies of DPKQDFMRFamide) and 2 additional peptides that terminate in FMHFamide or an FVRSamide. One receptor has been identified[71,72] that appears to be activated by all extended RFamides with varying potencies. Surprisingly, this receptor is also activated by myosuppressins and short NPFs. Three of the *Drosophila* extended FMRFamides decrease heartbeat.[73,74] *Drosophila* allatostatin-C that is unrelated in sequence to the FaRPs also is a potent inhibitor of heartbeat.[75] Seven extended RFamides stimulate contraction of body wall muscles in larval *Drosophila*.[76]

Myosuppressins were initially isolated from the cockroach *Leucophaea maderae* based on their ability to inhibit spontaneous contraction of the hindgut. Since that time myosuppressins have been found to inhibit contractions in foregut, midgut and hindgut. In *Drosophila*[73] and the locust *Schistocerca gregaria*[77] myosuppressins are cardioinhibitory. In locust, myosuppressins can stimulate skeletal muscle contraction but can also inhibit contraction of oviduct muscle.[78,79] This is further complicated as certain extended FLRFamides can stimulate oviduct muscle.[80] Myosuppressins can attenuate evoked transmitter release from the presynaptic membrane of excitatory motor neurons terminating on skeletal muscle of the mealworm, *Tenebrio molitor*.[81] Myosuppressins also stimulate indirect flight muscle in *Manduca*.[82] Two myosuppressin receptors have been identified in *Drosophila*[83] and one in mosquito.[84]

Myosuppressin action on cockroach hindgut contraction, mealworm neuromuscular junction and locust oviduct contraction can be mimicked by benzethonium chloride.[85,86] Benzethonium chloride is a stable nonpeptide and can activate the myosuppressin receptor or competitively displace myosuppressin from its receptor. The use of benzethonium chloride as a pest control agent would have to be approached with caution as the compound induces programmed cell death (apoptosis) and caspase activity in mammalian cells[87] and is a potent anti-microbial.[88] Understanding the mechanism of interaction of benzethonium chloride with the myosuppressin receptor may have value in the design of future compounds.

Several neuropeptides were purified on the basis of being stimulators of muscle contraction. These include proctolin, crustacean cardioactive peptide and corazonin.

Proctolin functions as a myostimulator with insect specific actions on foregut, midgut, hindgut; skeletal muscle, heart and muscles of the oviduct. As with many insect neuropeptides, crustacean cardioactive peptide and corazonin have varied functions in different insect Orders and hence offer interesting future targets for control.

Crustacean cardioactive peptide (CCAP) is a cyclic nonapeptide that was first isolated from the pericardial organs of the crab *Carcinus maenas* as a peptide that stimulated heartbeat.[89] The name assigned to this peptide was unfortunate as it was later found that the identical peptide can be isolated from numerous insect Orders. CCAP is identical to *Manduca* $Cap2_a$[20] that functions as a cardioaccelerator after adult emergence.[90,91] In *Manduca* $Cap2_a$ facilitates wing inflation and increases hemolymph circulation during flight and stimulates contraction of larval gut[90] and adult oviducts. *Locusta* CCAP/$Cap2_a$ stimulate both hindgut[92] and oviduct contraction.[93] Immunocytochemistry demonstrates that there is no CCAP/$Cap2_a$ innervation to hindgut or oviducts suggesting that this neuropeptide functions as a neurohormone.[93,94] CCAP plays a pivotal role in ecdysis and eclosion (See section pertaining to growth and development). As with CCAP, corazonin functions also relate to ecdysis and eclosion. Corazonin is synthesized by both neurons and neurosecretory cells. Corazonin was originally isolated from the corpora cardiaca of *Periplaneta* and was named based on potent cardioacceleratory activity.[95] An identical molecule was later found in the cockroach *Nauphoeta cinerea*, cricket *Gryllus bimaculatus*, moths *Manduca, Bombyx, Galleria mellonella* and fly *Drosophila melanogaster*.[96] Corazonin regulation of heartbeat is not found in these other insects. In locusts, a corazonin ortholog with one amino acid difference R7 to H7 appears to induce melanization or pigmentation changes.[97] In *Manduca*, corazonin immunoreactivity colocalizes with the PER protein in Type-$1a_1$ lateral neurosecretory cells. Ablation of these cells can impede the induction of pupal diapause under short-day photo-period conditions.[98] This may imply that corazonin participates in circadian rhythm and diapause. Corazonin appears multifunctional in *Manduca* as it appears in the hemolymph prior to the onset of ecdysis and may control initiation of ecdysis behaviours.[99] (See section pertaining to growth and development.) A single corazonin receptor that conforms to a GPCR has been identified in *Drosophila*,[68,100] *Anopheles*[101] and *Manduca*.[102]

Water Balance and Feeding Behaviour

Ion and water homeostasis have been viewed as an excellent target for insect control that could be manipulated through diuretic and anti-diuretic neuropeptides. Excretory water loss (diuresis) occurs in response to feeding and after eclosion. Given the timing of diuresis, hyperstimulation of a diuretic hormone could result in insect dehydration and death or blocking of the receptor could result in water retention resulting in decreased feeding, locomotion and/or higher susceptibility to predators. This idea was supported by feeding diuretic hormone-coated leaves to *Manduca* larvae. Larvae exhibited symptoms of reduced food intake, slower growth and higher mortality.[13,27]

Currently, the molecular/physiological basis of neuropeptide-stimulated diuresis is one of the more highly characterized as compared to other known insect neuropeptides. Secretion of primary urine occurs through regulation of two cell types in *Drosophila*, principal and stellate cells of the main segment of Malpighian tubules (MTs). Fluid secretion is coupled in these cells to ion transport. The basolateral surface of principal cells contains three distinct G-protein coupled receptors. Calcitonin-like diuretic hormone (e.g., DH_{31}) and corticotropin-releasing factor-like (CRF-like) diuretic peptide and Cap_{2b} (CAPA family of peptides related to periviscerokinins, PVKs) each bind to their respective receptor. Receptor binding by calcitonin and/or CRF-like peptides stimulates cAMP production which in turn activates apical membrane V-ATPase which translocates protons across the principal cell apical membrane. Secreted protons are returned to the cell by a cation/H^+ antiporter. Diuretic hormones stimulate KCl/NaCl transport through the basolateral membrane into the lumen. In *Drosophila*, several potassium channel types have been identified. Cap_{2b}-receptor interaction in principal cells serves the same purpose as the other two hormones, the activation of V-ATPase. The route of V-ATPase activation is different in that inositol-triphosphate (IP_3) is used as a second messenger. IP_3 increases internal Ca^{2+} levels and external Ca^{2+} influx through voltage dependent Ca^{2+} -channels.[103,104] The net increase in intracellular Ca^{2+} activates Ca^{2+}/calmodulin-sensitive nitric oxide (NO) synthase. NO activates guanylate cyclase that increases cGMP levels which then activates V-ATPase. The response to Cap_{2b} may not be the same in all insects as no cGMP increase was noted in *Schistocerca*.[105] cGMP increase due to Cap_{2b} (= MasCAPA1) is also diuretic in blood-feeding Diptera such as mosquitoes but anti-diuretic in blood-feeding Heteroptera such as *Rhodnius prolixus*.[106,107] Kinins (typified by leucokinin) are also diuretic peptides. Kinins interact with a basolateral G-protein coupled receptor of stellate cells and open Ca^{2+}-activated chloride channels that increases chloride availability for KCl/NaCl secretion.[108] Water exit is facilitated through aquaporin channels in stellate cells. Of the seven aquaporin channels in *Drosophila* only one is localized to MTs.[109]

Much less is known about anti-diuretic peptides (ADFs) with very few peptide examples isolated from insects. Anti-diuretic peptides prevent water loss by stimulating fluid reabsorption by the hindgut (ileum and rectum) or by reducing MT secretion. Two ADF peptides have been isolated in *Tenebrio* that may act by stimulating cGMP production in MTs which may activate a phosphodiesterase specific for cAMP.[110] These peptides have not been found in other insects. In *Locusta*, a large 72 aa peptide termed ion transport peptide increases electrogenic chloride transport across the ileum which then stimulates fluid readsorbtion from the insect hindgut.[111,112] A similar ITP peptide has been purified from *Bombyx* and has been identified as a sequence in the *Drosophila* genome.

Microarray analysis of the Malpighian tubule in *Drosophila* has revealed a large number of organic solute transporters whose purpose may be to detoxify substances (xenobiotics/insecticides) through excretion.[25] Examining what role neuropeptides may play in regulating these transporters will be of future interest.

Sulfakinins share sequence similarity with the vertebrate gastrin/CCK peptides that function in feeding regulation. Sulfakinins when injected in *Schistocerca* fifth instar larvae[113] or the German cockroach, *Blattella germanica*,[114] inhibit food uptake. Movement of food from the crop to the midgut appears to be the point of inhibition and the sulfatated tyrosine group in the peptide appears critical for this activity. Analogs made with chemical substitutions for the hydrolysis susceptible Tyr-sulfate may lead to improved biostability, greater hemolymph retention and extended negative effects on feeding.[115] A *Drosophila* sulfakinin receptor has been identified,[116] however, no similar receptor has yet been identified in other insects.

Leucomyosuppressin, an extended FLRFamide, also inhibits food intake when injected in the cockroach, *Blattella* with food accumulating in the foregut. Yet in the cockroach, *Diploptera punctata*, leucomyosuppressin or cockroach-type allatostatin stimulate midgut alpha-amylase and invertase activity.[117]

Behavioral aspects of feeding can also be regulated in insects by neuropeptide F (NPF). NPF has two forms a long and a short. The long 36 amino acid dNPF has been identified in the *Drosophila*,[118] *Aedes aegypti*[119]and *Anopheles gambiae*[120] and shares structural similarity with the vertebrate NPY. NPY has been implicated in the regulation of a variety of behaviors including feeding, anxiety, fear and responsiveness to stress.[121] The *Drosophila* and *Anopheles* dNPF receptors[122,123] share sequence identity with the vertebrate NPY receptor. dNPF neural network in the larval CNS changes in response to gustatory sugar stimulation, indicating that dNPF is an integral part of the chemosensory system that regulates eating behavior.[124] dNPF is expressed in the larval brain while the animal is in a feeding stage. As larvae age, dNPF expression ceases at a time when larvae stop feeding and wandering behaviour begins in preparation for pupation. Overexpressing dNPF in wandering larval brain leads to continuous feeding.[125] Well fed larvae will normally not accept noxious food. In *Drosophila* mutants where the dNPF receptor was overexpressed, well fed larvae would accept noxious food whereas loss of the dNPF receptor enhanced their discriminatory behaviour. The activity of the NPY receptor appears to be linked to insulin-like receptor signaling.[126] As a control strategy, finding a means of upregulating dNPF receptors may help eliminate an insects avoidance of chemical food baits.

Short NPF (sNPF) or "head peptides" have been isolated from several insects including three peptides from *Aedes*,[127] two peptides from *Helicoverpa zea*,[128] two from the colorodo potato beetle *Leptinotarsa decemlineata*[129] and one peptide from *Schistocerca*.[130] The *Drosophila* receptor has been identified and resembles the vertebrate Type 2 neuropeptide Y receptors.[131,132] *Drosophila* mutants that overexpress sNPY had greater appetites as shown through greater food intake which resulted in heavier flies. Loss of function sNPF mutants showed suppressed appetite and limited food intake.[133]

Growth and Reproduction

Williams suggested that endocrine imbalance might be an effective means of controlling pest insects.[134] His suggestion was proven correct as chemical insecticides that mimic the actions of the insect steroid hormone 20-hydroxyecdysone and the sesquiterpenoid juvenile hormone are effective insecticides.[10] 20-hydroxyecdysone and juvenile hormone are important regulators of insect growth, development, metamorphosis and reproduction.

The steroid ecdysone is synthesized and released by the paired prothoracic glands in response to a cerebral peptide termed prothoracicotropic hormone (PTTH). Before each molt and during early metamorphosis, dependent on nutritional state, ecdysone is released and converted to the active 20-hydroxyecdysone by a monooxygenase at target tissues. In *Bombyx* four extended FMRF-related (FaRPs) peptides expressed in neurosecretory cells of the thoracic ganglia directly innervate the prothoracic gland inhibiting ecdysteroidogenesis. The FaRPs decrease prothoracic gland cAMP via binding to a *Bombyx* myosuppressin (extended FLRFamide) receptor.[43,135] Circulating *Bombyx* myosuppressin and myoinhibitory peptide/prothoracicostatic peptide (MIP/PTSP) inhibit basal and PTTH induced secretion of ecdysone and thus function as additional prothoracicostatic factors. MIP/PTSP share a high degree of sequence identity with the cricket allatostatin B1.[136] Bombyxin, a *Bombyx* insulin-related peptide stimulates larval prothoracic gland ecdysteroidogenesis. Insulin-like peptides appear to regulate nutrition dependent cell and tissue growth. Insulin-like peptide action on growth of the prothoracic gland regulates the initiation of metamorphosis. In *Drosophila* body size can be manipulated. Activation of the insulin signaling pathway in prothoracic glands leads to overgrown prothoracic glands that initiate metamorphosis at smaller body weights. Reduced prothoracic gland size achieve larger body sizes prior to initiation of metamorphosis.[137]

The regulation of ecdysteroidogenesis by neuropeptides is a relatively new and growing field that still has not been explored yet in most pest insect species.

Changes in juvenile hormone titer, which regulates the growth and development of immature insects and reproduction in adult insects, is controlled at the levels of biosynthesis, secretion, sequestration and degradation. Juvenile hormone is synthesized and released from the corpora allata (CA) into the hemolymph. Allatostatins are neuropeptides that rapidly and reversibly inhibit juvenile hormone biosynthesis whereas allatotropin(s) stimulate juvenile hormone biosynthesis. Three distinct peptide sequences have allatostatic activity.[19] The first, cockroach-type or AST-As is a large family of peptides, with the conserved C-terminal sequence Y/FXFGLamide, isolated from *Diploptera*.[138,139] This family of peptides has been isolated in most invertebrates, however, allatostatic activity appears limited to cockroaches and crickets.

The second family (AST-B) of allatostatins was isolated from the cricket, *Gryllus bimaculatus* and has the general structure $W(X)_6W$amide. Allatostatic activity appears limited to the cricket.[140]

The third family (AST-C) of allatostatin isolated from Lepidoptera is a single 15 amino acid peptide with a non-amidated C-terminal sequence PISCF. AST-C inhibits JH biosynthesis in Lepidoptera[141,142] as well as the dipteran mosquito *Aedes*.[143,144] A similar peptide exists in *Drosophila* that may also function as an allatostatin as well as a potent cardioinhibitor.[75]

There is only one known allatotropin that stimulates JH biosynthesis and its action appears to be specific for Lepidoptera.[145-147] In *Drosophila* there is evidence that insulin action may affect JH biosynthesis through neuropeptides such as allatotropin,[148] however, an allatotropin sequence has yet to be found in the *Drosophila* genome.

AST analogs or pseudopeptides of cockroach-type or AST-A have been made with the aim of increasing resistance to degradation of ASTs by hemolymph peptidases. A methyleneamino analogue was less active as an inhibitor of vitellogenin production in vitro by the fat body of *Blattella*, but was more active in vivo in terms of both inhibition of JH biosynthesis and as an inhibitor of vitellogenin production by the fat body.[149] Further analogues have been made that also protect against membrane-bound peptidases that would inactivate the core sequence required for activity. These injected analogues are effective in inhibiting JH biosynthesis and oocyte growth without toxic effects on the insects physiology.[150,151] An impediment to the successful use of AST analogues in the field is the problem of delivery of the compounds to the animal. Although potent analogues resistant to catabolism have been produced, these compounds do not penetrate the cuticle or gut wall. Addition of a hydrophobic moiety to an active portion of these analogues may provide amphiphilic compounds capable of penetrating the cuticle while still retaining significant biological activity.[150] Recent success with creating pheromone biosynthesis activating neuropeptide (PBAN) analogues that are effective when applied topically[152] or orally[153] may allow similar strategies to be applied to the allatostatins.

Growth and Development

Insect growth is constrained by an external skeleton and as a consequence insects shed their old exoskeleton and create a new one in a process termed molting. Ecdysis is an innate behaviour necessary for shedding the old cuticle that is orchestrated by a precise cascade of neuropeptide action.[29,30] This behaviour includes rhythmic compression and relaxation movements that loosen the old cuticle (pre-ecdysis behaviour) and anterior directed peristaltic movement that sheds the old cuticle (ecdysis behaviour). Prior to ecdysis cerebral neuropeptide corazonin is released into the bloodstream which activates low level release of pre-ecdysis triggering hormone (PETH) and ecdysis triggering hormone (ETH) from Inka cells.[102] ETH acts via two G-protein coupled receptors (ETHR-A and ETHR-B) that are produced by alternative splicing of a single ETH receptor gene. ETHR-A is found localized in the abdominal central nervous system in known peptidergic neurons. These include neurons expressing both stimulatory and inhibitory neuropeptides including eclosion hormone (already known to respond to ETH), CCAP, allatostatins, calcitonin-like diuretic hormone, CRF-like diuretic hormones (DHs) 41 and 30, myoinhibitory peptide (MIP), bursicon (involved in wing expansion and cuticle hardening), short and long neuropeptide F, FaRPs and kinins.[99]

Although perturbation in the expression or the timing of expression of any of these neuropeptides could be effective in compromising the ecdysis process, the central role of the ETH receptor is attractive as a pest control target.

Future Developments

The identification of molecular/physiological targets for insect pest control is an ongoing process. Drawing on examples of mammalian targets for drug action may facilitate this process. If every neuropeptide gene has unique sequences that control specific neuronal cell and/or timing of expression, then specific transcription factors may regulate expression.Targetting specific transcription factors for neuropeptide genes may allow selectivity in gene silencing.

The ability to mutate/knockout gene expression in the fruitfly has advanced our understanding of gene expression. However, as noted above, neuropeptide physiology in *Drosophila* may not be representative of important pest insects. Gene expression inhibition in "pest insects" has recently been achieved. Through a technique known as RNA interference (RNAi) targetted genes can be reduced in expression. Inhibiting the expression of a protein target may pinpoint species-specific targets for pest control or may reduce proteins that confer resistance to insecticide action.

Methods of delivery of neuropeptides and neuropeptide genes into an insect have also been an area of intense research. Finally, introducing foreign substances into an insect will be met by the insect immune response. We have a limited understanding of the insect immune response and recent evidence indicates that neuropeptides may play a role in this response.

Targeting Transcription Factors

Transcription of a specific insect neuropeptide gene is activated or repressed in specific sets of neurons in the nervous system. Regulation may be in response to developmental or environmental cues. Transcription factors are typically proteins composed of a DNA binding domain responsible for specific contacts with DNA bases and an effecter domain (ED) that mediates activation or repression of targeted genes. Specific transcription factors recognize DNA elements (enhancers) in the promoter/5' flanking regions of neuropeptide genes and direct expression in individual neurons. In *Drosophila,* a single enhancer sequence directs the expression in only one of the 17 neurons that express the FMRFamide gene. Another enhancer in the 5' flanking region of the gene directs expression in a subset of six neurons.[154] These enhancer sequences are recognized by specific transcription factors. The *Drosophila* genes for Dromyosuppressin, FMRFamide-related and leucokinins appear to require a helix-loop-helix transcription factor known as DIMM for activation. Similarly, ETH, a peptide made in endocrine Inka cells, that controls a cascade of neuropeptides involved in ecdysis, requires a basic leucine zipper DNA binding protein, cryptocephal (crc).[155] Mutants of crc cause pleiotropic defects in ecdysone which in turn perturbs events through molting and metamorphosis.[156] With insect genomes, mining of transcription factors may lead to the discovery of neuropeptide gene specific transcription factors. In vertebrate systems, understanding transcription factor selectivity has led the creation of designer transcription factors[157,158] that serve as drug/therapeutic targets to up or down regulate genes of interest. In future, targeting of transcription factors for neuropeptide genes, neuropeptide receptors and/or genes activated via specific neuropeptide action as a means of pest control may be possible.

Inhibiting Gene Expression through RNA Interference

Genetics is a powerful tool available to researchers studying organisms such as *Drosophila* or *C. elegans*. The ability to mutate or "knockout" a gene has vastly accelerated our knowledge of development and biochemical pathways. RNAi, a technique that was discovered using *C. elegans* allows specific post-transcriptional silencing of genes.[159] The technique involves introduction of a gene-specific double-stranded (ds)RNA. dsRNA is recognized and processed, by an RNAse III enzyme known as Dicer into 21 to 23 nucleotide duplexes (siRNAs). A protein complex (RNA induced silencing complex) unwinds the siRNA duplex and becomes activated with single-stranded siRNA that guides the RISC complex to it complementary target RNA. The RISC complex then catalyzes the endonucleolytic cleavage of the mRNA. The net result is a reduction in specific mRNA levels and hence reduction in the gene product.

RNA interference has been shown to cause targeted disruption in a wide range of insects. Typically dsRNA is injected into the insect; however feeding appears to be effective with some insects.[160,161] Only a few examples exist for RNAi inhibition of neuropeptide genes. In *Drosophila*, dsRNA for CCAP was targeted to CCAP neurons and reduced expression of CCAP. The reduction was not sufficient to compromise the CCAP effects on ecdysis or heartbeat.[162] Similarly, dsRNA targeting the allatostatin gene in *Blattella germanica* females resulted in a reduction in mRNA and AST peptide (up to 70-80%) with no increase in JH biosynthesis for all age groups tested.[163] This contrasts with dsRNA targeting of the allatostatin gene in *Gryllus*. In crickets, JH titer increased with AST RNAi, however, the egg and testes development was reduced. This is opposite to what would be expected for the action of JH. Injection of the cockroach-type AST dsRNA into the moth *Spodoptera frugiperda* also elevated JH titers. This is again unexpected as the cockroach-type of AST doesn't affect JH biosynthesis in moths.[164] These results may address how there is still a lack of understanding as to how the allatostatin peptides function. Caution in interpretations should be noted due to a prevalence of off-target effects.[165]

In *B. mori*, PBAN interacts with its cognate receptor to stimulate an influx of extracellular calcium that activates an enzymatic cascade, which in turn regulates the terminal reductive step in pheromone production. Reducing the PBAN GPCR by approximately 75% accurately resulted in loss of lipase activity that liberates precursors for pheromone biosynthesis.[166] Dissecting the action of neuropeptides through their receptor(s) and receptor inhibition through RNAi may prove a useful route to understanding biosynthetic and signal transduction pathways[167] required for neuropeptide function.

Crossing the Cuticle Barrier

Insect neuropeptides are chemically incapable of penetrating, or being absorbed by, insect cuticle and are susceptible to degradation in the hemolymph or gut. To overcome the penetration obstacle, properties of a snowdrop lectin, *Galanthus nivalis* agglutinin (GNA) have been exploited. GNA orally ingested by insects is resistant to gut proteolytic enzymes and crosses the midgut epithelium barrier into the hemolymph. When tomato moth larvae were fed a fusion protein combining GNA and *Manduca* allatostatin, Mas-AST was stably delivered into the hemolymph. In line with the action of Mas-AST on gut, significant reduction in feeding and larval weight was found.[168] To date this is the only GNA-insect neuropeptide fusion protein that has been tested. Fusions of GNA with spider toxin[169,170] and red scorpion toxin[171] have since proven to be highly toxic to pest insects. The range of insects affected needs to be addressed before consideration for use as a general insecticide.

Crossing the cuticle barrier via viral and transposable element vectors has also been achieved. Baculoviruses are naturally-occurring viruses that infect the insect orders Lepidoptera, Hymenoptera and Diptera. Most baculoviruses that are used as biological control agents are Nucleopolyhedrosis viruses (NPVs) based on the occlusion bodies they produce after infection. These viruses exhibit species-specific infectivity. This makes them excellent candidates for limited host range insecticidal application. Twenty-seven NPV genomes that specifically infect Lepidoptera pests have been sequenced (Genbank), three NPVs specific for the Hymenoptera, *Neodiprion sertifer* (European pine sawfly), *Neodiprion lecontei* (red-headed pine sawfly) and *Neodiprion abietis* (balsam fir sawfly), one that infects the dipteran mosquito *Culex nigripalpus*.[172]

Baculovirus occlusion bodies are ingested by insect larvae and are dissolved by the high pH of their midguts. Sawfly and mosquito NPV infection is limited to gut cells whereas lepidopteran NPV infection can spread to other tissues. Infecting the gut causes the insect to stop feeding which leads to death. The drawback is that the infection process is slow, taking up to two weeks to disable the insect.[173] During this time, the insect continues to feed. With the aim of enhancing insecticidal efficacy, lepidoptera baculovirus genomes have been genetically modified to express either neuropeptides to alter the physiology of the insect or toxins to accelerate death. The first neuropeptide gene introduced into *Bombyx* baculovirus (BmNPV) expressed diuretic hormone that accelerated water loss in larval silkworms. The expression of diuretic hormone offered a promising results as larval death was accelerated by 20% in comparison to unmodified BmNPV infection.[174]

Less successful were the baculovirus driven expression of eclosion hormone or PTTH as no difference in mortality was noted in comparison to unmodified virus. A moth *Pseudeletia unipuncta* FLRFamide cDNA under the control of the polyhedrin promoter produced a significant increase in the levels of RFamide-like material in the blood of infected insects, however, inaccurate processing of the precursor occurred.[175] Inaccurate precursor processing appeared as a problem in the expression of *Helicoverpa* PBAN[176] and *Schistocerca* ion transport peptide.[177]

Although still offering potential, neuropeptide expression enhancement of baculovirus has fallen behind the introduction of insect specific toxin genes such as bacterial *Bacillus thuringiensis* (Bt) endotoxin,[178] spider[179] and scorpion toxins.[180] Recently, an engineered baculovirus contained DNA sequences (the promoter region of the gene) necessary for the distinct neural cell-specific expression pattern of the neuropeptide genes bombyxin and PTTH fused to a reporter gene. The reporter protein was specifically targeted into the correct cells of the nervous system.[181] This ability to direct native expression of the neuropeptide within the insect nervous system may allow for proper processing.

As an alternative to baculovirus, Densoviruses are invertebrate parvoviruses that infect a wider range of insect orders than do baculoviruses. Densoviruses are single-stranded DNA viruses and have been isolated from Lepidoptera, Diptera, Dictyoptera, Odonata, Hemiptera, Homoptera and Orthoptera. Densoviruses can take up to 20 days to kill their host and hence, like Baculovirus, might benefit from gene insertions/modifications through genetic engineering of the viruses to accelerate death.[182,183]

Germ-line transformation of genes into insects is possible through transposable elements such as piggybac, minos and hermes.[184] Germline transformation allows in vivo functional tests of selected gene product expression, insertional mutagenesis and discerning the contribution of cis-regulatory sequences.

Used in the latter context, the control of neuropeptide gene expression would be greatly facilitated. Such studies have currently been limited to developmental processes. Transposable elements should be viewed as a tool that will bring sophistication in manipulating insect genomes just as the transposable P-element has already accomplished in *Drosophila*.[185]

Insect Immunity

As bacteria, entomopathogenic fungi, parasites, parasitoids and viruses all have been proposed as agents of insect control, understanding how neuropeptides may activate or repress the insect immune response may be critical for effective application. In response to invading pathogens, insects have both humoral and cellular defenses. Humoral defense includes an increase in the reactive oxygen and nitrogen intermediates, the proteolytic conversion of prophenoloxidase to phenyloxidase that catalyzes the conversion of monophenols to diphenols and quinones toxic to micro-organisms and regulates coagulation of hemolymph. Later in humoral defense anti-microbial peptide (e.g., cecropins) production is stimulated for release from the fat body.[186] Cellular defense in insects involves entrapment of foreign material by hemocytes. Small foreign materials may be eliminated by phagocytosis, whereas hemocytes will coat invading microbes to form nodules and yet larger particles such as parasitoids will be coated in a process know as encapsulation.

When *Bacillus* bacteria, bacterial lipopolysaccharide (LPS), entomopathogenic fungus or fungal cell wall component laminarin (a β-1,3 glucan) are injected into adult locusts activation of the prophenoloxidase cascade occurs. If the same foreign substances are coinjected with locust AKH 1, the prophenoloxidase cascade is enhanced and prolonged. Nodules are formed in response to LPS and laminarin injection. Coinjection of AKH 1 with LPS or laminarin increases the number of nodules formed. If locusts are injected with AKH 1 when infected with fungus they die more rapidly in comparison to fungal infection alone.[187-189] The mechanism responsible for these enhanced effects on immune components is still unclear. In the cockroach, the neuropeptide allatostatin was shown to be produced by a population of granulated hemocytes that may contribute to the immune response.[190] In the fleshfly, *Neobellieria bullata,* 20-hydroxyecdysone application induces nodulation in response to laminarin whereas juvenile hormone application has the opposite effect[191] again supporting a role for neuropeptide regulation. The participation of neuropeptides in the immune response is not without precedent as neuropeptides are produced in the vertebrates during the immune response.[192] In the rat, vasoactive intestinal peptide (VIP) occurs in the nervous system, in endocrine gut cells and in mast cell.[193] Human peripheral lymphocytes -infected with virus produce endorphin-like material,[194] and mouse eosinophils of granulomas resulting from *Schistosomiasis* infection synthesize substance P.[195]

As more insect genomes are sequenced, genome-wide screens and proteonomic analysis of genes and products involved in immune responses may facilitate how neuropeptides may be participating. Such screens have been made in *Drosophila* and evidence suggests that genes induced during the anti-bacterial immune response differ from genes that are induced to fight a parasitoid attack.[196-198] A similar screen of hemocyte and fat body transcripts was made in the moth, *Spodoptera*, in response to polydna virus infection.[199]

Conclusion

Currently, no individual insect neuropeptide can serve as a panacea in pest control. Progress toward neuropeptide-based pest control is hampered by the large number of insects and our limited understanding of the diverse physiology that each displays. Investigators are recognizing that molecular and physiology-based approaches need to merge to understand all the factors that are involved in neuropeptide synthesis, release and physiological action. Most neuropeptide receptors have now been identified in *Drosophila*; however, few orthologs have yet been identified in pest insects of economic importance to the agricultural and forestry sectors. *In vivo* signal transduction pathways used by an individual neuropeptide upon receptor binding are still largely unknown. This may be important as certain receptors (such as FaRPs) appear to be promiscuous in their binding. As genome sequences of pest insects become available, further unique protein targets will be identified for future pest control purposes.

Acknowledgements

Work is supported by the Natural Sciences and Engineering Research Council of Canada (NSERC). I am grateful to Dr. Ian Orchard, Dr. Angela Lange and Holly Mitchell for helpful comments.

References

1. Wetzel S, Duchesne LC, Laporte MF eds. Bioproducts from Canadian Forests: New Partnerships in the Bioeconomy. USA: Springer; 2006.
2. World malaria report 2005. Available at: http://www.who.int/malaria/includes_en/whomalariapublications19982004.htm#2006
3. Chalubinski M, Kowalski ML. Endocrine disrupters—potential modulators of the immune system and allergic response. Allergy 2006; 61:1326-35.
4. Ishaaya I, Kontsedalov S, Horowitz AR. Biorational insecticides: Mechanism and cross-resistance. Arch Insect Biochem Physiol 2005; 58:192-9.
5. Lynd A, Ranson H, McCall PJ et al. A simplified high-throughput method for pyrethroid knock-down resistance (kdr) detection in Anopheles gambiae. Malar J 2005; 4:16.
6. David JP, Strode C, Vontas J et al. The Anopheles gambiae detoxification chip: A highly specific micro-array to study metabolic-based insecticide resistance in malaria vectors. Proc Natl Acad Sci USA 2005; 102:4080-4.
7. Raymond-Delpech V, Matsuda K, Sattelle BM et al. Ion channels: Molecular targets of neuroactive insecticides. Invert Neurosci 2005; 5:119-33.
8. Mandavilli A. Health agency backs use of DDT against malaria. Nature 2006; 443:250-1.
9. Guinovart C, Navia MM, Tanner M et al. Malaria: Burden of disease. Curr Mol Med 2006; 6:137-40.

10. Dhadialla TS, Carlson GR, Le DP. New insecticides with ecdysteroidal and juvenile hormone activity. Annu Rev Entomol 1998; 43:545-69.
11. Ishaaya I. Introduction: Biorational insecticides-mechanism and application. Arch Insect Biochem Physiol 2003; 54:144.
12. Quistad GB, Adams ME, Scarborough RM et al. Metabolism of proctolin, a pentapeptide neurotransmitter in insects. Life Sci 1984; 34:569-76.
13. Gade G, Goldsworthy GJ. Insect peptide hormones: A selective review of their physiology and potential application for pest control. Pest Manag Sci 2003; 59:1063-75.
14. Nassel DR, Homberg U. Neuropeptides in interneurons of the insect brain. Cell Tissue Res 2006; 326:1-24.
15. Nassel DR. Neuropeptides in the nervous system of Drosophila and other insects: Multiple roles as neuromodulators and neurohormones. Prog Neurobiol 2002; 68:1-84.
16. Kastin A ed. The Handbook of Biologically Active Peptides. 1st ed. Elsevier; 2006.
17. Taghert PH, Veenstra JA. Drosophila neuropeptide signaling. Adv Genet 2003; 49:1-65.
18. Stay B, Tobe SS. The role of allatostatins in juvenile hormone synthesis in insects and crustaceans. Annu Rev Entomol 2007; 52:277-99.
19. Bendena WG, Donly BC, Tobe SS. Allatostatins: A growing family of neuropeptides with structural and functional diversity. Ann N Y Acad Sci 1999; 897:311-29.
20. Predel R, Wegener C. Biology of the CAPA peptides in insects. Cell Mol Life Sci 2006:1-14.
21. Isaac RE, Taylor CA, Hamasaka Y et al. Proctolin in the post-genomic era: New insights and challenges. Invert Neurosci 2004; 5:51-64.
22. Sokolowski MB. NPY and the regulation of behavioral development. Neuron 2003; 39:6-8.
23. Elekonich MM, Horodyski FM. Insect allatotropins belong to a family of structurally-related myoactive peptides present in several invertebrate phyla. Peptides 2003; 24:1623-32.
24. Wu Q, Brown MR. Signaling and function of insulin-like peptides in insects. Annu Rev Entomol 2006; 51:1-24.
25. Dow JA, Davies SA. The malpighian tubule: Rapid insights from post-genomic biology. J Insect Physiol 2006; 52:365-78.
26. Gade G. Regulation of intermediary metabolism and water balance of insects by neuropeptides. Annu Rev Entomol 2004; 49:93-113.
27. Coast GM, Orchard I, Phillips JE et al. Insect diuretic and antidiuretic hormones. Adv Insect Physiol 2002; 29:279-409.
28. Coast GM, Garside CS. Neuropeptide control of fluid balance in insects. Ann NY Acad Sci 2005; 1040:1-8.
29. Truman JW. Hormonal control of insect ecdysis: Endocrine cascades for coordinating behavior with physiology. Vitam Horm 2005; 73:1-30.
30. Ewer J. Behavioral actions of neuropeptides in invertebrates: Insights from Drosophila. Horm Behav 2005; 48:418-29.
31. Simonet G, Poels J, Claeys I et al. Neuroendocrinological and molecular aspects of insect reproduction. J Neuroendocrinol 2004; 16:649-59.
32. Veenstra JA. Mono- and dibasic proteolytic cleavage sites in insect neuroendocrine peptide precursors. Arch Insect Biochem Physiol 2000; 43:49-63.
33. Baggerman G, Liu F, Wets G et al. Bioinformatic analysis of peptide precursor proteins. Ann NY Acad Sci 2005; 1040:59-65.
34. Amare A, Hummon AB, Southey BR et al. Bridging neuropeptidomics and genomics with bioinformatics: Prediction of mammalian neuropeptide prohormone processing. J Proteome Res 2006; 5:1162-7.
35. Kolhekar AS, Roberts MS, Jiang N et al. Neuropeptide amidation in Drosophila: Separate genes encode the two enzymes catalyzing amidation. J Neurosci 1997; 17:1363-76.
36. Jiang N, Kolhekar AS, Jacobs PS et al. PHM is required for normal developmental transitions and for biosynthesis of secretory peptides in Drosophila. Dev Biol 2000; 226:118-36.
37. Sossin WS, Fisher JM, Scheller RH. Cellular and molecular biology of neuropeptide processing and packaging. Neuron 1989; 2:1407-17.
38. Sieburth D, Ch'ng Q, Dybbs M et al. Systematic analysis of genes required for synapse structure and function. Nature 2005; 436:510-7.
39. Hewes RS, Taghert PH. Neuropeptides and neuropeptide receptors in the Drosophila melanogaster genome. Genome Res 2001; 11:1126-42.
40. Vanden Broeck J. Neuropeptides and their precursors in the fruitfly, Drosophila melanogaster. Peptides 2001; 22:241-54.
41. Hauser F, Cazzamali G, Williamson M et al. A review of neurohormone GPCRs present in the fruitfly Drosophila melanogaster and the honey bee apis mellifera. Prog Neurobiol 2006; 80:1-19.

42. Claeys I, Poels J, Simonet G et al. Insect neuropeptide and peptide hormone receptors: Current knowledge and future directions. Vitam Horm 2005; 73:217-82.
43. Yamanaka N, Zitnan D, Kim YJ et al. Regulation of insect steroid hormone biosynthesis by innervating peptidergic neurons. Proc Natl Acad Sci USA 2006; 103:8622-7.
44. Wang Z, Lange AB, Orchard I. Coupling of a single receptor to two different G proteins in the signal transduction of FMRFamide-related peptides. Biochem Biophys Res Commun 1995; 212:531-8.
45. Satake S, Masumura M, Ishizaki H et al. Bombyxin, an insulin-related peptide of insects, reduces the major storage carbohydrates in the silkworm Bombyx mori. Comp Biochem Physiol B Biochem Mol Biol 1997; 118:349-57.
46. Sharrer B. Neuroendocrine physiology of insects. Pflugers Arch 1952; 255:154-63.
47. Scharrer B. The neurosecretory neuron in neuroendocrine regulatory mechanisms. Am Zool 1967; 7:161-9.
48. Clark L, Zhang JR, Tobe S et al. Proctolin: A possible releasing factor in the corpus cardiacum/corpus allatum of the locust. Peptides 2006; 27:559-66.
49. Clark L, Agricola HJ, Lange AB. Proctolin-like immunoreactivity in the central and peripheral nervous systems of the locust, Locusta migratoria. Peptides 2006; 27:549-58.
50. Veelaert D, Passier P, Devreese B et al. Isolation and characterization of an adipokinetic hormone release-inducing factor in locusts: The crustacean cardioactive peptide. Endocrinology 1997; 138:138-42.
51. Nassel DR, Passier PC, Elekes K et al. Evidence that Locustatachykinin I is involved in release of adipokinetic hormone from locust corpora cardiaca. Regul Pept 1995; 57:297-310.
52. Nassel DR, Vullings HG, Passier PC et al. Several isoforms of Locustatachykinins may be involved in cyclic AMP-mediated release of adipokinetic hormones from the locust corpora cardiaca. Gen Comp Endocrinol 1999; 113:401-12.
53. Vullings HG, Ten Voorde SE, Passier PC et al. A possible role of SchistoFLRFamide in inhibition of adipokinetic hormone release from locust corpora cardiaca. J Neurocytol 1998; 27:901-13.
54. Van der Horst DJ, Van Marrewijk WJ, Diederen JH. Adipokinetic hormones of insect: Release, signal transduction and responses. Int Rev Cytol 2001; 211:179-240.
55. Gade G, Auerswald L. Mode of action of neuropeptides from the adipokinetic hormone family. Gen Comp Endocrinol 2003; 132:10-20.
56. Patel RT, Soulages JL, Arrese EL. Adipokinetic hormone-induced mobilization of fat body triglyceride stores in Manduca sexta: Role of TG-lipase and lipid droplets. Arch Insect Biochem Physiol 2006; 63:73-81.
57. Isabel G, Martin JR, Chidami S et al. AKH-producing neuroendocrine cell ablation decreases trehalose and induces behavioral changes in Drosophila. Am J Physiol Regul Integr Comp Physiol 2005; 288:R531-8.
58. Lee G, Park JH. Hemolymph sugar homeostasis and starvation-induced hyperactivity affected by genetic manipulations of the adipokinetic hormone-encoding gene in Drosophila melanogaster. Genetics 2004; 167:311-23.
59. Kaufmann C, Briegel H. Flight performance of the malaria vectors Anopheles gambiae and Anopheles atroparvus. J Vector Ecol 2004; 29:140-53.
60. Oguri E, Steele JE. Lipid metabolism in the cockroach, Periplaneta americana, is activated by the hypertrehalosemic peptide, HTH-I. Peptides 2003; 24:1545-51.
61. Kodrik D, Socha R, Simek P et al. A new member of the AKH/RPCH family that stimulates locomotory activity in the firebug, Pyrrhocoris apterus (Heteroptera). Insect Biochem Mol Biol 2000; 30:489-98.
62. Milde JJ, Ziegler R, Wallstein M. Adipokinetic hormone stimulates neurones in the insect central nervous system. J Exp Biol 1995; 198:1307-11.
63. Nassel DR. Neuropeptides, amines and amino acids in an elementary insect ganglion: Functional and chemical anatomy of the unfused abdominal ganglion. Prog Neurobiol 1996; 48:325-420.
64. Roeder T. Tyramine and octopamine: Ruling behavior and metabolism. Annu Rev Entomol 2005; 50:447-77.
65. Wicher D, Agricola HJ, Sohler S et al. Differential receptor activation by cockroach adipokinetic hormones produces differential effects on ion currents, neuronal activity and locomotion. J Neurophysiol 2006; 95:2314-25.
66. Hansen KK, Hauser F, Cazzamali G et al. Cloning and characterization of the adipokinetic hormone receptor from the cockroach Periplaneta americana. Biochem Biophys Res Commun 2006; 343:638-43.
67. Staubli F, Jorgensen TJ, Cazzamali G et al. Molecular identification of the insect adipokinetic hormone receptors. Proc Natl Acad Sci USA 2002; 99:3446-51.
68. Park Y, Kim YJ, Adams ME. Identification of G protein-coupled receptors for Drosophila PRXamide peptides, CCAP, corazonin and AKH supports a theory of ligand-receptor coevolution. Proc Natl Acad Sci USA 2002; 99:11423-8.

69. Kaufmann C, Brown MR. Adipokinetic hormones in the african malaria mosquito, Anopheles gambiae: Identification and expression of genes for two peptides and a putative receptor. Insect Biochem Mol Biol 2006; 36:466-81.
70. Orchard I, Lange AB, Bendena WG. FMRFamide-related peptides: A multifunctional family of structurally related neuropeptides in insects. Advances in insect physiology London: Academic Press, 2001; 28:267-329.
71. Meeusen T, Mertens I, Clynen E et al. Identification in Drosophila melanogaster of the invertebrate G protein-coupled FMRFamide receptor. Proc Natl Acad Sci USA 2002; 99:15363-8.
72. Cazzamali G, Grimmelikhuijzen CJ. Molecular cloning and functional expression of the first insect FMRFamide receptor. Proc Natl Acad Sci USA 2002; 99:12073-8.
73. Johnson E, Ringo J, Dowse H. Native and heterologous neuropeptides are cardioactive in Drosophila melanogaster. J Insect Physiol 2000; 46:1229-36.
74. Merte J, Nichols R. Drosophila melanogaster FMRFamide-containing peptides: Redundant or diverse functions? Peptides 2002; 23:209-20.
75. Price MD, Merte J, Nichols R et al. Drosophila melanogaster flatline encodes a myotropin orthologue to Manduca sexta allatostatin. Peptides 2002; 23:787-94.
76. Hewes RS, Snowdeal EC, 3rd, Saitoe M et al. Functional redundancy of FMRFamide-related peptides at the Drosophila larval neuromuscular junction. J Neurosci 1998; 18:7138-51.
77. Robb S, Packman LC, Evans PD. Isolation, primary structure and bioactivity of schistoflrf-amide, a FMRF-amide-like neuropeptide from the locust, Schistocerca gregaria. Biochem Biophys Res Commun 1989; 160:850-6.
78. Peeff NM, Orchard I, Lange AB. Isolation, sequence and bioactivity of PDVDHVFLRFamide and ADVGHVFLRFamide peptides from the locust central nervous system. Peptides 1994; 15:387-92.
79. Lange AB, Cheung IL. The modulation of skeletal muscle contraction by FMRFamide-related peptides of the locust. Peptides 1999; 20:1411-8.
80. Lange AB, Peeff NM, Orchard I. Isolation, sequence and bioactivity of FMRFamide-related peptides from the locust ventral nerve cord. Peptides 1994; 15:1089-94.
81. Yamamoto D, Ishikawa S, Holman GM et al. Leucomyosuppressin, a novel insect neuropeptide, inhibits evoked transmitter release at the mealworm neuromuscular junction. Neurosci Lett 1988; 95:137-42.
82. Kingan TG, Teplow DB, Phillips JM et al. A new peptide in the FMRFamide family isolated from the CNS of the hawkmoth, Manduca sexta. Peptides 1990; 11:849-56.
83. Egerod K, Reynisson E, Hauser F et al. Molecular cloning and functional expression of the first two specific insect myosuppressin receptors. Proc Natl Acad Sci USA 2003; 100:9808-13.
84. Scholler S, Belmont M, Cazzamali G et al. Molecular identification of a myosuppressin receptor from the malaria mosquito Anopheles gambiae. Biochem Biophys Res Commun 2005; 327:29-34.
85. Lange AB, Orchard I, Wang Z et al. A nonpeptide agonist of the invertebrate receptor for SchistoFLRFamide (PDVDHVFLRFamide), a member of a subfamily of insect FMRFamide-related peptides. Proc Natl Acad Sci USA 1995; 92:9250-3.
86. Nachman RJ, Olender EH, Roberts VA et al. A nonpeptidal peptidomimetic agonist of the insect FLRFamide myosuppressin family. Peptides 1996; 17:313-20.
87. Yip KW, Mao X, Au PY et al. Benzethonium chloride: A novel anticancer agent identified by using a cell-based small-molecule screen. Clin Cancer Res 2006; 12:5557-69.
88. Takeoka G, Dao L, Wong RY et al. Identification of benzethonium chloride in commercial grapefruit seed extracts. J Agric Food Chem 2001; 49:3316-20.
89. Stangier J, Hilbich C, Beyreuther K et al. Unusual cardioactive peptide (CCAP) from pericardial organs of the shore crab carcinus maenas. Proc Natl Acad Sci USA 1987; 84:575-9.
90. Tublitz N, Brink D, Broadie KS et al. From behavior to molecules: An integrated approach to the study of neuropeptides. Trends Neurosci 1991; 14:254-9.
91. Cheung CC, Loi PK, Sylwester AW et al. Primary structure of a cardioactive neuropeptide from the tobacco hawkmoth, Manduca sexta. FEBS Lett 1992; 313:165-8.
92. Stangier J, Hilbich C, Keller J. Occurance of crustacean cardioactive peptide (CCAP) in the nervous-system of an insect, Locusta migratoria. J Comp Physiol [B] 1989; 159:5-11.
93. Donini A, Agricola H, Lange AB. Crustacean cardioactive peptide is a modulator of oviduct contractions in Locusta migratoria. J Insect Physiol 2001; 47:277-85.
94. Donini A, Ngo C, Lange AB. Evidence for crustacean cardioactive peptide-like innervation of the gut in Locusta migratoria. Peptides 2002; 23:1915-23.
95. Veenstra JA. Isolation and structure of corazonin, a cardioactive peptide from the american cockroach. FEBS Lett 1989; 250:231-4.
96. Hua Y, Ishibashi J, Saito H et al. Identification of [Arg^7]corazonin in the silkworm, Bombyx mori and the cricket, Gryllus bimaculatus, as a factor inducing dark color in an albino strain of the locust, Locusta migratoria. J Insect Physiol 2000; 46:853-60.

97. Tanaka S. Endocrine mechanisms controlling body-color polymorphism in locusts. Arch Insect Biochem Physiol 2001; 47:139-49.
98. Shiga S, Davis NT, Hildebrand JG. Role of neurosecretory cells in the photoperiodic induction of pupal diapause of the tobacco hornworm Manduca sexta. J Comp Neurol 2003; 462:275-85.
99. Kim YJ, Zitnan D, Cho KH et al. Central peptidergic ensembles associated with organization of an innate behavior. Proc Natl Acad Sci USA 2006; 103:14211-6.
100. Cazzamali G, Saxild N, Grimmelikhuijzen C. Molecular cloning and functional expression of a Drosophila corazonin receptor. Biochem Biophys Res Commun 2002; 298:31-6.
101. Belmont M, Cazzamali G, Williamson M et al. Identification of four evolutionarily related G protein-coupled receptors from the malaria mosquito Anopheles gambiae. Biochem Biophys Res Commun 2006; 344:160-5.
102. Kim YJ, Spalovska-Valachova I, Cho KH et al. Corazonin receptor signaling in ecdysis initiation. Proc Natl Acad Sci USA 2004; 101:6704-9.
103. Rosay P, Davies SA, Yu Y et al. Cell-type specific calcium signalling in a Drosophila epithelium. J Cell Sci 1997; 110(Pt 15):1683-92.
104. MacPherson MR, Pollock VP, Broderick KE et al. Model organisms: New insights into ion channel and transporter function. L-type calcium channels regulate epithelial fluid transport in Drosophila melanogaster. Am J Physiol Cell Physiol 2001; 280:C394-407.
105. Pollock VP, McGettigan J, Cabrero P et al. Conservation of capa peptide-induced nitric oxide signalling in diptera. J Exp Biol 2004; 207:4135-45.
106. Quinlan MC, Tublitz NJ, O'Donnell MJ. Anti-diuresis in the blood-feeding insect Rhodnius prolixus stal: The peptide CAP(2b) and cyclic GMP inhibit malpighian tubule fluid secretion. J Exp Biol 1997; 200:2363-7.
107. Paluzzi JP, Orchard I. Distribution, activity and evidence for the release of an anti-diuretic peptide in the kissing bug Rhodnius prolixus. J Exp Biol 2006; 209:907-15.
108. Radford JC, Davies SA, Dow JA. Systematic G-protein-coupled receptor analysis in Drosophila melanogaster identifies a leucokinin receptor with novel roles. J Biol Chem 2002; 277:38810-7.
109. Kaufmann N, Mathai JC, Hill WG et al. Developmental expression and biophysical characterization of a Drosophila melanogaster aquaporin. Am J Physiol Cell Physiol 2005; 289:C397-407.
110. Eigenheer RA, Nicolson SW, Schegg KM et al. Identification of a potent antidiuretic factor acting on beetle malpighian tubules. Proc Natl Acad Sci USA 2002; 99:84-9.
111. King DS, Meredith J, Wang YJ et al. Biological actions of synthetic locust ion transport peptide (ITP). Insect Biochem Mol Biol 1999; 29:11-8.
112. Meredith J, Ring M, Macins A et al. Locust ion transport peptide (ITP): Primary structure, cDNA and expression in a baculovirus system. J Exp Biol 1996; 199:1053-61.
113. Wei Z, Baggerman G, J Nachman R et al. Sulfakinins reduce food intake in the desert locust, Schistocerca gregaria. J Insect Physiol 2000; 46:1259-65.
114. Maestro JL, Aguilar R, Pascual N et al. Screening of antifeedant activity in brain extracts led to the identification of sulfakinin as a satiety promoter in the german cockroach are arthropod sulfakinins homologous to vertebrate gastrins-cholecystokinins? Eur J Biochem 2001; 268:5824-30.
115. Nachman RJ, Vercammen T, Williams H et al. Aliphatic amino diacid asu functions as an effective mimic of tyr(SO3H) in sulfakinins for myotropic and food intake-inhibition activity in insects. Peptides 2005; 26:115-20.
116. Kubiak TM, Larsen MJ, Burton KJ et al. Cloning and functional expression of the first Drosophila melanogaster sulfakinin receptor DSK-R1. Biochem Biophys Res Commun 2002; 291:313-20.
117. Fuse M, Zhang JR, Partridge E et al. Effects of an allatostatin and a myosuppressin on midgut carbohydrate enzyme activity in the cockroach Diploptera punctata. Peptides 1999; 20:1285-93.
118. Brown MR, Crim JW, Arata RC et al. Identification of a Drosophila brain-gut peptide related to the neuropeptide Y family. Peptides 1999; 20:1035-42.
119. Stanek DM, Pohl J, Crim JW et al. Neuropeptide F and its expression in the yellow fever mosquito, Aedes aegypti. Peptides 2002; 23:1367-78.
120. Riehle MA, Garczynski SF, Crim JW et al. Neuropeptides and peptide hormones in Anopheles gambiae. Science 2002; 298:172-5.
121. Pedrazzini T. Importance of NPY Y1 receptor-mediated pathways: Assessment using NPY Y1 receptor knockouts. Neuropeptides 2004; 38:267-75.
122. Garczynski SF, Brown MR, Shen P et al. Characterization of a functional neuropeptide F receptor from Drosophila melanogaster. Peptides 2002; 23:773-80.
123. Garczynski SF, Crim JW, Brown MR. Characterization of neuropeptide F and its receptor from the african malaria mosquito, Anopheles gambiae. Peptides 2005; 26:99-107.
124. Shen P, Cai HN. Drosophila neuropeptide F mediates integration of chemosensory stimulation and conditioning of the nervous system by food. J Neurobiol 2001; 47:16-25.

125. Wu Q, Wen T, Lee G et al. Developmental control of foraging and social behavior by the Drosophila neuropeptide Y-like system. Neuron 2003; 39:147-61.
126. Wu Q, Zhao Z, Shen P. Regulation of aversion to noxious food by Drosophila neuropeptide Y- and insulin-like systems. Nat Neurosci 2005; 8:1350-5.
127. Veenstra JA. Isolation and identification of three RFamide-immunoreactive peptides from the mosquito Aedes aegypti. Peptides 1999; 20:31-8.
128. Huang Y, Brown MR, Lee TD et al. RF-amide peptides isolated from the midgut of the corn earworm, helicoverpa zea, resemble pancreatic polypeptide. Insect Biochem Mol Biol 1998; 28:345-56.
129. Spittaels K, Verhaert P, Shaw C et al. Insect neuropeptide F (NPF)-related peptides: Isolation from colorado potato beetle (Leptinotarsa decemlineata) brain. Insect Biochem Mol Biol 1996; 26:375-82.
130. Schoofs L, Clynen E, Cerstiaens A et al. Newly discovered functions for some myotropic neuropeptides in locusts. Peptides 2001; 22:219-27.
131. Mertens I, Meeusen T, Huybrechts R et al. Characterization of the short neuropeptide F receptor from Drosophila melanogaster. Biochem Biophys Res Commun 2002; 297:1140-8.
132. Garczynski SF, Brown MR, Crim JW. Structural studies of Drosophila short neuropeptide F: Occurrence and receptor binding activity. Peptides 2006; 27:575-82.
133. Lee KS, You KH, Choo JK et al. Drosophila short neuropeptide F regulates food intake and body size. J Biol Chem 2004; 279:50781-9.
134. Williams CM. Third-generation pesticides. Sci Am 1967; 217:13-7.
135. Truman JW. Steroid hormone secretion in insects comes of age. Proc Natl Acad Sci USA 2006; 103:8909-10.
136. Hua YJ, Tanaka Y, Nakamura K et al. Identification of a prothoracicostatic peptide in the larval brain of the silkworm, Bombyx mori. J Biol Chem 1999; 274:31169-73.
137. Mirth C, Truman JW, Riddiford LM. The role of the prothoracic gland in determining critical weight for metamorphosis in Drosophila melanogaster. Curr Biol 2005; 15:1796-807.
138. Woodhead AP, Stay B, Seidel SL et al. Primary structure of four allatostatins: Neuropeptide inhibitors of juvenile hormone synthesis. Proc Natl Acad Sci USA 1989; 86:5997-6001.
139. Pratt GE, Farnsworth DE, Siegel NR et al. Identification of an allatostatin from adult Diploptera punctata. Biochem Biophys Res Commun 1989; 163:1243-7.
140. Lorenz MW, Kellner R, Hoffmann KH. A family of neuropeptides that inhibit juvenile hormone biosynthesis in the cricket, Gryllus bimaculatus. J Biol Chem 1995; 270:21103-8.
141. Jansons IS, Cusson M, McNeil JN et al. Molecular characterization of a cDNA from Pseudaletia unipuncta encoding the Manduca sexta allatostatin peptide (Mas-AST). Insect Biochem Mol Biol 1996; 26:767-73.
142. Kramer SJ, Toschi A, Miller CA et al. Identification of an allatostatin from the tobacco hornworm Manduca sexta. Proc Natl Acad Sci USA 1991; 88:9458-62.
143. Li Y, Hernandez-Martinez S, Fernandez F et al. Biochemical, molecular and functional characterization of PISCF-allatostatin, a regulator of juvenile hormone biosynthesis in the mosquito Aedes aegypti. J Biol Chem 2006; 281:34048-55.
144. Li Y, Hernandez-Martinez S, Noriega FG. Inhibition of juvenile hormone biosynthesis in mosquitoes: Effect of allatostatic head factors, PISCF- and YXFGL-amide-allatostatins. Regul Pept 2004; 118:175-82.
145. Kataoka H, Toshi A, Li JP et al. Identification of an allatotropin from adult Manduca sexta. Science 1989; 243:1481-3.
146. Taylor PA 3rd, Bhatt TR, Horodyski FM. Molecular characterization and expression analysis of Manduca sexta allatotropin. Eur J Biochem 1996; 239:588-96.
147. Truesdell PF, Koladich PM, Kataoka H et al. Molecular characterization of a cDNA from the true armyworm Pseudaletia unipuncta encoding Manduca sexta allatotropin peptide(1). Insect Biochem Mol Biol 2000; 30:691-702.
148. Tu MP, Yin CM, Tatar M. Mutations in insulin signaling pathway alter juvenile hormone synthesis in Drosophila melanogaster. Gen Comp Endocrinol 2005; 142:347-56.
149. Piulachs MD, Vilaplana L, Bartolome JM et al. Ketomethylene and methyleneamino pseudopeptide analogues of insect allatostatins inhibit juvenile hormone and vitellogenin production in the cockroach Blattella germanica. Insect Biochem Mol Biol 1997; 27:851-8.
150. Garside CS, Nachman RJ, Tobe SS. Injection of dip-allatostatin or dip-allatostatin pseudopeptides into mated female Diploptera punctata inhibits endogenous rates of JH biosynthesis and basal oocyte growth. Insect Biochem Mol Biol 2000; 30:703-10.
151. Garside CS, Hayes TK, Tobe SS. Degradation of dip-allatostatins by hemolymph from the cockroach, Diploptera punctata. Peptides 1997; 18:17-25.

152. Abernathy RL, Teal PE, Meredith JA et al. Induction of pheromone production in a moth by topical application of a pseudopeptide mimic of a pheromonotropic neuropeptide. Proc Natl Acad Sci USA 1996; 93:12621-5.
153. Nachman RJ, Teal PE, Strey A. Enhanced oral availability/pheromonotropic activity of peptidase-resistant topical amphiphilic analogs of pyrokinin/PBAN insect neuropeptides. Peptides 2002; 23:2035-43.
154. Benveniste RJ, Taghert PH. Cell type-specific regulatory sequences control expression of the Drosophila FMRF-NH2 neuropeptide gene. J Neurobiol 1999; 38:507-20.
155. Gauthier SA, Hewes RS. Transcriptional regulation of neuropeptide and peptide hormone expression by the Drosophila dimmed and cryptocephal genes. J Exp Biol 2006; 209:1803-15.
156. Hewes RS, Schaefer AM, Taghert PH. The cryptocephal gene (ATF4) encodes multiple basic-leucine zipper proteins controlling molting and metamorphosis in Drosophila. Genetics 2000; 155:1711-23.
157. Blancafort P, Segal DJ, Barbas CF 3rd. Designing transcription factor architectures for drug discovery. Mol Pharmacol 2004; 66:1361-71.
158. Urnov FD, Rebar EJ. Designed transcription factors as tools for therapeutics and functional genomics. Biochem Pharmacol 2002; 64:919-23.
159. Fire A, Xu S, Montgomery MK et al. Potent and specific genetic interference by double-stranded RNA in Caenorhabditis elegans. Nature 1998; 391:806-11.
160. Turner CT, Davy MW, MacDiarmid RM et al. RNA interference in the light brown apple moth, epiphyas postvittana (walker) induced by double-stranded RNA feeding. Insect Mol Biol 2006; 15:383-91.
161. Araujo RN, Santos A, Pinto FS et al. RNA interference of the salivary gland nitrophorin 2 in the triatomine bug Rhodnius prolixus (Hemiptera: Reduviidae) by dsRNA ingestion or injection. Insect Biochem Mol Biol 2006; 36:683-93.
162. Dulcis D, Levine RB, Ewer J. Role of the neuropeptide CCAP in Drosophila cardiac function. J Neurobiol 2005; 64:259-74.
163. Maestro JL, Belles X. Silencing allatostatin expression using double-stranded RNA targeted to preproallatostatin mRNA in the german cockroach. Arch Insect Biochem Physiol 2006; 62:73-9.
164. Meyering-Vos M, Merz S, Sertkol M et al. Functional analysis of the allatostatin-A type gene in the cricket Gryllus bimaculatus and the armyworm Spodoptera frugiperda. Insect Biochem Mol Biol 2006; 36:492-504.
165. Ma Y, Creanga A, Lum L et al. Prevalence of off-target effects in Drosophila RNA interference screens. Nature 2006; 443:359-63.
166. Ohnishi A, Hull JJ, Matsumoto S. Targeted disruption of genes in the Bombyx mori sex pheromone biosynthetic pathway. Proc Natl Acad Sci USA 2006; 103:4398-403.
167. Nybakken K, Vokes SA, Lin TY et al. A genome-wide RNA interference screen in Drosophila melanogaster cells for new components of the hh signaling pathway. Nat Genet 2005; 37:1323-32.
168. Fitches E, Audsley N, Gatehouse JA et al. Fusion proteins containing neuropeptides as novel insect contol agents: Snowdrop lectin delivers fused allatostatin to insect haemolymph following oral ingestion. Insect Biochem Mol Biol 2002; 32:1653-61.
169. Fitches E, Edwards MG, Mee C et al. Fusion proteins containing insect-specific toxins as pest control agents: Snowdrop lectin delivers fused insecticidal spider venom toxin to insect haemolymph following oral ingestion. J Insect Physiol 2004; 50:61-71.
170. Down RE, Fitches EC, Wiles DP et al. Insecticidal spider venom toxin fused to snowdrop lectin is toxic to the peach-potato aphid, myzus persicae (Hemiptera: Aphididae) and the rice brown planthopper, nilaparvata lugens (Hemiptera: Delphacidae). Pest Manag Sci 2006; 62:77-85.
171. Pham Trung N, Fitches E, Gatehouse JA. A fusion protein containing a lepidopteran-specific toxin from the south indian red scorpion (Mesobuthus tamulus) and snowdrop lectin shows oral toxicity to target insects. BMC Biotechnol 2006; 6:18.
172. Duffy SP, Young AM, Morin B et al. Sequence analysis and organization of the Neodiprion abietis nucleopolyhedrovirus genome. J Virol 2006; 80:6952-63.
173. Possee RD, Barnett AL, Hawtin RE et al. Engineering baculoviruses for pest control. Pest Manag Sci 1999; 51:462-70.
174. Maeda S. Increased insecticidal effect by a recombinant baculovirus carrying a synthetic diuretic hormone gene. Biochem Biophys Res Commun 1989; 165:1177-83.
175. Lee E, Lange A, Orchard I et al. Characterization and baculovirus-directed expression of a myosuppressin encoding cDNA from the true armyworm, Pseudaletia unipuncta. Peptides 2002; 23:747-56.
176. O'Reilly DR, Kelly TJ, Masler EP et al. Overexpression of Bombyx mori prothoracicotropic hormone using baculovirus vectors. Insect Biochem Mol Biol 1995; 25:475-85.
177. Pfeifer TA, Hegedus D, Wang YJ et al. Analysis of an insect neuropeptide, Schistocerca gregaria ion transport peptide (ITP), expressed in insect cell systems. Arch Insect Biochem Physiol 1999; 42:245-52.
178. Chang JH, Choi JY, Jin BR et al. An improved baculovirus insecticide producing occlusion bodies that contain bacillus thuringiensis insect toxin. J Invertebr Pathol 2003; 84:30-7.

179. Johnson JH, Bloomquist JR, Krapcho KJ et al. Novel insecticidal peptides from Tegenaria agrestis spider venom may have a direct effect on the insect central nervous system. Arch Insect Biochem Physiol 1998; 38:19-31.
180. Rajendra W, Hackett KJ, Buckley E et al. Functional expression of lepidopteran-selective neurotoxin in baculovirus: Potential for effective pest management. Biochim Biophys Acta 2006; 1760:158-63.
181. Moto K, Kojima H, Kurihara M et al. Cell-specific expression of enhanced green fluorescence protein under the control of neuropeptide gene promoters in the brain of the silkworm, Bombyx mori, using Bombyx mori nucleopolyhedrovirus-derived vectors. Insect Biochem Mol Biol 2003; 33:7-12.
182. Jiang H, Zhang JM, Wang JP et al. Genetic engineering of Periplaneta fuliginosa densovirus as an improved biopesticide. Arch Virol 2006.
183. Carlson J, Suchman E, Buchatsky L. Densoviruses for control and genetic manipulation of mosquitoes. Adv Virus Res 2006; 68:361-92.
184. Handler AM. Use of the piggyBac transposon for germ-line transformation of insects. Insect Biochem Mol Biol 2002; 32:1211-20.
185. Ryder E, Russell S. Transposable elements as tools for genomics and genetics in Drosophila. Brief Funct Genomic Proteomic 2003; 2:57-71.
186. Lavine MD, Strand MR. Insect hemocytes and their role in immunity. Insect Biochem Mol Biol 2002; 32:1295-309.
187. Mullen LM, Goldsworthy GJ. Immune responses of locusts to challenge with the pathogenic fungus metarhizium or high doses of laminarin. J Insect Physiol 2006; 52:389-98.
188. Goldsworthy G, Chandrakant S, Opoku-Ware K. Adipokinetic hormone enhances nodule formation and phenoloxidase activation in adult locusts injected with bacterial lipopolysaccharide. J Insect Physiol 2003; 49:795-803.
189. Mullen LM, Lightfoot ME, Goldsworthy GJ. Induced hyperlipaemia and immune challenge in locusts. J Insect Physiol 2004; 50:409-17.
190. Skinner JR, Fairbairn SE, Woodhead AP et al. Allatostatin in hemocytes of the cockroach Diploptera punctata. Cell Tissue Res 1997; 290:119-28.
191. Franssens V, Smagghe G, Simonet G et al 20-hydroxyecdysone and juvenile hormone regulate the laminarin-induced nodulation reaction in larvae of the flesh fly, Neobellieria bullata. Dev Comp Immunol 2006; 30:735-40.
192. Besedovsky HO, del Rey A. Immune-neuro-endocrine interactions: Facts and hypotheses. Endocr Rev 1996; 17:64-102.
193. Cutz E, Chan W, Track NS et al. Release of vasoactive intestinal polypeptide in mast cells by histamine liberators. Nature 1978; 275:661-2.
194. Smith EM, Blalock JE. A molecular basis for interactions between the immune and neuroendocrine systems. Int J Neurosci 1988; 38:455-64.
195. Weinstock JV. Production of neuropeptides by inflammatory cells within the granulomas of murine Schistosomiasis mansoni. Eur J Clin Invest 1991; 21:145-53.
196. Wertheim B, Kraaijeveld AR, Schuster E et al. Genome-wide gene expression in response to parasitoid attack in Drosophila. Genome Biol 2005; 6:R94.
197. Irving P, Ubeda JM, Doucet D et al. New insights into Drosophila larval haemocyte functions through genome-wide analysis. Cell Microbiol 2005; 7:335-50.
198. Johansson KC, Metzendorf C, Soderhall K. Microarray analysis of immune challenged Drosophila hemocytes. Exp Cell Res 2005; 305:145-55.
199. Barat-Houari M, Hilliou F, Jousset FX et al. Gene expression profiling of Spodoptera frugiperda hemocytes and fat body using cDNA microarray reveals polydnavirus-associated variations in lepidopteran host genes transcript levels. BMC Genomics 2006; 7:160.

CHAPTER 10

Neuropeptide Biology in *Drosophila*

Elke Clynen, Ank Reumer, Geert Baggerman, Inge Mertens and Liliane Schoofs*

Abstract

Drosophila melanogaster is since decades the most important invertebrate model. With the publishing of the genome sequence, *Drosophila* also became a pioneer in (neuro)peptide research. Neuropeptides represent a major group of signaling molecules that outnumber all other types of neurotransmitters/modulators and hormones. By means of bioinformatics 119 (neuro)peptide precursor genes have been predicted from the *Drosophila* genome. Using the neuropeptidomics technology 46 neuropeptides derived from 19 of these precursors could be biochemically characterized. At the cellular level, neuropeptides usually exert their action by binding to membrane receptors, many of which belong to the family of G-protein coupled receptors or GPCRs. Such receptors are the major target for many contemporary drugs. In this chapter, we will describe the identification, localization and functional characterization of neuropeptide-receptor pairs in *Drosophila melanogaster*.

Introduction: *Drosophila* as a Model to Study Neuropeptide Signaling

Drosophila has revolutionized biology more than any other organism. The entire genus contains about 1,500 species and is very diverse in appearance, behavior and breeding habitat. One species in particular, *Drosophila melanogaster*, has been heavily used in research in genetics and developmental biology. Also for neuropeptide research *Drosophila* is a very suitable model organism, especially since its genome has been nearly fully sequenced and is publicly accessible.[1]

Neuropeptides form the largest class of signaling molecules in animals. They transmit and regulate bio-information in the circulatory as well as the neuronal system and exert their role mostly by acting on G-protein coupled receptors or GPCRs. As such, neuropeptides play critical roles in regulating most biological processes. Neuropeptides are diverse in structure, localization and function. Their only common feature is that they are all synthesized as peptide precursor proteins, also called preproproteins. Besides the neuropeptides themselves, also the receptors they act on are structurally diverse and the resulting signaling cascades are also highly varied, so there is a tremendous potential of different effects on living cells. Therefore, peptides are attractive for pharmaceutical and agro-industrial companies because they represent (lead) compounds that can be further exploited for diverse practical applications. Peptides as such cannot be used as therapeutics or as insecticides, because they are usually broken down before they reach their target. Therefore, small compounds called peptidomimetics that mimic or block the interaction of the peptide with its receptor are being developed. As more genomes become available, the findings in *Drosophila* can be readily expanded to other species, including those of economic interest.

*Corresponding Author: Liliane Schoofs—Department of Animal Physiology and Neurobiology, K.U. Leuven, Naamsestraat 59, 3000 Leuven, Belgium. Email: liliane.schoofs@bio.kuleuven.be

Neuropeptide Systems as Targets for Parasite and Pest Control, edited by Timothy G. Geary and Aaron G. Maule.

In this chapter we will successively discuss the prediction of neuropeptides from the genome by specialized bioinformatics programs, their biochemical characterization by neuropeptidomics, deorphanization of neuropeptide GPCRs and localization and functional characterization of neuropeptides in *Drosophila.*

Bioinformatics

The *Drosophila melanogaster* genome sequence was published in 2000.[1] Currently, twenty-two other *Drosophila* species are undergoing or have completed whole genome shotgun sequencing. These genome sequencing projects gave a new impulse to (neuro)peptide research, as putative peptides can now be mined from the genome. This, however, presents a major challenge as neuropeptide precursors share little common features. Precursors encoding multiple structurally related peptides, as well as precursors encoding multiple, unrelated peptides and precursors encoding just a single bioactive peptide occur. The only common feature is the presence of an amino-terminal signal peptide that directs the ribosomes synthesizing neuropeptide precursors to the endoplasmic reticulum (ER). However, this is not a unique criterion as all proteins synthesized and sorted in the secretory pathway possess such a signal peptide. In the ER, the precursor is posttranslationally processed into bioactive peptides by a series of enzymatic steps. A typical feature for neuropeptide precursors is the presence of cleavage sites. Neuropeptides are typically cleaved from the precursor at dibasic sites. However, not all dibasic sites are actually used as a cleavage site and other, unconventional sites also occur.

Two motif-finding programs, MEME and Pratt, were used to search for common motifs in all known neuropeptide precursor proteins from *Drosophila*.[2] No general pattern or motif was found, only a very degenerative one in the area of the signal peptide sequence. Only in smaller subpopulations of datasets, common motifs could be found, corresponding to the conserved sequences of various peptide families. These conserved sequences mostly correspond to the biologically active core that interacts with the receptor proteins that mediate their action.

Based on sequence similarity to known peptide genes from other organisms, the *Drosophila melanogaster* genome was screened by means of BLAST (Basic Local Alignment Search Tool) analysis. This way, 43 peptide precursors could be annotated.[3,4] Peptide precursor genes are, however, poorly predicted by the BLAST algorithm as it is not very efficient for finding similarity to short sequences when they are scanned against the whole genome sequence. Most peptide precursors are between 50 and 500 amino acids in length and in general only a small part of the precursor consists of the actual active peptide(s). Also, putative peptide sequences for which no orthologous peptide has been identified will not be revealed this way.

Recently, an alternative searching program was developed to scan predicted proteins for the structural hallmarks of a neuropeptide precursor.[5] This program started from a protein database of *D. melanogaster* and selected all proteins less than 500 amino acids in length that contain an amino-terminal signal peptide. The resulting 5096 proteins were, after removal of the signal sequence, in silico split into short subsequences at cleavage sites typical for neuropeptide precursors. A second database comprised all known peptide precursor subsequences from Metazoa known to date. These were also split into subsequences. Next, a BLAST analysis was conducted on these two databases. Because similarity not necessarily implies homology, the output was further screened. The resulting proteins had to comply to one of following criteria: or the proteins had to contain at least two similar subsequences (based on the principle that multiple peptides encoded by a single invertebrate peptide precursor gene are often highly related) or they should contain a well-conserved motif (these putative peptide precursor genes encode multiple nonrelated peptides or only a single putative peptide). In addition, the motif should be close to a cleavage site. Motifs for neuropeptides in Metazoans have recently been catalogued.[6] In this way, 76 additional putative secretory peptide genes were predicted, which brings the total of predicted peptide precursor genes in *Drosophila melanogaster* to a total of 119.

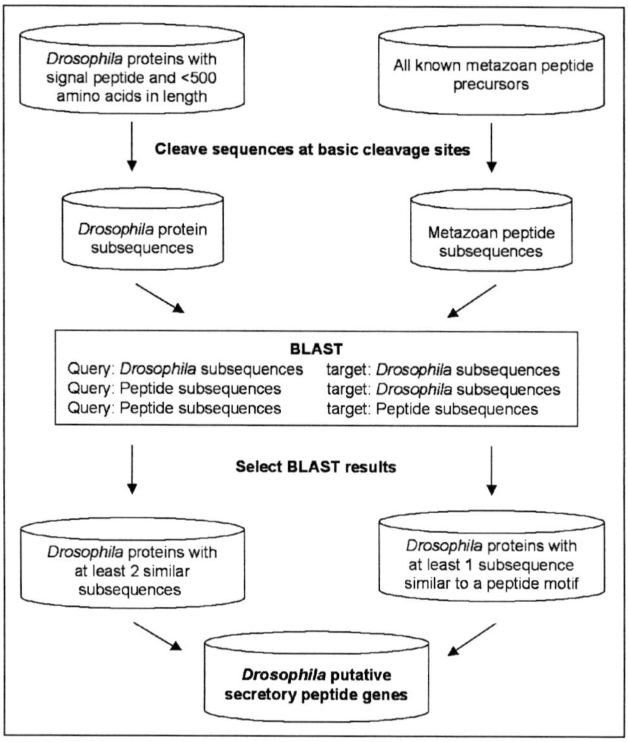

Figure 1. Schematic representation of the in silico neuropeptide searching program developed by Liu et al.[5]

Neuropeptidomics

Bioinformatic predictions do not reveal which peptides are ultimately expressed. Conventional cleaving sites are not always used and there have also been reports on unconventional cleaving sites, meaning that it is hard to reveal the sequences corresponding to the biologically active peptides starting from the precursor sequence. Moreover, the processing of a common precursor can differ during development or between tissues. And, the nucleotide sequence does not give information about posttranslational modifications, which are often essential for neuropeptide stability and activity. Therefore, a biochemical characterization of neuropeptides remains necessary. This sequence information used to arrive slowly, due to the huge efforts required for tissue collection and purification to ultimately isolate and sequence a peptide, until in 2001 the concept of peptidomics was introduced. Peptidomics presents a global strategy, by which all peptides present in a biological sample that can be derived from a cell, tissue, body liquid or even the whole organism, are simultaneously visualized and identified.[7-9] Peptidomics complements proteomics, the study of proteins, which are long chains of amino acids. There are several possible peptidomic methods, all based on mass spectrometry. The most common tool is a combination of nanoscale liquid chromatography, tandem mass spectrometry and database mining, which allows the detection and sequencing of low concentrations of peptides from complex mixtures with a high degree of automation. This way, the *Drosophila* neuropeptidome was investigated starting from only 50 larval CNS, allowing the identification of 47 neuropeptides derived from 19 different precursors (Table 1).[10-12] Only 7 of these peptides had been biochemically characterized before. Four of the precursors were not identified or predicted as neuropeptide precursor before.

***Table 1. List of neuropeptides identified in* Drosophila melanogaster[10,11]**

Peptide	Sequence
FMRFamide	
FMRFamide 1	DPKQDFMRFa
FMRFamide 2	TPAEDFMRFa
FMRFamide 3	SDNFMRFa
FMRFamide 4	PDNFMRFa
FMRFamide 5	SVQDNFMHFa
FMRFamide 6	MDSNFIRFa
MS (myosuppressin)	
MS	TDVDHVFLRFa
MS^{2-9}	DVDHVFLRFa
SK (sulfakinin)	
SK 1	FDDY(SO_3)GHMRFa
SK 2	GGDDQFDDY(SO_3)GHMRFa
sNPF (short neuropeptide F)	
sNPF 1	SPSLRLRFa
sNPF-AP	SDPDMLNSIVE
sNPF 2^{1-10}	WFGDVNQKPI
AKH (adipokinetic hormone)	
AKH	pQLTFSPDWa
CRZ (corazonin)	
Crz	pQTFQYSRGWTNa
Crz^{3-11}	FQYSRGWTNa
AST (allatostatin)	
AST 2^{1-11}	AYMYTNGGPGM
AST 3	SRPYSFGLa
AST 4	TTRPQPFNFGLa
MIP (myoinhibiting peptide)	
MIP 2	AWKSMNVAW
MIP 5	DQWQKLHGGWa
DIM 2 (immune induced peptide 2)	
Dim 2	GNVVINGDCKYCVNGa
DIM 4 (immune induced peptide 4)	
Dim 4	GTVLIQTDNTQYIRTa
CAPA (cardio acceleratory peptide)	
Cap 1	GANMGLYAFPRVa

continued on next page

Figure 2, viewed on following page. Flowchart of the neuropeptidomics setup used to analyze the peptidome of the *Drosophila* central nervous system. Nervous tissue was homogenized and the extract was separated on a nanoLC system directly coupled to a Q-TOF mass spectrometer. At each moment in the LC run, the mass spectrometer software uses the MS spectrum to decide on peak selection for fragmentation. Typically, ions with 2-4 positive charges are automatically selected for MS/MS analysis through collision induced dissociation (CID). The obtained fragmentation spectra are converted to peak list files, which are then used for querying a database in order to identify the peptides.[8]

Table 1. Continued

Peptide	Sequence
Cap 2	ASGLVAFPRVa
MT	TGPSASSGLWFGPRLa
MT^{2-15}	GPSASSGLWFGPRLa
LK (leucokinin)	
LK	NSVVLGKKQRFHSWGa
LK-AP	SPEPPILPDY
TK (tachykinin)	
TK 1	APTSSFIGMRa
TK 2	APLAFVGLRa
TK 3	APTGFTGMRa
TK 4	APVNSFVGMRa
TK 5	APNGFLGMRa
HUG (hugin)	
MT 2	SVPFKPRLa
IFamide	
IFamide	AYRKPPFNGSIFa
NPLP1 (neuropeptide-like precursor 1)	
NAP	SVAALAAQGLLNAP
MTYamide	YIGSLARAGGLMTYa
VQQ	NLGALKSSPVHGVQQ
IPNamide	NVGTLARDFQLPIPNa
GVQ	GALKSSPVHGVQ
NPLP2 (neuropeptide-like precursor 2)	
NEF	TKAQGDFNEF
LTK	EESNPAQEFLTK
KLK	AQGDFNEFIEKLK
NPLP3 (neuropeptide-like precursor 3)	
SHA	VVSVVPGAISHA
VVIamide	SVHGLGVVIa
NPLP4 (neuropeptide-like precursor 4)	
YSY	pQYYYGASGGYYDSPYSY

The lack of identification of more (predicted) peptides may be due to several reasons. First, their concentration in the conditions used may be below the sensitivity of the instrumental setup. In this respect one has to consider that the concentration of a peptide may vary during development and is dependent on the physiological condition of the organism(s). Second, not all peptides are extracted

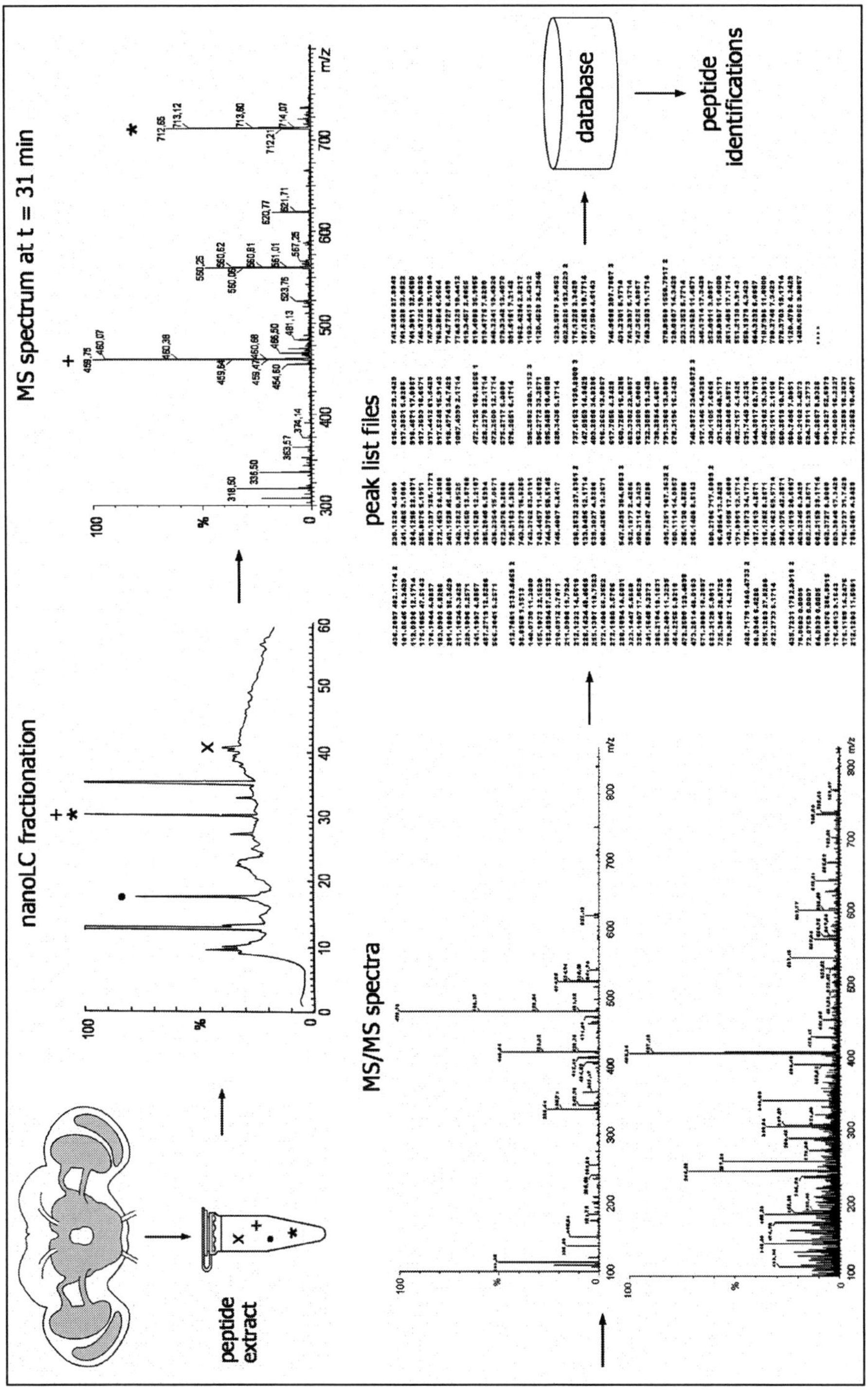

Figure 2. Please see legend on previous page.

or ionized with the same efficiency. Third, in contrast to peptides obtained after tryptic digestion of a protein, endogenous peptides in most cases do not contain a basic amino acid at their C-terminus and will therefore yield fragmentation spectra that are hard to interpret. Fourth, the presence of a putative peptide sequence derived from the genome sequence does not necessarily mean that this peptide is present in vivo. Fifth, the predicted peptides may be present in other tissues than the ones analyzed so far.

Deorphanization of *Drosophila* Neuropeptide GPCRs

Most neuropeptides interact with G-protein coupled receptors (GPCRs), thereby generating an intracellular response.[13] G protein-coupled receptors constitute the largest family of cell surface proteins. They have a typical 7 transmembrane spanning structure and upon ligand activation, they signal via G proteins, composed of an α, β and γ subunit. The α- and $\beta\gamma$-subunits dissociate from each other and separately activate several classical effectors, including adenylyl cyclases and phospholipases and regulate the activity of ion transporters, several kinases and ion channels. GPCRs can be stimulated by a diverse array of external stimuli, including bioactive peptides, chemoattractants, neurotransmitters, hormones, phospholipids, photons, odorants and taste ligands.[14] Based on shared sequence motifs, all GPCRs are categorized into six subfamilies or classes.[15] Neuropeptide GPCRs all belong to the rhodopsin-like (class A) or the secretin-like (class B) subfamily (Table 2).

In 2003, seven of the twenty most prescribed drugs interact with GPCRs, representing a total sales number of 14.3 billion US dollars.[16] Overall, GPCRs are the molecular target for over 30% of all currently marketed drugs, making the GPCR superfamily one of the most valuable target molecules for drug development.[17] Likewise, GPCRs can also be used as pesticide targets in agricultural applications.

Because the structure and function of most neuropeptide GPCRs is conserved in the phylum of the Arthropoda, *Drosophila* neuropeptide GPCRs can be used to find new lead compounds that bind the receptor, thereby inhibiting their function. The identified lead compounds can then be further engineered to real pesticides. However, to reach that goal, it is important that all *Drosophila* GPCRs are linked with their naturally occurring bioactive ligand and that the function(s) of this receptor-ligand pair is revealed. Only then, neuropeptide GPCRs can be selected as candidates for pest control.

Since the publication of the *Drosophila* genome, a large number of orphan GPCRs were characterized (Table 2), but about half of the neuropeptide GPCRs in *Drosophila* remain orphan to date because deorphanization is not straightforward for some GPCRs.[1] When a cellular assay platform is used, several difficulties can hinder the deorphanization process. These can be situated on the level of cloning, expression or signal transduction pathway. An overview of all cellular assays that can be used in receptor deorphanization is given in Mertens et al.[18] Although cell-based assays are adequate for many GPCRs and even new GPCR biosensors and imaging technologies have recently been developed that hold promise for the development of functional GPCR screens in living cells, it is likely that these cell-based formats will limit the development of higher density GPCR assays.[19] Therefore it is not surprising that recently the focus is on further miniaturized assays, ultra-high throughput assays and, eventually microarray/bioship assay formats. Stable, robust, cell-free signaling assemblies comprising receptor and appropriate molecular switching components will form the basis for such future GPCR assay platforms.[20]

Functional Role of Neuropeptides: Localization, Reverse Genetics and Bioassays

Localization

The CNS of *Drosophila* larvae contains approximately 10000 neurons, about 200 of which are peptidergic. The majority of neuropeptides has been demonstrated in brain and VNC interneurons of various types and/or neurosecretory cells. The latter are cells that have axon terminations in neurohemal release sites in contact with the circulation system. In addition, they commonly have varicose processes in the brain, which could act as interneuronal segments.

Table 2. List of characterized and/or predicted G-protein coupled receptors and corresponding ligands in* Drosophila*: original classification by Hewes and Taghert and update.[4] The receptors depicted in italic were annotated incorrectly and the ones depicted in bold were not annotated at all.

Gene	Receptor	Ligand(s)	Year	Reference
Family A/Group II-B: gastrin/cholecystokinin receptors				
CG6881/CG6894	DSK-R1	Drm-SK-1 and -2	2002	57
CG6857	orphan			
CG14593	orphan			
CG30340	orphan			
Family A/Group III-B: neurokinin receptors				
CG10626	DLKR	Dromyokinin/drosokinin	2002	58
CG6515	DTK-R	DTK	2003	59
CG7887	DTK-R	DTK	2003,2005	59,60
CG8784	Drm-PK-2 receptor	Drm-PK-2	2003	61
CG8795	Drm-PK-2 receptor	Drm-PK-2	2003	61
CG10823	SIFa-R	Drm-SIFa	2006	62
Family A/Group III-B: neuropeptide Y receptors				
CG1147	NPFR-1	NPF	2002	63
CG7395	sNPF-R	Drm-sNPF-1, -2, -3 and -4	2002	64,65
CG5811	orphan			
CG12610	orphan			
CG13995	orphan			
Family A/Group III-B: growth hormone secretagogue, neurotensin, neuromedin U and thyrotropin releasing hormone receptors				
CG2114	FMRFaR	Drm-FMRFamide 1-8	2002	66
CG14575	Capa-1 and -2 receptor	Drm-capa-1 and -2	2002	67,68
CG5911-A and -B	ETH receptors	Drm-ETH-1 and -2	2002	69
CG6986	*Proctolin receptor*	*Proctolin*	*2003*	70
CG8985	*DMS-R*	*Drome-MS*	*2003*	71,59
CG13803	*DMS-R*	*Drome-MS*	*2003*	71,59
CG9918	*pyrokinin-1*	*PK-1*	*2005*	72
CG33639 (CG5936)	orphan			
CG13229	orphan			
CG13575	orphan			
CG14003	orphan			

continued on next page

Table 2. Continued

Gene	Receptor	Ligand(s)	Year	Reference
CG16726	orphan			
Family A/Group V: galanin/allatostatin and opoid/somatostatin receptors				
CG2872	AlstR-2/DAR-1	Drostatin-1	1999	73,74
CG10001	AlstR-2/DAR-2	Drostatin-2	2001	73,74
CG7285	Drostar-1	Drostatin-C	2002	75
CG13702	Drostar-2	Drostatin-C	2002	75
CG14484 (CG30106)	AstB-R	Ast-B	2003	59
CG4313	orphan			
Family A/Group V: gonadotropin releasing hormone, vasopressin and oxytonin receptors				
CG11325	AKH-R1	Drm-AKH	2002	68,76
CG10698	Cor-R	Drm-COR	2002	68,77
CG6111/ CG14547	CCAP-R	Drm-CCAP	2003	78
Family A/Group V (Type Ic): glycoprotein hormone receptors				
CG8930 (DLGR-2)	LGR-2 (rk)	bursicon	2005	79,80
CG7665	DLGR-1	GPA2/GPB5	2005	81
CG4187	orphan			
CG5042 (CG31096/ DLGR-3)	orphan			
Family A: unclassified orphan receptors				
CG3171	**orphan**	**neuropeptide?**		
CG4322	**orphan**	**neuropeptide?**		
CG12290	**orphan**	**neuropeptide?**		
Family B/Group I: calcitonin and diuretic hormone receptors				
CG8422	CG8422	DH44	2004	82
CG13758	PDF-R	PDF	2005	83,84,85
CG17415	DH31-R	DH31	2005	86
CG4395	orphan			
CG12370	orphan			
Family B: methuselah-like receptors				
CG6936	**Mth receptor**	**Sun A and B**	**2004**	87
Family B: unclassified orphan receptors				
CG11318	**orphan**	**neuropeptide?**		
CG8639 (CIRL)	**orphan**	**neuropeptide?**		
CG15556	**orphan**	**neuropeptide?**		

The localization of a neuropeptide reveals whether it plays a role as circulatory hormone (present at neurohemal release sites) or as neuromodulator (present in interneurons) and possibly cotransmitter (colocalized with a classical neurotransmitter). Double-immunolabeling experiments have shown that neuropeptides are often colocalized with conventional fast-acting transmitters, like GABA, biogenic amines and nitric oxide that act on an ion-channel-type of receptor. The localization is also often essential for the design of experimental approaches to determine the function.

Representatives of most of the characterized neuropeptide families have been localized in *D. melanogaster*. Each neuropeptide precursor displays a unique neuronal distribution pattern.[21] Some neuropeptide precursor genes encode multiple peptides, which mostly seem to be co-expressed. Commonly, each type of neuropeptide is localized to a relatively small number of neurons, typically a specific subset of 6 to 20 cells. Two or more different neuropeptides can be present in partly overlapping cell populations.

For a precise and comparable morphological description of peptidergic neurons, one needs an anatomical reference system. The segmental nerves and regularly distributed transverse and longitudinal fasciclin-2 expressing fibers provide a convenient 3D-coordinate system in which peptide neurons can be mapped using peptide-specific GAL4-driven expression of GFP markers or immunostaining.[22] This system also enables to identify presynaptic neurons providing inputs onto peptidergic interneurons and neurosecretory cells.

Neuropeptides are produced by a series of enzymatic steps that sequentially cleave and further modify larger precursor molecules. The synthesis and secretion of neuropeptides has to be strictly regulated in order to properly execute a complex behavior. All factors that are involved in the regulation hereof represent possible targets for parasite and pest control. Therefore, (co)localization studies that map proteins critical for neuropeptide signaling, including transcription factors regulating cell-specific neuropeptide expression, peptide-processing enzymes, G-protein coupled receptors and neurosecretory proteins that are required for exocytosis, are important.

There are several possible tools to study the localization of a neuropeptide. The most popular being immunocytochemistry and in situ hybridization. The latter localizes the precursor mRNA and hence gives no information on the translated peptides.

In *Drosophila*, peptidergic neurons can also be easily identified without staining methods based on the GAL4/UAS binary system. The GAL4 protein activates transcription of only those genes containing GAL4 binding sites or upstream activating sequences (UAS). When a certain promoter (or enhancer) directs expression of the transcriptional activator GAL4 in a particular pattern, GAL4 in turn directs transcription of the GAL4-responsive (UAS) target gene in an identical pattern. This system can be used to visualize neuropeptide neurons, using the promoter of a certain neuropeptide precursor gene to drive GAL4 expression and a UAS—reporter gene to reveal the pattern. The green fluorescent protein (GFP) is often used as reporter, as its expression can be visualized by virtue of its natural fluorescence in live specimens throughout development. The GAL4/UAS system is often used in combination with immunostaining to perform colocalization studies. It does, however, not reveal posttranslational (differential) processing of the precursor.

Also mass spectrometry has become an important tool to study the localization of neuropeptides. It enables the rapid and accurate identification of the almost complete neuropeptide identity profile from small numbers of tissues, cell groups or even single cells. Thereby it can confirm, refine and extend data from immunostaining. While using immunostaining only one (or a few) peptide(s) can be studied at a time, mass spectrometric profiling identifies the complete neuropeptide profile of a single neuron, nerve or neurohemal organ at a given moment. This gives us more insight in the coordinated action of neuropeptides. Also the problem of cross-reactivity between structurally related peptides inherent to immunostaining is not posed by MS methods. Here, neuropeptides that differ by only one amino acid (peptide isoforms) can be easily distinguished based on their exact mass and also posttranslational modifications can be identified. This way, one can uncover differential peptide processing and get new insights into the posttranslational processing of peptide precursors.

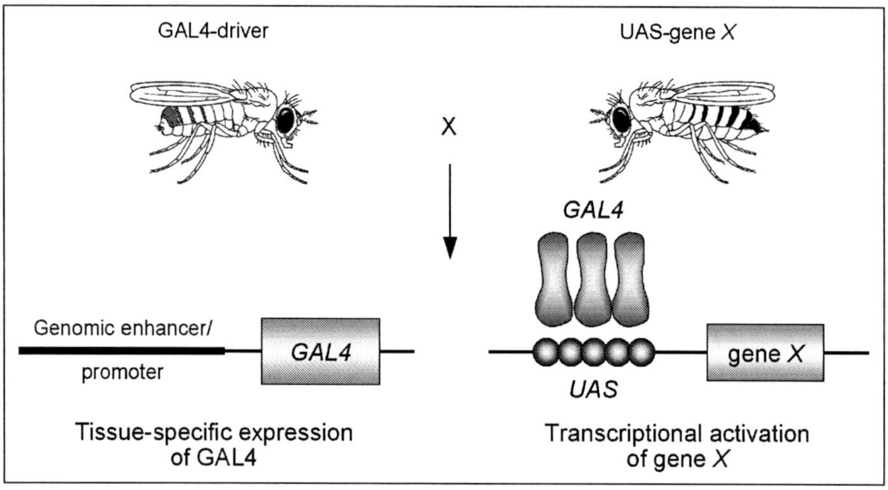

Figure 3. GAL4/UAS system: flylines are generated in which the expression of a transcriptional activator of yeast, the GAL4 protein, is under the influence of different genomic enhancers or promoters. In this way, a plethora of different spatial and/or temporal patterns of GAL4 expression is available. This transcription factor acts on a GAL4 responsive promoter or 'upstream activating sequence' UAS. Flylines can be generated in which any DNA sequence is placed downstream of a UAS sequence. The cross between a transgenic UAS line and a GAL4 line can thus result in spatio-temporal controlled GAL4 driven expression of this sequence.

The peptidome of the major neurohemal organs, i.e., specialized organs where the hormonal release of neuropeptides takes place, of *D. melanogaster* was characterized by direct MALDI-TOF mass spectrometry.[23,24] Peptides present at these sites might classify as (putative) hormones. The ring gland, dorsal sheath of the ventral nerve cord, the epitracheal organs, thoracic perisympathetic organs and abdominal transverse nerves 1-3 were studied. All detected masses up to 2.5 kDa in the neurohemal organs could be assigned to bioactive neurohormones or intermediates of prohormone processing.

Recently, the first single-cell mass analysis in *Drosophila* has been reported. The large (± 20 μm) lateral ventral neurons of the optic lobes (LNvs), known to express the neuropeptide pigment-dispersing factor (PDF), were visualized by (UAS) GFP-expression using pdf-GAL4 flies and subsequently isolated from adult flies and directly analyzed by MALDI-TOF MS.[25] The resulting spectra showed strong mass signals for PDF, but no evidence for other colocalized neuropeptides. The same single cell analyses were performed in larvae for the (smaller) hugin-neurons (± 15 μm), which are expressed in a group of neurosecretory cells of the suboesophageal ganglion.[26]

Reverse Genetics and Bioassays

The advantage of *Drosophila* as a model system for neuropeptide research is that a lot of genetic tools are already available and continuously new techniques are being developed that can be used to probe the in vivo function of neuropeptide precursor genes. For many years the emphasis has been on a forward genetic approach to unravel the function of genes that were discovered on the basis of their mutant phenotype.[27] However, since the publication of the genome sequence of *Drosophila*, the development of reverse genetic methods to search for a gene's function, starting from its DNA sequence and its location in the genome, has boosted. These reverse genetic approaches can be divided into two groups.

The first group focuses on the acquisition of mutations in specific genes of interest by means of modified forward genetic screens. Chemical mutagenesis, making use of DNA-damaging agents like ethyl methanesulphonate (EMS) and ethylnitrosourea (ENU), as well as transposable element mutagenesis are often the workhorse of this class.[28] Since the development of a *P*-element transgenesis method to restore the wild type function of the *rosy* gene by Rubin and Spradling in 1982, the *P*-element has become the most utilized transposable element for studying gene function in *Drosophila melanogaster*.[29,30] This enormously powerful tool unfortunately suffers from some constraints, like the *P*-elements target site specificity that makes it difficult to obtain insertions in some regions of the genome called "cold spots". As an alternative, one can either make use of the *Drosophila melanogaster hobo* elements, although they are a rarely used transposon, or one can utilize transposable elements from other species that have been adapted for use in *Drosophila*, like the Lepidopteran derived *piggyback* elements, the housefly-derived *Hermes* element and the *Minos* and *mariner* elements from *Drosophila hydei* and *Drosophila mauritiana* respectively.[31]

The second group of reverse genetic approaches emphasizes on the altering of the function of a gene of interest. The main strategies in this group are site-specific recombination and gene targeting.

The principle of site-specific recombination relies on the discovery of a class of enzymes that recognize specific DNA sequences and carry out reciprocal recombination between two copies of that sequence. They were subsequently called site-specific recombinases.[32] In *Drosophila*, the FLP recombinase from the yeast 2μ plasmid and the Cre recombinase from the P1 bacteriophage are the two most used site-specific recombinase enzymes.

Gene targeting can be divided into two main strategies: targeted gene replacement by homologous recombination and RNA interference. For targeted gene replacement, the homing endonuclease enzyme strategy is used. These enzymes recognize and cut at a specific and long sequence in double stranded DNA leaving behind a double stranded break (DSB). This method makes it possible to precisely modify an endogenous gene sequence by homologous recombination between an introduced DNA fragment and the homologous target gene. In this way, mutations that reduce, destroy or alter a gene's function in a defined manner can be made or even a complete replacement of the endogenous gene with for example a marker gene can be obtained. In *Drosophila*, the I-*Sce*I and the I-*Cre*I endonucleases are used.[33,34]

Andrew Fire and Craig Mello (1998) were the first to describe their observation of the endogenous RNA interference (RNAi) mechanism for which they were awarded (2006) with 'The Nobel Prize in Physiology or Medicine'.[35] The RNA interference process is initiated by the presence of double stranded RNA (dsRNA) which induces an efficient sequence-specific silencing of gene expression. In *Drosophila*, RNAi can be used for cultured *Drosophila* cells as well as for in vivo experiments. The first in vivo experiments indicated that upon injection dsRNA could be a strong antagonist of gene function in precellular embryos.[36] To overcome the limitations associated with dsRNA injection e.g., the repetition of the injection regarding the quantity, the location and timing of the injection; methods were developed to stably express dsRNA in vivo. Hereby, the place and time of expression can be controlled using the GAL4/UAS binary system. A general overview can be found in '*Drosophila*: a laboratory handbook' written by Scott Hawley, Kent Golic and Michael Ashburner.[37] Although most methods have proven to be effective, they all have their limitations owing to their specific experimental approach.

In a short overview we will demonstrate the use of these reverse genetic approaches in *Drosophila* neuropeptide research.

Transposon mutagenesis can be seen as the umbrella upon which the use of most of the reverse genetic approaches relies, chemical mutagenesis excluded. The hunt for the in vivo functional characterization of a gene mostly starts with a survey of the extensive collections of transposable element insertions in search for hits into or near a gene of interest. Flybase maintains links to the websites with information about these collections. When this survey does not give the desired outcome, one can make use of a collection of *P*-element (or other transposon) transformation vectors, which make it possible to bring any desired piece of DNA into the fly's genome.

Mobilization and imprecise excision of existing *P*-element insertions have already proven to be useful for the elucidation of the function of the pigment dispersing factor or *pdf* and the ecdysis-triggering hormone or *eth* neuropeptide genes. Mobilization of a *P*-element residing in the

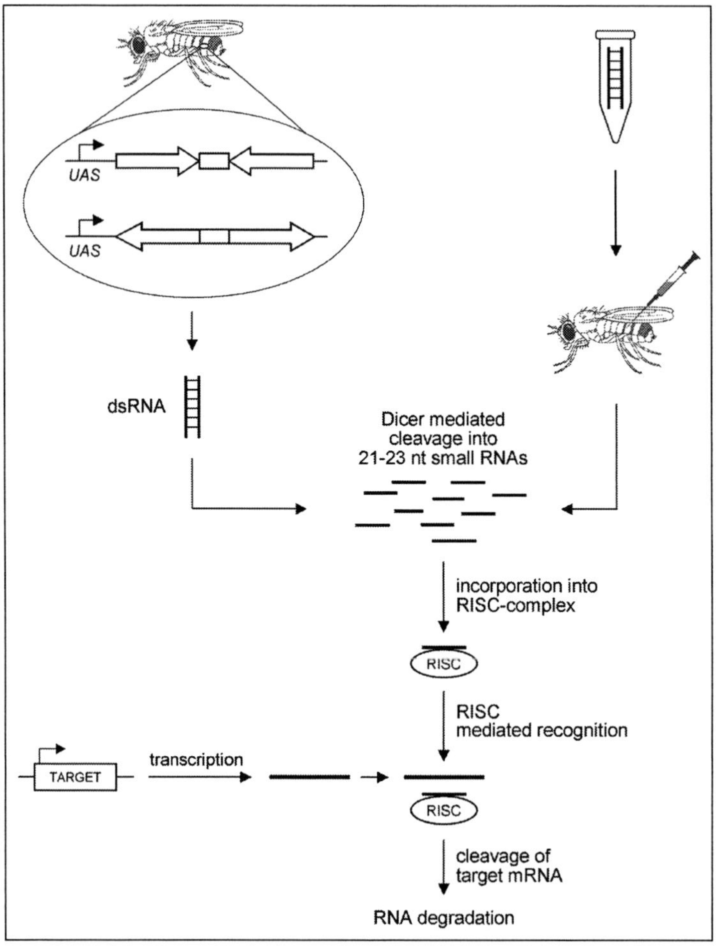

Figure 4. The RNA interference mechanism in the fruitfly. Double stranded RNA can be delivered into *Drosophila* by injection of in vitro transcribed RNA (upper right), or by stable expression of dsRNA in vivo through GAL4 driven expression of inverted repeats (upper left). The ribonuclease Dicer acts upon this dsRNA trigger and cleaves it into 21-23 nt small RNAs.[52] RISC is a multiprotein complex in which a single stranded part of a small RNA is incorporated. This then guides the sequence specific cleavage of the target mRNA.[53]

pdf gene gave rise to a fly strain bearing a nonsense mutation (pfd^{01}). Video recording of locomotor events of this *pfd*-null mutant showed that these flies are arrhythmic in constant darkness and this was one reason for Renn et al to suggest that pdf acts as the principal transmitter in circadian behavior.[38] A deletion of *eth*, the gene encoding ecdysis-triggering hormone (ETH) in *Drosophila*, was obtained by imprecise excision of a *P*-element. In video recorded ecdysis behavior assays, these null mutants (eth^-) failed to inflate the new respiratory system on schedule, did not perform the ecdysis behavioral sequence and exhibited the phenotype *buttoned-up*, which is characterized by incomplete ecdysis and 98% mortality at the transition from first to second larval instar.[39]

Targeted ablation of neurons expressing a neuropeptide precursor of interest is generally used to investigate the phenotypical, behavioral and physiological consequence of the absence of its peptides. Neuropeptide enhancer driven GAL4 expression combined with UAS-linked cell death

Table 3. P-element insertion stocks available from stock centers, Bloomington—(http://flystocks.bio.indiana.edu/; Szeged...) Author's Note: The Szeged stock center unfortunately had to close doors on June 30th 2009.

P-Element Insertions	Symbol	Transposon	Reference
Bloomington	EP	P{EP}	88
		P{EPg}	89
	EY	P{EPgy2}	90,91
	BG	P{GT1}	92
	PlacW	P{lacW}	93,94
	LA	P{Mae-UAS.6.11}	95
	PZ	P{PZ}	96
	KG and KV	P{SUPor-P}	90
	XP	P{XP}	97
	DG	P{wHy}	98
Szeged	RS	P{RS3} and P{RS5}	99
	EP	P{EP}	88
	l(3)Sxxxxxx, Sxxxxxx, l(2)SHxxxx	P{lacW}	100,101,102
Exelixis	XP	P{XP}	97
DGRC	NP	P{GawB}	103
	GS	P{GS}, P{Mae-UAS.GS}	104
	LA	P{Mae-UAS.6.11}	95

genes is the most used experimental strategy. Therefore, the first step is transposon transformation of the desired GAL4 and UAS constructs. In this respect, *dAkh* (Drosophila *adipokinetic hormone*)-GAL4,[40] *CCAP* (*crustacean cardioactive peptide*)-GAL4,[41,42] *EHups* (*eclosion hormone upstream sequence*)-GAL4[43] and *pdf*-GAL4[38,44] were used in combination with UAS-*reaper* (UAS-*rpr*), *npf* (*neuropeptide F*)-GAL4 combined with UAS-*DTI* (an attenuated *diphteria toxin* gene)[45,46] and *pfd*-GAL4 combined with UAS-*head involution defective* (UAS-*hid*),[38,44] for the targeted ablation of the respective neuropeptide precursor expressing neurons.

Spatio-temporal controlled or ectopic expression of a neuropeptide gene of interest also has been proven to be a useful approach. Ectopic expression of the *hugin* gene, for example, resulted in larval death predominantly at or shortly after ecdysis from second to third instar, a pattern reminiscent of *ETH* mutants, suggesting that at least one of the posttranslational cleavage products affects molting of the larvae by interfering with the regulation of ecdysis.[47] *Actin5C*-GAL4 driven expression of the *proctolin* gene in the CNS and midgut resulted in a 14% increase in the heart rate in pupae, providing evidence in support of a cardioacceleratory endocrine function for proctolin in *Drosophila*.[48]

Although recently developed, gene targeting and transgenic expression of dsRNA in particular, has already demonstrated to be an excellent tool in the quest for a gene's function.

Winther et al specifically eliminated tachykinin related peptides (TKRPs) in the nervous system of *Drosophila* using targeted RNAi of the *dtk* (Drosophila *tachykinin*) gene to examine odor perception with a larval olfactory test and locomotor activity with the 'Buridan's paradigm'.[49,50] They found that the gene silencing of these peptides resulted in a loss of sensitivity towards specific odorants and concentrations and also in hyperactivity.

In most cases however, more insight into the function(s) of neuropeptides is gained by using a combination of different reverse genetic approaches and multiple bioassays.

With a food response assay, a glucose sensing and motivational feeding assay and a social behavior and burrowing assay Wu et al demonstrated that transgenic larvae deficient in *Drosophila* neuropeptide F (*d*NPF) signaling through targeted gene silencing of the *dNPF* gene precociously exhibited the phenotypes of food aversion and social behavior normally displayed by older nonfeeding larvae.[46] Conversely, *dNPF* overexpression in the larval CNS prolonged the feeding activity and suppressed the social behavior of older larvae. Whereas Wen et al used a behavioral assay and an ethanol content assay to demonstrate that these *d*NPF (or its receptor *d*NPFR1) RNAi flies showed a decreased sensitivity to ethanol sedation and that overexpression of *d*NPF increased alcohol sensitivity.[45] Also the controlled functional disruption (*npf-GAL4* driven *diphteria toxin* (*DTI*) expression) of *d*NPF and *d*NPFR1 neurons rapidly triggered acute resistance to ethanol sedation, suggesting that the NPF pathway tonically controls acute alcohol response.

The use of gain-of-function and loss-of-function transgenic flies for the short form of NPF, *d*sNPF (Drosophila *short neuropeptide F*), in a feeding assay led to the suggestion that *d*sNPF regulates food intake and body size. Overexpression of *d*sNPF in the CNS of *Drosophila* larvae promots food intake and results in bigger and heavier flies whereas the targeted knock down results in a suppression of food intake. In contrast to *d*NPF, the *d*sNPF did not prolong the feeding behavior suggesting that they are involved in different aspects of the mechanisms controlling feeding.[51]

As proven above, a clever use of reverse genetics combined with a wise selection of bioassays greatly enhances our knowledge about neuropeptides and their functions.

Other Neuronal Molecules as Potential Targets for Insecticides

Neurotransmitter-receptors can be used as targets for insecticides, as they combine the extreme specificity of the ligand-receptor recognition process and fast mode of signaling with a great physiological relevance. Mainly the nicotinic acetylcholine receptor and the GABA-receptor are considered as important targets for insecticides because of their importance in the neurotransmission within the insect nervous tissue, acetylcholine being the most important excitatory and GABA the most important inhibitory neurotransmitter.[54,55] Unfortunately, the pharmacology of these two receptors is closely related to that of the homologous vertebrate receptors, which results in a high toxicity for vertebrates of the respective insecticides. Lindane for instance, is an insecticide that is currently widely used (e.g., in shampoos for lice). Lindane acts through a GABA-A receptor-chloride channel complex. The problem with Lindane, however, is that insects have become resistant and that it has toxic neurologic effects. Therefore, these receptors are not ideally suited to serve as targets for the development of new insecticides. For a review on ion channels as molecular targets for neuroactive insecticides see ref. [56]

GPCRs with insect-specific ligands are a good alternative for the development of new biodegradable, safe, specific and nontoxic insecticides. Besides the receptors, also enzymes that are involved in the formation of bioactive peptides from their inactive precursor proteins represent possible targets for insecticides.

Conclusion

The publishing of the *Drosophila* genome has revolutionized its use for studying a plethora of often conserved physiological systems, and for exploring neuropeptide biology in particular. Despite all this, little still is known about the function(s) of many annotated (neuro-)peptides as well as the receptor(s) they act on. In addition, the use of novel bioinformatic tools has recently led to the prediction of even more peptide precursor genes of which the in vivo peptide expression was explored using neuropeptidomic techniques and other localization studies. The genetic methods that have been used for unraveling the function of some neuropeptides, however, are most promising. In the near future, quite a lot of functions thus presumably will be elucidated when adequate bioassays are available. Since most of the genetic tools used for examining *Drosophila* genes are not (yet) at one's disposal in other insects of which the genome has been sequenced, *Drosophila* thus is by far the most suitable organism for investigating neuropeptide, as well as other peptide/protein functioning. Before generalizing neuropeptide function(s), one naturally first has to explore there possible conservation since some peptides do not occur in *Drosophila*, or in other insects, and are therefore regarded to be insect family specific. Some of the latter thus can not be examined in the fruit fly. *Drosophila* neuropeptide research nonetheless has enhanced, and will further increase the knowledge of important conserved signaling systems in the future.

Acknowledgements

E. Clynen, G. Baggerman and I. Mertens are postdoctoral researchers of the Fund for Scientific Research-Flanders (FWO-Vlaanderen). Special thanks to J. Puttemans for figure design.

References

1. Adams MD, Celniker SE, Holt RA et al. The genome sequence of Drosophila melanogaster. Science 2000; 287:2185-2195.
2. Baggerman G, Liu F, Wets G et al. Bioinformatic analysis of peptide precursor proteins. Ann NY Acad Sci 2005; 1040:59-65.
3. Vanden Broeck J. Neuropeptides and their precursors in the fruitfly, Drosophila melanogaster. Peptides 2001; 22:241-254.
4. Hewes RS, Taghert PH. Neuropeptides and neuropeptide receptors in the Drosophila melanogaster genome. Gen Res 2001; 11:1126-1142.
5. Liu F, Baggerman G, D'Hertog W et al. In silico identification of new secretory peptide genes in Drosophila melanogaster. Mol Cell Proteomics 2006; 5:510-522.
6. Liu F, Baggerman G, Schoofs L et al. Uncovering conserved patterns in bioactive peptides in Metazoa. Peptides 2006; 27:3137-3153.
7. Clynen E, De Loof A, Schoofs L. The use of peptidomics in endocrine research. Minireview. Gen Comp Endocrinol 2003; 132:1-9.
8. Baggerman G, Verleyen P, Clynen E et al. Peptidomics J Chromatogr B Analyt Technol Biomed Life Sci 2004; 803:3-16.
9. Schulz-Knappe P, Zucht HD, Heine C et al. Peptidomics: the comprehensive analysis of peptides in complex biological mixtures. Comb Chem High T Scr 2001; 4:207-217.
10. Baggerman G, Cerstiaens A, De Loof A et al. Peptidomics of the larval Drosophila melanogaster central nervous system. J Biol Chem 2002; 277:40368-40374.
11. Baggerman G, Boonen K, Verleyen P et al. Peptidomic analysis of the larval Drosophila melanogaster central nervous system by two-dimensional capillary liquid chromatography quadrupole time-of-flight mass spectrometry. J Mass Spectrom 2005; 40:250-260.
12. Schoofs L, Baggerman G. Peptidomics in Drosophila melanogaster. Briefings in functional genomics and proteomics 2003; 2:114-120.
13. Zupanc GKH. Peptidergic transmission: from morphological correlates to functional implications. Micron 1996; 27:35-91.
14. Bockaert J, Pin JP. Molecular tinkering of G protein-coupled receptors: an evolutionary success. EMBO J 1999; 18:1723-1729.
15. Gether U. Uncovering molecular mechanisms involved in activation of G protein-coupled receptors. Endocrine Reviews 2000; 21:90-113.
16. Schlyer S, Horuk R. I want a new drug: G-protein-coupled receptors in drug development. Drug Discovery Today 2006; 11:481-493.
17. Drews J. Drug discovery: a historical perspective. Science 2000; 287:1960-1964.
18. Mertens I, Vandingenen A, Meeusen T et al. Postgenomic characterization of G-protein-coupled receptors. Pharmacogenomics 2004; 5:657-672.
19. Thomsen W, Frazer J, Unett D. Functional assays for screening GPCR targets. Current Opinion in Biotechnology 2005; 16:655-665.
20. Leifert WR, Aloia AL, Bucco O et al. G-protein-coupled receptors in drug discovery: nanosizing using cell-free technologies and molecular biology approaches. J Biomol Screen 2005; 10:765-779.
21. Nässel DR, Homberg U. Neuropeptides in interneurons of the insect brain. Cell Tissue Res 2006; 326:1-24.
22. Santos JS, Voemel M, Homberg U et al. Three-dimensional charting of peptidergic neurons in the larval ventral nerve cord of the fruit fly, Drosophila melanogaster. J Neurogenet 2006; 20:214-215.
23. Predel R, Wegener C, Russell WK et al. Peptidomics of CNS-associated neurohemal systems of adult Drosophila melanogaster: a mass spectrometric survey of peptides from individual flies. J Comp Neurol 2004; 474:379-392.

24. Wegener C, Reinl T, Jansch L et al. Direct mass spectrometric peptide profiling and fragmentation of larval peptide hormone release sites in Drosophila melanogaster reveals tagma-specific peptide expression and differential processing. J Neurochem 2006; 96:1362-1374.
25. Neupert S, Johard HAD, Nässel DR et al. Mass spectrometric analysis of peptide content in single circadian pacemaker neurons (large LNVS) in Drosophila melanogaster. J Neurogenet 2006;20:186-187.
26. Neupert S, Predel R. Analysis of neuropeptide expression in single hugin-neurons using MALDI-TOF MS. J Neurogenet 2006;20:187-187.
27. St Johnston D. The art and design of genetic screens: Drosophila melanogaster. Nat Rev Genet 2002; 3:176-188.
28. Adams MD, Sekelsky JJ. From sequence to phenotype: reverse genetics in Drosophila melanogaster. Nat Rev Genet 2002; 3:189-198.
29. Rubin GM, Spradling AC. Genetic transformation of Drosophila with transposable element vectors. Science 1982; 218:348-353.
30. Spradling AC, Rubin GM. Transposition of cloned P elements into Drosophila germ line chromosomes. Science 1982; 218:341-347.
31. Ryder E, Russell S. Transposable elements as tools for genomics and genetics in Drosophila. Brief Funct Genomic Proteomic 2003; 2:57-71.
32. Echols H, Gingery R, Moore L. Integrative recombination function of bacteriophage lambda: evidence for a site-specific recombination enzyme. J Mol Biol 1968; 34:251-260.
33. Gong WJ, Golic KG. Ends-out, or replacement, gene targeting in Drosophila. Proc Natl Acad Sci USA 2003; 100:2556-2561.
34. Rong YS, Golic KG. Gene targeting by homologous recombination in Drosophila. Science 2000; 288:2013-2018.
35. Fire A, Xu S, Montgomery MK et al. Potent and specific genetic interference by double-stranded RNA in Caenorhabditis elegans. Nature 1998; 391:806-811.
36. Kennerdell JR, Carthew RW. Use of dsRNA-mediated genetic interference to demonstrate that frizzled and frizzled 2 act in the wingless pathway. Cell 1998; 95:1017-1026.
37. Ashburner M, Golic KG, Hawley RS. Drosophila: a laboratory handbook. 2005; second edition.
38. Renn SC, Park JH, Rosbash M et al. A pdf neuropeptide gene mutation and ablation of PDF neurons each cause severe abnormalities of behavioral circadian rhythms in Drosophila. Cell 1999; 99:791-802.
39. Park Y, Filippov V, Gill SS et al. Deletion of the ecdysis-triggering hormone gene leads to lethal ecdysis deficiency. Development 2002; 129:493-503.
40. Lee G, Park JH. Hemolymph sugar homeostasis and starvation-induced hyperactivity affected by genetic manipulations of the adipokinetic hormone-encoding gene in Drosophila melanogaster. Genetics 2004; 167:311-323.
41. Dulcis D, Levine RB, Ewer J. Role of the neuropeptide CCAP in Drosophila cardiac function. J Neurobiol 2005; 64:259-274.
42. Park JH, Schroeder AJ, Helfrich-Forster C et al. Targeted ablation of CCAP neuropeptide-containing neurons of Drosophila causes specific defects in execution and circadian timing of ecdysis behavior. Development 2003; 130:2645-2656.
43. Baker JD, McNabb SL, Truman JW. The hormonal coordination of behavior and physiology at adult ecdysis in Drosophila melanogaster. J Exp Biol 1999; 202:3037-3048.
44. Yoshii T, Heshiki Y, Ibuki-Ishibashi T et al. Temperature cycles drive Drosophila circadian oscillation in constant light that otherwise induces behavioural arrhythmicity. Eur J Neurosci 2005; 22:1176-1184.
45. Wen T, Parrish CA, Xu D et al. Drosophila neuropeptide F and its receptor, NPFR1, define a signaling pathway that acutely modulates alcohol sensitivity. Proc Natl Acad Sci USA 2005; 102:2141-2146.
46. Wu Q, Wen T, Lee G et al. Developmental control of foraging and social behavior by the Drosophila neuropeptide Y-like system. Neuron 2003; 39:147-161.
47. Meng X, Wahlstrom G, Immonen T et al. The Drosophila hugin gene codes for myostimulatory and ecdysis-modifying neuropeptides. Mech Dev 2002; 117:5-13.
48. Taylor CA, Winther AM, Siviter RJ et al. Identification of a proctolin preprohormone gene (Proct) of Drosophila melanogaster: expression and predicted prohormone processing. J Neurobiol 2004; 58:379-391.
49. Winther AM, Acebes A, Ferrus A. Tachykinin-related peptides modulate odor perception and locomotor activity in Drosophila. Mol Cell Neurosci 2006; 31:399-406.
50. Gotz KG. Visual guidance in Drosophila. Basic Life Sci 1980; 16:391-407.
51. Lee KS, You KH, Choo JK et al. Drosophila short neuropeptide F regulates food intake and body size. J Biol Chem 2004; 279:50781-50789.
52. Hammond SM, Bernstein E, Beach D et al. An RNA-directed nuclease mediates posttranscriptional gene silencing in Drosophila cells. Nature 2000; 404:293-296.

53. Bernstein E, Caudy AA, Hammond SM et al. Role for a bidentate ribonuclease in the initiation step of RNA interference. Nature 2001; 409:363-366.
54. Breer H, Sattelle DB. Molecular properties and functions of insect acetylcholine receptors. J Insect Physiol 1987; 33:771-790.
55. Rauh JJ, Lummis SC, Sattelle DB. Pharmacological and biochemical properties of insect GABA receptors. Trends Pharmacol Sci 1990; 11:325-329.
56. Raymond-Delpech V, Matsuda K, Sattelle BM et al. Ion channels: molecular targets of neuroactive insecticides. Invert Neurosci 2005; 5:119-133.
57. Kubiak TM, Larsen MJ, Burton KJ et al. Cloning and functional expression of the first Drosophila melanogaster sulfakinin receptor DSK-R1. Biochem Biophys Res Commun 2002; 291:313-320.
58. Radford JC, Davies SA, Dow JAT. Systematic G-protein-coupled receptor analysis in Drosophila melanogaster identifies a leucokinin receptor with novel roles. J Biol Chem 2002; 277:38810-38817.
59. Johnson EC, Bohn LM, Barak L et al. Identification of Drosophila neuropeptide receptors by G protein-coupled receptors-bèta-arrestin2 interactions. J Biol Chem 2003; 278:52172-52178.
60. Birse RT, Johnson EC, Taghert PH et al. Widely distributed Drosophila G-protein-coupled receptor (CG7887) is activated by endogenous tachykinin-related peptides. J Neurobiol 2006; 66:33-46.
61. Rosenkilde C, Cazzamali G, Williamson M et al. Molecular cloning, functional expression and gene silencing of two Drosophila receptors for the Drosophila neuropeptide pyrokinin-2. Biochem Biophys Res Commun 2003; 309:485-494.
62. Jorgensen LM, Hauser F, Cazzamali G et al. Molecular identification of the first SIFamide receptor. Biochem Biophys Res Commun 2006; 340:696-701.
63. Garczynski SF, Brown MR, Shen P et al. Characterization of a functional neuropeptide F receptor from Drosophila melanogaster. Peptides 2002; 23:773-780.
64. Feng GP, Reale V, Chatwin H et al. Functional characterization of a neuropeptide F-like receptor from Drosophila melanogaster. Eur J Neurosci 2003; 18:227-238.
65. Mertens I, Meeusen T, Huybrechts R et al. Characterization of the short neuropeptide F receptor from Drosophila melanogaster. Biochem Biophys Res Commun 2002; 297:1140-1148.
66. Meeusen T, Mertens I, Clynen E et al. Identification in Drosophila melanogaster of the invertebrate G protein-coupled FMRFamide receptor. Proc Natl Acad Sci USA 2002; 99:15363-15368.
67. Iversen A, Cazzamali G, Williamson M et al. Molecular cloning and functional expression of a Drosophila receptor for the neuropeptides capa-1 and -2. Biochem Biophys Res Commun 2002; 299:628-633.
68. Park Y, Kim YJ, Adams ME. Identification of G protein-coupled receptors for Drosophila PRXamide peptides, CCAP, corazonin and AKH supports a theory of ligand-receptor co-evolution. Proc Natl Acad Sci USA 2002; 99:11423-11428.
69. Iversen A, Cazzamali G, Williamson M et al. Molecular identification of the first insect ecdysis triggering hormone receptors. Biochem Biophys Res Commun 2002; 299:924-931.
70. Johnson EC, Garczynski SF, Park D et al. Identification and characterization of a G protein-coupled receptor for the neuropeptide proctolin in Drosophila melanogaster. Proc Natl Acad Sci USA 2003; 100:6198-6203.
71. Egerod K, Reynisson E, Hauser F et al. Molecular cloning and functional expression of the first two specific insect myosuppressin receptors. Proc Natl Acad Sci USA 2003; 100:9808-9813.
72. Cazzamali G, Torp M, Hauser F et al. The Drosophila gene CG9918 codes for a pyrokinin-1 receptor. Biochem Biophys Res Commun 2005; 335:14-19.
73. Birgul N, Weise C, Kreienkamp HJ et al. Reverse physiology in Drosophila: identification of a novel allatostatin-like neuropeptide and its cognate receptor structurally related to the mammalian somatostatin/galanin/opioid receptor family. EMBO J 1999; 18:5892-5900.
74. Larsen MJ, Burton KJ, Zantello MR et al. Type A allatostatins from Drosophila melanogaster and Diplotera puncata activate two Drosophila allatostatin receptors, DAR-1 and DAR-2, expressed in CHO cells. Biochem Biophys Res Commun 2001; 286:895-901.
75. Kreienkamp HJ, Larusson HJ, Witte I et al. Functional annotation of two orphan G-protein-coupled receptors, Drostar1 and -2, from Drosophila melanogaster and their ligands by reverse pharmacology. Journal of Biological Chemistry 2002; 277:39937-39943.
76. Staubli F, Jorgensen TJD, Cazzamali G et al. Molecular identification of the insect adipokinetic hormone receptors. Proc Natl Acad Sci USA 2002; 99:3446-3451.
77. Cazzamali G, Saxild NPE, Grimmelikhuijzen CJP. Molecular cloning and functional expression of a Drosophila corazonin receptor. Biochem Biophys Res Commun 2002; 298:31-36.
78. Cazzamali G, Hauser F, Kobberup S et al. Molecular identification of a Drosophila G protein-coupled receptor specific for crustacean cardioactive peptide. Biochem Biophys Res Commun 2003; 303:146-152.
79. Mendive FM, Van Loy T, Claeysen S et al. Drosophila molting neurohormone bursicon is a heterodimer and the natural agonist of the orphan receptor DLGR2. Febs Letters 2005; 579:2171-2176.

80. Luo CW, Dewey EM, Sudo S et al. Bursicon, the insect cuticle-hardening hormone, is a heterodimeric cystine knot protein that activates G protein-coupled receptor LGR2. Proc Natl Acad Sci USA 2005; 102:2820-2825.
81. Sudo S, Kuwabara Y, Park JI et al. Heterodimeric fly glycoprotein hormone-alpha 2 (GPA2) and glycoprotein hormone-beta 5 (GPB5) activate fly leucine-rich repeat-containing G protein-coupled receptor-1 (DLGR1) and stimulation of human thyrotropin receptors by chimeric fly GPA2 and human GPB5. Endocrinology 2005; 146:3596-3604.
82. Johnson EC, Bohn LM, Taghert PH. Drosophila CG8422 encodes a functional diuretic hormone receptor. J Exp Biol 2004; 207:743-748.
83. Mertens I, Vandingenen A, Johnson EC et al. PDF receptor signalling in Drosophila contributes to both circadian and geotaxic behaviors. Neuron 2005.
84. Lear BC, Merrill CE, Lin JM et al. A G protein-coupled receptor, groom-of-PDF, is required for PDF neuron action in circadian behavior. Neuron 2005; 48:221-227.
85. Hyun S, Lee Y, Hong ST et al. Drosophila GPCR Han is a receptor for the circadian clock neuropeptide PDF. Neuron 2005; 48:267-278.
86. Johnson EC, Shafer OT, Trigg JS et al. A novel diuretic hormone receptor in Drosophila: evidence for conservation of CGRP signaling. J Exp Biol 2005; 208:1239-1246.
87. Cvejic S, Zhu Z, Felice SJ et al. The endogenous ligand Stunted of the GPCR Methuselah extends lifespan in Drosophila. Nat Cell Biol 2004; Epub.
88. Rorth P. A modular misexpression screen in Drosophila detecting tissue-specific phenotypes. Proc Natl Acad Sci USA 1996; 93:12418-12422.
89. Staudt N, Molitor A, Somogyi K et al. Gain-of-function screen for genes that affect Drosophila muscle pattern formation. PLoS Genet 2005; 1:e55.
90. Roseman RR, Johnson EA, Rodesch CK et al. A P element containing suppressor of hairy-wing binding regions has novel properties for mutagenesis in Drosophila melanogaster. Genetics 1995; 141:1061-1074.
91. Bellen HJ, Levis RW, Liao G et al. The BDGP gene disruption project: single transposon insertions associated with 40% of Drosophila genes. Genetics 2004; 167:761-781.
92. Lukacsovich T, Asztalos Z, Awano W et al. Dual-tagging gene trap of novel genes in Drosophila melanogaster. Genetics 2001; 157:727-742.
93. Bier E, Vaessin H, Shepherd S et al. Searching for pattern and mutation in the Drosophila genome with a P-lacZ vector. Genes Dev 1989; 3:1273-1287.
94. Torok T, Tick G, Alvarado M et al. P-lacW insertional mutagenesis on the second chromosome of Drosophila melanogaster: isolation of lethals with different overgrowth phenotypes. Genetics 1993; 135:71-80.
95. Beinert N, Werner M, Dowe G et al. Systematic gene targeting on the X chromosome of Drosophila melanogaster. Chromosoma 2004; 113:271-275.
96. Mlodzik M, Hiromi Y. Enhancer trap method in Drosophila: its application to neurobiology. 1992; 397-414.
97. Thibault ST, Singer MA, Miyazaki WY et al. A complementary transposon tool kit for Drosophila melanogaster using P and piggyBac. Nat Genet 2004; 36:283-287.
98. Huet F, Lu JT, Myrick KV et al. A deletion-generator compound element allows deletion saturation analysis for genomewide phenotypic annotation. Proc Natl Acad Sci USA 2002; 99:9948-9953.
99. Golic KG, Golic MM. Engineering the Drosophila genome: chromosome rearrangements by design. Genetics 1996; 144:1693-1711.
100. Deak P, Omar MM, Saunders RD et al. P-element insertion alleles of essential genes on the third chromosome of Drosophila melanogaster: correlation of physical and cytogenetic maps in chromosomal region 86E-87F. Genetics 1997; 147:1697-1722.
101. Bellotto M, Bopp D, Senti KA et al. Maternal-effect loci involved in Drosophila oogenesis and embryogenesis: P element-induced mutations on the third chromosome. Int J Dev Biol 2002; 46:149-157.
102. Oh SW, Kingsley T, Shin HH et al. A P-element insertion screen identified mutations in 455 novel essential genes in Drosophila. Genetics 2003; 163:195-201.
103. Brand AH, Perrimon N. Targeted gene expression as a means of altering cell fates and generating dominant phenotypes. Development 1993; 118:401-415.
104. Toba G, Ohsako T, Miyata N et al. The gene search system. A method for efficient detection and rapid molecular identification of genes in Drosophila melanogaster. Genetics 1999; 151:725-737.

Chapter 11

Neuropeptide Receptors as Possible Targets for Development of Insect Pest Control Agents

Matthias B. Van Hiel, Tom Van Loy, Jeroen Poels, Hans Peter Vandersmissen, Heleen Verlinden, Liesbeth Badisco and Jozef Vanden Broeck*

Abstract

Various insect species have a severe impact on human welfare and environment and thus force us to continuously develop novel agents for pest control. Neuropeptides constitute a very versatile class of bioactive messenger molecules that initiate and/or regulate a wide array of vital biological processes in insects by acting on their respective receptors in the plasmamembrane of target cells. These receptors belong to two distinct categories of signal transducing proteins, i.e., heptahelical or G protein-coupled receptors (7TM, GPCR) and single transmembrane containing receptors. An increasing amount of evidence indicates that insect neuropeptide-receptor couples play crucial roles in processes as diverse as development, metabolism, ecdysis and reproduction. As such, they gain growing interest as promising candidate targets for the development of a new generation of species- and receptor-specific insect control agents that may generate fewer side effects. In this chapter, we will present some examples of insect neuropeptide receptors and aim to demonstrate their fundamental importance in insect biology.

Introduction

Insects represent the most diverse animal group on earth, occupying almost all terrestrial ecological niches. With nearly 1 million documented species and an estimated total of 10 times more, they account for at least 70% of animal diversity. Due to their unequalled adaptive radiation and their incredible diversity, they fulfil a crucial role in numerous ecosystems. They also produce economically interesting consumables (e.g., honey, silk or wax), are the pollinators of many flowering plants and are recyclers of dead organic material. Perhaps their main and most overlooked benefit lies in their insectivorous nature. Many insects, such as dragonflies, wasps and beetles feed on other insects and in this way control the populations of potential pest species. On the other hand, numerous insects have harmful effects on humanity, as vectors for diseases or as competitors for our nutrition. For instance, the spreading of malaria by mosquitoes costs billions for prevention of infection and still these insects infect ca. 650 million people annually, from which at least 1 million succumb. Other illnesses transmitted through insects include the bubonic plague (fleas), dengue and yellow fever (mosquitoes), sleeping sickness (tsetse fly), Chagas' disease (assassin bug) and leishmaniasis (sandfly). In addition, outbreaks of herbivorous insects, such as locusts, can cause massive damage to harvests, endangering the food supply for millions of people.

*Corresponding Author: Jozef Vanden Broeck—Department of Animal Physiology and Neurobiology, Zoological Institute K.U.Leuven, Naamsestraat 59, B-3000 Leuven, Belgium. Email: Jozef.VandenBroeck@bio.kuleuven.be

Neuropeptide Systems as Targets for Parasite and Pest Control, edited by Timothy G. Geary and Aaron G. Maule. ©2010 Landes Bioscience and Springer Science+Business Media.

Clearly, there is a continuous need to develop novel insect control agents to prevent outbreaks and to protect our crops. Most of the currently used insecticides are either detrimental to the environment or are toxic to nontarget organisms, including mankind. Another problem is the emergence of growing insect populations that are resistant to the noxious agents. These problems cause an incessant search for new, selective targets to alleviate the constant threat. Currently, novel approaches direct the attention to neuropeptides for the development of so-called fourth generation insecticides.[1,2] Neuropeptides constitute a very versatile class of extracellular signalling molecules primarily, but not exclusively, found in the nervous system. They regulate many developmental, metabolic, behavioral and reproductive processes. To exploit neuropeptidergic systems in the search for potential control agents, it is of vital importance to characterize the receptors, given that potential insecticides might modulate receptor activity as superagonists, antagonists or inverse agonists. Insect neuropeptide receptors can be classified into two major categories: single transmembrane (TM) receptors and heptahelical (7TM) or G protein-coupled receptors (GPCRs), the latter group comprising the majority of insect peptide and protein hormone receptors. So far, no insect neuropeptide receptors have been discovered within the third class of membrane receptors, the ligand-gated ion channels.

Since the annotation of the *Drosophila melanogaster* genome, more than half of the 45 predicted GPCRs that display similarity with mammalian peptide or protein hormone receptors[3,4] have been "deorphanized", meaning that their endogenous ligands have been identified (for a recent review see ref. 5). Other insect genome projects revealed that the majority of these receptors, as well as the corresponding neuropeptides, have (co)evolved in diverse insect orders. This short review will not give a complete overview of all known neuropeptide receptors of insects, but aims to suggest the potential use of some receptors as targets for the development of a novel generation of pest controlling agents (Table 1). We will briefly discuss the significance of peptide receptor mediated signalling in processes as diverse as growth, development, reproduction, water and salt balance, energy metabolism and behavior. These receptors may represent examples of candidate targets for novel insect control agents, based on their involvement in processes that are likely to be crucial for the fitness and success of insect populations.

Receptors Controlling Insect Development

Allatostatin Receptors

Before the emergence as an adult, several moults take place in an insect. This is necessary since the cuticle, which functions as an exoskeleton, is a very rigid structure. Two 'classic' hormones cooperate in regulating these moults: 20-hydroxyecdysone (20E) triggers the actual moult, whereas juvenile hormone (JH) defines which type of moult is taking place. Two neuropeptide-mediated activities work antagonistically to control the production of JH. Allatotropins (AT) stimulate JH production in the *corpora allata* (CA), while allatostatins (AST) inhibit the synthesis of JH.

At present, several receptors for AST have been identified, though allatotropin receptors have yet to be discovered. The existence of nervous control of the CA had already been postulated by Scharrer in the late 1940s[6] and later it was shown that neurosecretory cells in the brain of *Locusta migratoria* nymphs produced both stimulatory and inhibitory factors affecting the CA.[7,8] The isolation of the first of these neuropeptides, however, had to wait until two decades later.[9,10] Three types of AST exist: the A-type, first discovered in the cockroach,[10] the B-type or "cricket" allatostatins[11,12] and the C-type, which was first characterized in the moth *Manduca sexta*.[13] Nowadays, ASTs have been identified from over 30 species of insects and crustaceans.[14] In addition to their inhibitory effect on JH production, they still induce some other effects as well. For instance, in certain cockroaches, A-type ASTs reduce the heartbeat or increase the activity of amylase and invertase.[15] B-type ASTs are also known as "myoinhibiting peptides" (MIP) for their inhibitory effects on visceral muscle motility and inhibit ecdysteroid synthesis in the prothoracic glands of *Manduca* and *Bombyx*.[16] Seven allatostatin receptors have been characterized, so far. Two type A receptors have been identified in *Drosophila*: DAR-1 and DAR-2, which are activated by *Drm*-AstA.[17-19]

Table 1. Deorphanized neuropeptide receptors discussed in this chapter

Species	Name	Synonyms	Agonist	Ref	Species	Name	Synonyms	Agonist	Ref
GPCRs with developmental functions					**GPCRs with behavioral functions**				
D mel	DAR-1	CG2872, DalstR1	Allatostatin-A (*Drm*-AstA 3)	17	*D mel*	NKD	*CG6515, Takr86C*	*Lom*-TK II (Locustatachykinin)	138
D mel	DAR-1	CG2872, DAlstR1	Allatostatin-A (*Drm*-AstA)	18	*D mel*	NKD	*CG6515, Takr86C*	*Drm*-TK 1	22
D mel	DAR-2	CG10001, DAlstR2	Allatostatin-A (*Drm*-AstA)	18	*D mel*	DTKR	CG7887, *Takr99D*	Vertebrate TK (substance P)	139
D mel	DAR-2	CG10001, DAlstR2	Allatostatin-A (drostatins) *Drm*-AstA-4	19	*D mel*	DTKR	CG7887, *Takr99D*	*Drm*-TK 1	22
P ame	AlstR		Allatostatin-A (*Dip*-AstA 8)	20	*S cal*	STKR	Stomoxytachykinin receptor	Stomoxytachykinin (*Stc*-TK)	140,141
B mori	BAR		*Bom*-AstA (bostatins) *Bom*-AstA-2,3	21	**NPF/NPY receptors**				
D mel	AstB-R	CG30106, CG14484, MIPR	*Drm*-AstB 1 (*Drm*-MIP 1)	22	*D mel*	NPFR1	CG1147, *NPFR1*	*Drm*-NPF	102
					D mel	NPFR1	CG1147, NPFR1	*Drm*-NPF	22
D mel	Drostar1	CG7285	Gln-AstC; pyroGlu-AstC	23	*D mel*	*Drm*-sNPF-R	CG7395, CG18639, *NPFR76F*	*Drm*-sNPF-1 *Drm*-sNPF-2	109,110
D mel	Drostar1	CG7285	AstC	22	*D mel*	*Drm*-sNPF-R	CG7395, CG18639, *NPFR76F*	*Drm*-sNPF-1 *Drm*-sNPF-2 *Drm*-sNPF-3 *Drm*-sNPF-4	108

continued on next page

Table 1. Continued

Species	Name	Synonyms	Agonist	Ref	Species	Name	Synonyms	Agonist	Ref
D mel	Drostar2	CG13702, ALCR2	Gln-AstC; pyroGlu-AstC	23	**GPCRs with functions in ecdysis**				
D mel	Drostar2	CG13702, ALCR2	AstC	22	*D mel*	ETHR	CG5911a and CG5911b	ETH-1 ETH-2	125,124
D mel	DLGR-1	*Fsh*, CG7665	Fly GPA/GPB	25,27	*D mel*	DCRZR	CG10698, DGHRH II	Corazonin (*Drm*-CRZ)	22,71,130
D mel	DLGR-2	Rickets (rk), CG8930	Bursicon	32,142	*M sexta*	MasCRZR		Corazonin	129
GPCRs with metabolic functions					*D mel*	Putative CCAP-R	CG6111	CCAP AKH	71
D mel		CG17415	Diuretic hormone 31 (DH31)	64	*D mel*	CCAP-R	CG6111 (corrected sequence)	CCAP	143
D mel	*Drm*-DH-R	CG8422	DH 44	63	**GPCRs for PK/PBAN peptides**				
M sexta	*Mas*-DH-R		Mas-DH, Pea-DH Acd-DH, Lom-DH Mas-DPII, Mas-DH	60	*D mel*	Putative Cap2b-R	CG14575; capaR	Cap2b-1 Cap2b-2	71,125
A dom	*Acd*-DH-R		Acd-DH, Mas-DH Pea-DH, Acd-DH	61	*D mel*	Putative PRXa R, DPKR1	CG8795	Hugγ, *Drm*-PK-2 *Drm*-PK-1/Cap2b-3 ETH-1 Hugγ,*Lem*-PK	71,133
D mel	DAKHR	CG11325, DGRHR	*Drm*-AKH	72	*D mel*	Putative PRXa R	CG9918	*Drm*-PK-1/Cap2b-3	71
D mel	DAKHR	CG11325, DGRHR	*Drm*-AKH	124	*D mel*	DPK-R2	CG8784	Drm-PK-2, Hugγ	133
B mori	BAKHR		*Hez*-HrTH, *Mas*-AKH	72	*B mori*	*Bom*-PBANR		PBAN	135
					H zea	Hez-PBANR		PBAN	136
					S lit	PBANR		*Hez*-PBAN	137

Bombyx mori and *Periplaneta americana* both have a characterized AstA receptor.[20,21] The other characterized AST receptors are: the AstB-R, activated by *Drm*-AstB1 and Drostar1 and 2, for which AstC is the ligand.[22,23]

Leucine-Rich Repeat Containing G Protein-Coupled Receptors

Leucine-rich repeat containing G protein-coupled receptors (LGRs) form a unique GPCR subgroup that has been very well preserved, both structurally and at the amino acid sequence level, throughout metazoan evolution. They serve essential physiological and developmental functions in diverse animal phyla.[24] The most typical structural feature of LGRs is their large N-terminal extracellular domain involved in ligand binding.

Genome sequencing projects triggered the comparative study of LGR-encoding genes and revealed the existence of three distinct LGR subtypes that likely predate the divergence between vertebrates and invertebrates.[24] The first insect LGR ever cloned is the fruit fly DLGR1 (*Fsh*) that is homologous to mammalian gonadotropin [Follicle stimulating hormone (FSH)/Luteinizing hormone (LH)/human Choriogonadotropin (hCG)] and thyroid stimulating hormone (TSH) receptors (referred to as Type A LGRs).[25,26] The expression of this fly receptor starts 8-16 hours after oviposition and remains high until after pupariation, suggesting a possible role during development.[25] Recently, it was demonstrated that a fly dimeric hormone, composed of a glycoprotein hormone α (GPA)- and β (GPB)-subunit related protein respectively, was capable of activating DLGR1 resulting in increased intracellular cAMP-levels.[27] Furthermore, genes coding for Type A LGRs and putative glycoprotein hormone-like subunits were predicted in silico in the genomes of other insects, such as the malaria mosquito, *Anopheles gambiae*,[24,28] as well as in nematodes.[29,30] This supports the idea that at least some hormone-LGR couples originated before the radiation of the major animal phyla.

The first Type B LGR to be fully characterized was the fruit fly bursicon receptor (DLGR2 or *rickets*)[31,32] that exhibits structural homology with three mammalian orphan receptors, i.e., LGRs 4, 5 and 6, as well as with predicted GPCRs from other insects, such as the mosquito, *Anopheles gambiae*[28] and the honeybee, *Apis mellifera*.[33] In insects, bursicon steers the remodeling, i.e., sclerotization and melanization of the newly synthesized exoskeleton (cuticle), after each moulting cycle. In addition, phenotypic analysis of bursicon deficient *Drosophila* mutants revealed its regulatory role in wing expansion behavior and in wing development in adult insects escaping from their puparium.[34,35] Known bursicon bioactivity seems to be restricted to very short periods of time during the insect's lifespan and, therefore, this important pathway may be considered as a potentially promising target for novel pesticides. For instance, a substance that interferes with the bursicon receptor would likely not kill indiscriminately the entire population, but would mainly act as a regulatory agent, for example during an epidemic expansion of a pest. In addition, the relatively low similarity between the hormone-interacting receptor domain of the bursicon receptor from diverse insects (*cf.* there is ca. 50% identity between *Drosophila* and *Anopheles*), may offer the opportunity to develop low molecular weight, species-specific receptor antagonists or modulators.

In several insect species, a third category of LGRs (known as Type C LGRs) exists. These insect receptors are structurally related to mammalian LGR7 and LGR8. Whereas the ligands for insect Type C LGRs remain to be identified, their human counterparts mediate the bioactivity of relaxin-related peptides, thereby steering important developmental processes, such as testicular descent.[36,37] In several invertebrate genomes, orthologous LGR Type C receptors are encoded.[3,4,28,33]

Insulin Receptors

Members of the insulin superfamily can also initiate a highly conserved signalling pathway by activating specific receptor tyrosine kinases (RTKs), a category of single membrane-spanning proteins. In different insect species, insulin receptors (IRs) have been characterized. Like its mammalian homolog, the *Drosophila* insulin receptor (DIR) is composed of two α-subunits and two β-subunits linked by disulfide bonds, but displays amino- and carboxyl-terminal extensions.[38] Insulin receptors have also been cloned from the ovaries of the yellow fever mosquito

(*Aedes aegypti*)[39] and of the silkmoth, *Bombyx mori* (Lindstrom-Dinnetz and Iatrou, unpublished, database accession number AF025542).

Upon binding of insulin or insulin-related peptides to RTK, specific tyrosine residues of the latter become phosphorylated and are able to recruit the insulin receptor substrate(s) (IRS), which consequently can initiate two signalling pathways,[40,41] i.e., the mitogen-activated protein kinase (MAPK) pathway and the phosphatidylinositol-3-OH kinase/protein kinase B (PI3K/PKB) pathway.[42,43] Various studies have demonstrated the importance of the insulin/IRP signalling pathway in a variety of physiological processes, including growth and reproduction (for a review see ref. 44).

Loss-of-function mutations in various components of the insulin/PI3K signalling pathway all appeared to have a negative effect on growth in *D. melanogaster.*[45-47] Moreover, insulin/IRP signalling was suggested to play a role in regulating the link between growth and nutritional status. For *D. melanogaster* and *B. mori*, IRP expression was shown to vary with hemolymph carbohydrate levels, which are suggested to be indicators of nutritional status.[48-50] Inhibition of the insulin/PI3K signalling pathway mimics the effect of starvation, whereas activation of the pathway bypasses the nutritional requirements for cell growth.[51] It appears that both the central nervous system and the fat body are capable of adjusting the secretion of growth regulators in response to the level of nutrients in the haemolymph.[50,52,53] Hence, insulin-related peptides in insects were suggested to act as similar growth regulators.

Various studies demonstrated that insulin-related peptides are likely to play a role in insect reproduction. Putative insulin receptors have been observed in ovarian cells of three different lepidopteran species.[54] In these species, insulin-like substances are probably involved in ovarian development. In *Ae. aegypti*, a mosquito insulin receptor (MIR) has been demonstrated in follicular cells. In adult female insects, these cells are the primary source of ecdysteroids. Interestingly, treatment with bovine insulin caused the *Ae. aegypti* follicular cells to produce ecdysteroids,[55] which further stimulate egg maturation and vitellogenesis in dipteran species.[56,57] Moreover, MIR appeared to be expressed in function of the reproductive cycle.[58] Female *D. melanogaster* require an intact insulin/IRP signalling pathway in the ovaries to regulate egg production as response to changes in the diet.[59] In general, animals (Metazoa) with reduced insulin/IRP signalling also show reduced fertility.

Diuretic Hormone Receptors as Regulators of Water and Salt Balances

In insects, a tight regulation of water and salt balance is maintained by a number of diuretic peptide hormones (DHs), which display sequence similarity to corticotrophin releasing factor (CRF) and calcitonin. The first insect receptor for this category of peptides was identified in the tobacco hornworm, *Manduca sexta*. It transduces the DH signal generating cAMP as a second messenger.[60] In 1996, the orthologous receptor in the cricket, *Acheta domesticus*, was identified,[61] a few years later followed by the cloning of a related receptor from the Malpighian tubules of *Bombyx mori.*[62]

In the genome of *Drosophila melanogaster*, five related receptors were identified—CG13758, CG17415, CG4395, CG8422, CG12370.[3,4] Two of these receptors have been characterized as functional diuretic hormone receptors. CG8422 was shown to be the receptor for the CRF-like peptide DH44,[63] while CG17415 recognizes DH31,[64] a calcitonin-like peptide.

The receptors mentioned above are members of the class B G-protein coupled receptors (GPCRs), also known as the secretin receptor family, most of which are activated by medium-sized peptides. Not all of these are involved in controlling water and salts homeostasis. Another subfamily within this class of GPCRs, are the Methuselah-like receptors, of which a dozen lie encoded in the *Drosophila* genome. The GPCR, Methuselah, is of great importance in the fruit fly's stress response and biological ageing as heterozygous mutants lived longer and showed an increased resistance to several sources of stress.[65] The same phenotype was observed in flies that were mutated in the *sun* gene, which encodes the ligands for Mth, Sun A and B.[66]

Recently, antagonists for Mth were developed by mRNA display selection which also induce the effects described above. Binding studies using these antagonists revealed that they interact with the receptor at the same site as the Sun peptides. Based on X-ray diffraction analysis of the receptor-antagonist complex, this site was identified as an interface between the ectodomain and the extracellular loops of Mth.[67]

AKH Receptors Are Involved in the Control of Energy Metabolism

An important group of metabolic peptides are the adipokinetic hormones (AKHs). By activating lipases and glycogen phosphorylase in the fat body, they set free energy rich substrates, such as carbohydrates, lipids and proline, in the haemolymph and enable the insect to perform intense and exigent muscular tasks, such as locomotor activity or flight. In addition to this primary function, these pleiotropic hormones are also capable of inhibiting synthesis of RNA, fatty acids and proteins and of stimulating muscle contractions. They are found in nearly all insect orders and most insects appear to have at least one of these peptides. AKH's are produced in glandular cells of the *corpora cardiaca* (CC).[68,69] Several receptors for AKH have been characterized in insects. The first was found in *Manduca sexta*, whereby membrane fractions from the fat body were used to determine the optimal binding conditions for tritium-labelled Manduca-AKH.[70] Park et al deorphanized the *Drosophila* AKHR,[71] a discovery that was confirmed by another research group, which also identified the *Bombyx* AKHR by applying the AKH-like peptide of the moth *Helicoverpa zea*.[72] Recently categorized AKHR's include the *Periplaneta americana* AKHR that can be activated by Pea-AKH-I, as well as -II and has a slightly higher affinity for AKH-II[73,74] and the *Anopheles gambiae* AKH receptor.[75]

Receptors with a Role in Behavior

If it were possible to alter the behavior of a specific insect, this would form a powerful tool for pest control. Impairing their feeding behavior, for example, not only would hamper growth, but also would reduce damage to crops. A drastic inhibition of the insects' ecdysis behavior would reduce their capacity for further growth.

Tachykinin Receptors

Tachykinins (or neurokinins) exert an extensive spectrum of functions; they play an important neuromodulatory role in the central nervous system and exhibit a broad range of peripheral activities. They have been implicated in immunomodulation, cancer growth, pain transmission, smooth muscle contraction, etc.[76,77] Tachykinin-like peptides are also present in invertebrate species and can be separated into two distinct groups based on their respective sequence characteristics. All known vertebrate and a few invertebrate tachykinins share a common C-terminal amino acid motif, -FXGLMa. The "insectatachykinins" (also referred to as "tachykinin related peptides", TRPs), display structural similarity to the first group and possess a C-terminal -FX_1G/AX_2Ra consensus sequence. Peptides belonging to this group have not (yet) been found in vertebrates.[5,78]

In invertebrates, a limited number of tachykinin receptors have been fully cloned and characterized.[79-85] These are G protein-coupled receptors (GPCRs) and display sequence similarity to the vertebrate tachykinin receptors $NK_{1\text{-}3}$. Insect tachykinins are encoded as multiple isoforms on large precursor proteins; e.g., in *Drosophila*, 6 TRPs (*Drm*-TK1-6) are encoded by the *Tk* gene product, while only 2 TRP receptors (DTKR and NKD) have been found. Because there is no evidence for alternative splicing of the *Tk* gene, it is expected that *Drm*-TKs are colocalized and released together in equimolar concentrations.[86,87] This raises the question whether *Drosophila* tachykinins are redundant or show some specificity towards their receptors. While *Drm*-TKs are displaying nearly equipotent activities on DTKR, this seems not the case for NKD (Poels J et al, personal communication).[84,88]

STKR, a tachykinin receptor from the stable fly *Stomoxys calcitrans* shows a remarkable differentiation between common TRPs (C-terminal -FX*G*XRa) and Ala-containing TRPs (-FX*A*XRa; the endogenous *Stomoxys* tachykinin belongs to this group). This apparently small structural change leads to reduced efficacy (partial agonism) and increased potency of Ala-containing TRPs when STKR mediated Ca^{2+}-changes are investigated. Remarkably, the efficacy of Gly- or Ala-containing isoforms is not influenced when STKR-induced modifications of cAMP are concerned. Consequently, in analogy to the vertebrate NK receptors, a receptor model was postulated that takes into account the existence of 2 active states for STKR. When compared to the normal Gly-analogs, the Ala-analogs may stabilize a receptor conformation which less effectively activates the PLC-pathway or which is less abundant.[89] This discrimination of highly related TRPs could prove to be an important mechanism of physiological fine-tuning for this class of receptors in insects, where the amount of TRPs seems to exceed by far the number of putative receptors.

Interestingly, while insect tachykinins are not able to activate vertebrate receptors at physiological concentrations or vice versa, some invertebrates possess tachykinins with an -FXGLMa consensus sequence. Recently, a novel TRP receptor was cloned from *Octopus vulgaris*, an invertebrate that contains salivary gland specific vertebrate like tachykinins (oct-TK-I-II) as well as 7 TRPs (oct-TKRP). While the octopus TRP receptor was present in salivary glands, it could not be activated by oct-TK-I but was stimulated by oct-TRPs.[85] Whether -FXGLMa TKs exert physiological functions in invertebrates remains to be investigated, but since these tachykinins seem to reside in salivary glands of invertebrates that feed on vertebrate prey, a role in vasodilatation via vertebrate NK receptors is not unlikely.

Insect TRPs are mainly found in interneurons in the CNS and endocrine cells of the intestine, suggesting functions in neuromodulation and the gut. Myostimulatory in vitro actions of TRPs have been described several times, as well as regulation of fluid secretion by Malpighian tubules.[90]

Recently, in vivo roles for *Drm*-TKs in *Drosophila* were investigated by means of RNA interference of the *Tk* gene. Complete abolishment of *Drm*-TKs results in embryonic lethality, matching the presence of the two TRP receptors in *Drosophila*, which are differentially expressed throughout embryonic development. Neural *Tk* knock-down flies show subtle changes in behavioral responses to odorant-specific stimuli as well as increased locomotor activity. These behavioral changes correspond well to the wild type expression pattern of *Drm*-TKs in olfactory lobes and the central body complex.[91]

TRPs have also been detected in mushroom bodies of the honeybee *Apis mellifera*. The expression level of the *Apis* TRP precursor gene thereby seems to be differentially regulated in queens, workers and drones. Since mushroom bodies are associated with social behavior and because *Apis* TRPs are enriched in these specific brain regions, a possible role in the honeybee's social life seems possible.[92,93]

Given the vast array of physiological functions that tachykinins and most likely TRPs exert; the corresponding receptors, as well as tachykinin degrading enzymes, could prove to be valuable targets for insecticides. Several issues have to be addressed first; e.g., not much is known concerning the conformational constraints that TRPs undertake when activating their receptors. One report showed that analogues of the locustatachykinins and the mammalian tachykinin substance P, containing a sterically hindered Aib-NMePhe/Tyr residue block, share similar low-energy turn conformations.[94] Also, only very few peptide antagonists (Spantides I-III) have been tested on TRP receptors, while no reports exist on the action of nonpeptide antagonists.[84,95,96] In any case, the development of insect (species) specific TRP receptor antagonists would prove very useful in this field of research.

NPF/NPY Receptors

The neuropeptide F (NPF) family of peptides found in invertebrates is structurally related to the vertebrate neuropeptide Y (NPY) family.[97] NPY is one of the most abundant and widely distributed neuropeptides in the vertebrate central nervous system, which stimulates food consumption, affects blood pressure, induces anxiolysis, enhances memory retention and affects circadian rhythms. In insects, NPFs have been identified in the fruit fly, *Drosophila melanogaster* (*Drm*-NPF),[98] the yellow fever mosquito, *Aedes aegypti* (*Aea*-NPF)[99] and the African malaria mosquito, *Anopheles gambiae* (*Ang*-NPF).[100]

Receptors for NPY family members have been characterized in many vertebrates, but only a few invertebrates. All are seven-transmembrane G protein-coupled receptors (GPCR). In most mammals studied to date, five subtypes of NPY receptors appear to occur in each species.[101] Insect receptors for NPFs have been functionally identified in *D. melanogaster* and *A. gambiae.*[102,103] Sequences of invertebrate NPF receptors most closely resemble those of the Y_2 subtype. For insects, the presence of NPF in the central nervous system (CNS) and in the midgut[98,99] resembles the distribution of NPY family peptides in vertebrates. A role for NPF in feeding behavior was suggested by studies in *D. melanogaster*[104-106] and *A. aegypti.*[99] These findings suggest that the role of the NPY-like family in feeding behavior is highly conserved between mammals and insects.[5,106]

Invertebrate NPF-like peptides occur in both long and short forms. The long forms range in size from 36 to 40 amino acid residues and display sequence similarity to vertebrate NPY, others are much shorter but display some C-terminal sequence similarity with both NPY and NPF, as well as with some RFa-containing neuropeptides (e.g., prolactin-releasing peptide) of vertebrates.[107] These peptides are designated as "short NPFs" (sNPFs).

Receptors for sNPFs have been functionally identified in *D. melanogaster,*[108-110] the red imported fire ant, *Solenopsis invicta*[111] and *A. gambia.*[112] In insects, sNPFs seem to play a role as neurotransmitters or neuromodulators and probably function as regulators of feeding behavior and reproduction.[108,111-113]

Receptors Controlling Ecdysis Behavior

Because of its rigid structure the insect's exoskeleton or cuticle needs to be replaced periodically to allow growth and development, a process known as moulting. A rise in the 20-hydroxy-ecdysone (20E) titer initiates the moult that begins with apolysis, the detachment of the cuticle from the underlying epidermis. As the moult progresses, the old cuticle is partially degraded in preparation for the insect's escape, while a new one is being synthesized. Once the 20E titers fall below a treshold level, moulting culminates with the stereotyped behavioral program of ecdysis, i.e., the shedding of the remains of the old cuticle. Ecdysis requires two sets of endocrine cells: the peripherally located Inka cells that release pre-ecdysis triggering hormone (PETH) and ecdysis triggering hormone (ETH) [114,115] and the centrally located ventromedial (VM) neurons that release eclosion hormone (EH).[116,117] A positive feedback system allows both endocrine cells to reciprocally excite each other resulting in a massive EH/ETH surge in the haemolymph, as well as a release of EH within the central nervous system (CNS). Also, the undecapeptide hormone corazonin (CRZ) is likely implicated in ecdysis control. In the tobacco hawkmoth, *Manduca sexta,* CRZ receptor is expressed in the peripheral endocrine Inka cells and injections of this peptide hormone evoke premature ecdysis behavior by stimulating the release of PETH and ETH.[118] ETH triggers the first motor program of ecdysis, the pre-ecdysis behavior.[119,120] Subsequently, EH released within the CNS triggers the secretion of crustacean cardioactive peptide (CCAP), a peptide initially discovered through its acceleratory activity on the heart rate of the shore crab,[121] from a subset of neurons in the ventral nervous system.[122] CCAP plays a key role in ecdysis behavior. Apart from suppressing the pre-ecdysis phase, it initiates the ecdysis motor program.[123,124]

Although several insect genomes harbour putative ETH receptors, only the *Drosophila* ETH receptor (ETHR), for which two subtypes exist (ETHR-A and ETHR-B), has been characterized. Both receptor subtypes bind ETH-1 and ETH-2 and in both cases ETH-1 is the most potent agonist.[125,124] Interestingly, ETHR-A and ETHR-B are differently expressed in larval brain suggesting dissimilar functions. Also two CCAP receptors have been described, one in *Drosophila melanogaster,*[126] the other in the mosquito, *Anopheles gambiae.*[75] Apart from its function in ecdysis, CCAP exerts numerous other effects in vivo. Furthermore, this peptide is implicated in the intestinal digestion[127] and induces the release of adipokinetic hormone (AKH) in locusts.[128]

"Corazonin" is another peptide with a name that is linked to its primarily described activity. It was discovered in cockroaches as a cardio-acceleratory peptide. Several research groups have identified the endogenous ligand for the CRZ receptor of the fruitfly and the tobacco hornworm orthologue has been characterized as well.[22,71,129,130] Recently, the *Anopheles* receptor has also been laid bare.[75]

Receptors Controlling Insect Reproduction

Instead of impeding with processes to prevent insects from maturing into adulthood, another strategy focuses on directly interfering with reproduction. Obviously, the control of reproduction would offer an optimal means to control population growth. Several developmental peptides are involved in the regulation of reproductive physiology.[131] In addition, a very prominent role in insect reproduction is occupied by pheromones, since the localization of a suitable partner is crucial for the survival of a species.

PK/PBAN Receptors

The production of pheromones involved in sexual communication of (lepidopteran) insects is controlled by a neuropeptide produced in neuroendocrine cells of the suboesophageal ganglion. This peptide, called pheromone biosynthesis activating neuropeptide (PBAN), shares a conserved C-terminal FXPRLamide motif with pyrokinins (PK), diapause hormones (DH), myotropins (MT) and other peptides, that constitute the PK/PBAN family. Apparently, some members of this family can elicit pheromonotropic reactions, as PBAN. It is widely accepted that the target site of PBAN is the pheromone gland and several studies support this. Conversely, other reports claim an indirect control of PBAN via the terminal abdominal ganglion or the *bursa copulatrix* (reviewed by Altstein).[132] Other important activities are induced by this pleiotropic peptide family, but these will not be further discussed in detail here.

The first receptors for FXPRLa peptides were discovered in the fruitfly.[71] The receptor, CG14575, is activated by Cap2b-1 and Cap2b-2, peptides which were known for their cardio-acceleratory effects in both *Manduca* and *Drosophila*. In addition, CG8795 can be activated by Hugγ, *Drm*-PK-2 and *Drm*-PK-1/Cap2b-3. The latter peptide is also recognized by a third putative PK receptor (CG9918).[71] The activation of CG8795, also called DPK-R1, by *Drm*-PK-2 en Hugγ was confirmed by another group that also cloned and tested DPK-R2 (CG8784) recognizing the same ligands as DPK-R1.[133] CG9918 was later confirmed to be the *Drosophila* PK receptor.[134] In the silkworm *Bombyx mori*[135] and the moth *Helicoverpa zea*[136] the PBANR was activated by conspecific PBAN. Recently, *Hez*-PBAN was found to activate the MAPK pathway through the PBANR of *Spodoptera littoralis*.[137]

Conclusion and Future Prospects

This review clearly indicates that neuropeptides and their receptors are major players in the regulation of numerous vital processes. Therefore, it is easily understood that the scientific community has, for decades already, considered these proteins a top research subject, as potential targets for drug development or for the production of control agents to combat insect pests. For instance, the agrochemical industry is interested in the development of products that are not only very effective, but also highly selective for the targeted pest species, thus complying with the ever-growing awareness for the environment. At present, it has become possible to start developing synthetic ligands for neuropeptide receptors, which might lead to valuable products for pest control.

Previous insect genome projects already revealed the existence of many different genes that are believed to be involved in peptide ligand induced signalling functions. Initially, it is important to elucidate the biochemical pathways through which neuropeptides exert their function, as well as the biodynamic interplay of this large set of genes. A number of recent research techniques facilitate the study of specific genes and their encoded proteins. For instance, RNA interference can be employed to study the phenotypic effects of selective posttranscriptional gene silencing. Furthermore, specific changes in gene expression can be studied by real-time quantitative reverse transcriptase PCR or, at a genome-wide scale, microarray analyses can be performed. Both methods are complementary to each other and may prove to become very useful in the identification of peptide induced downstream signalling events leading to changes in gene expression at the transcriptional level. Another benefit is the advent of higly sensitive methods for the detection of the gene products themselves, at the peptide or protein level (proteomics).

These and other techniques will prove to be useful tools in the characterization of the many receptors, of which the exact function still eludes us today, as well as in the identification of ligands for remaining orphan receptors.

Note Added in Proof

Since the acceptance of this chapter a few papers, relevant to this issue, have been published. Rewitz et al. found that the neuropeptide PTTH (prothoracicotropic hormone) activates the RTK Torso in Drosophila, which then initiates metamorphosis [Rewitz et al. Science 2009; 326(5958):1403-5]. This receptor, which is also involved in embryonic terminal cell fate, could be an interesting target for pest control agents. Another research group deorphanised the inotocin GPCR from Tribolium, an oxytocin/vasopressin-like receptor in insects, and discovered there counterparts in numerous other arthropods [Stafflinger et al. Proc Natl Acad Sci USA 2008; 105(9):3262-7]. In addition, unpublished data mentioned in the text has since then been published. The neuropeptide Drosophila tachykinin-related peptide 6 (DTK6) activates the NKD receptor (neurokinin receptor from Drosophila; CG6515) [Poels et al. Peptides 2009; 30(3):545-56].

References

1. Hoffmann KH, Lorenz MW. Recent Advances in Hormones in Insect Pest Control. Phytoparasitica 1998; 26:323-330.
2. Masler EP, Kelly TJ, Menn JJ. Insect neuropeptides: discovery and application in insect management. Arch Insect Biochem Physiol 1993; 22:87-111.
3. Vanden Broeck J. Insect G protein-coupled receptors and signal transduction. Archives of Insect Biochemistry and Physiology 2001; 48:1-12.
4. Hewes RS, Taghert PH. Neuropeptides and neuropeptide receptors in the Drosophila melanogaster genome. Genome Res 2001; 11:1126-1142.
5. Claeys I, Poels J, Simonet G et al. Insect neuropeptide and peptide hormone receptors: current knowledge and future directions. Vitam Horm 2005; 73:217-282.
6. Scharrer B. Section of the nervi corpori cardiaci in Leucophaea maderae (Orthoptera). anatomical record 1946; 96:577.
7. Girardie A. Contribution a l'étude du controle de l'activité des corpora allata par la pars intercerebralis chez Locusta migratoria (L). Comptes rendus de l'Académie des sciences 1965; 261:4876-4878.
8. Girardie A. Controle neurohormonal de la metamorphose et de la pigmentation chez Locusta migratoria cinerascens. Bull Biol 1967; 101:79-114.
9. Kataoka H, Toschi A, Li JP et al. Identification of an Allatotropin from Adult Manduca Sexta. Science 1989; 243:1481-1483.
10. Woodhead AP, Stay B, Seidel SL et al. Primary structure of four allatostatins: neuropeptide inhibitors of juvenile hormone synthesis. Proc Natl Acad Sci USA 1989; 86:5997-6001.
11. Lorenz MW, Kellner R, Hoffmann KH. A family of neuropeptides that inhibit juvenile hormone biosynthesis in the cricket, Gryllus bimaculatus. J Biol Chem 1995; 270:21103-21108.
12. Schoofs L, Holman GM, Hayes TK et al. Isolation, identification and synthesis of locustamyoinhibiting peptide (LOM-MIP), a novel biologically active neuropeptide from Locusta migratoria. Regul Pept 1991; 36:111-119.
13. Kramer SJ, Toschi A, Miller CA et al. Identification of an allatostatin from the tobacco hornworm Manduca sexta. Proc Natl Acad Sci USA 1991; 88:9458-9462.
14. Stay B, Tobe SS. The role of allatostatins in juvenile hormone synthesis in insects and crustaceans. Annu Rev Entomol 2007; 52:277-299.
15. Gade G. Allatoregulatory peptides—molecules with multiple functions. Invertebrate Reproduction & Development 2002; 41:127-135.
16. Hua YJ, Tanaka Y, Nakamura K et al. Identification of a prothoracicostatic peptide in the larval brain of the silkworm, Bombyx mori. J Biol Chem 1999; 274:31169-31173.
17. Birgul N, Weise C, Kreienkamp HJ et al. Reverse physiology in drosophila: identification of a novel allatostatin-like neuropeptide and its cognate receptor structurally related to the mammalian somatostatin/galanin/opioid receptor family. EMBO J 1999; 18:5892-5900.
18. Larsen MJ, Burton KJ Zantello MR et al. Type A allatostatins from Drosophila melanogaster and Diplotera puncata activate two Drosophila allatostatin receptors, DAR-1 and DAR-2, expressed in CHO cells. Biochem Biophys Res Commun 2001; 286:895-901.

19. Lenz C, Williamson M, Hansen GN et al. Identification of four Drosophila allatostatins as the cognate ligands for the Drosophila orphan receptor DAR-2. Biochem Biophys Res Commun 2001; 286:1117-1122.
20. Auerswald L, Birgul N, Gade G et al. Structural, functional and evolutionary characterization of novel members of the allatostatin receptor family from insects. Biochem Biophys Res Commun 2001; 282:904-909.
21. Secher T, Lenz C, Cazzamali G et al. Molecular cloning of a functional allatostatin gut/brain receptor and an allatostatin preprohormone from the silkworm Bombyx mori. J Biol Chem 2001; 276:47052-47060.
22. Johnson EC, Bohn LM, Barak LS et al. Identification of Drosophila neuropeptide receptors by G protein-coupled receptors-beta-arrestin2 interactions. J Biol Chem 2003; 278:52172-52178.
23. Kreienkamp HJ, Larusson HJ, Witte I et al. Functional annotation of two orphan G-protein-coupled receptors, Drostar1 and -2, from Drosophila melanogaster and their ligands by reverse pharmacology. J Biol Chem 2002; 277:39937-39943.
24. Van Loy T, Vandersmissen HP, Van Hiel MB et al. Comparative genomics of leucine-rich repeats containing G protein-coupled receptors and their ligands. Gen Comp Endocrinol 2008;155:14-21.
25. Hauser F, Nothacker HP, Grimmelikhuijzen CJ. Molecular cloning, genomic organization and developmental regulation of a novel receptor from Drosophila melanogaster structurally related to members of the thyroid-stimulating hormone, follicle-stimulating hormone, luteinizing hormone/choriogonadotropin receptor family from mammals. J Biol Chem 1997; 272:1002-1010.
26. Nishi S, Hsu SY, Zell K et al. Characterization of two fly LGR (leucine-rich repeat-containing, G protein-coupled receptor) proteins homologous to vertebrate glycoprotein hormone receptors: constitutive activation of wild-type fly LGR1 but not LGR2 in transfected mammalian cells. Endocrinology 2000; 141:4081-4090.
27. Sudo S, Kuwabara Y, Park JI et al. Heterodimeric fly glycoprotein hormone-alpha2 (GPA2) and glycoprotein hormone-beta5 (GPB5) activate fly leucine-rich repeat-containing G protein-coupled receptor-1 (DLGR1) and stimulation of human thyrotropin receptors by chimeric fly GPA2 and human GPB5. Endocrinology 2005; 146:3596-3604.
28. Hill CA, Fox AN, Pitts RJ et al. G protein-coupled receptors in Anopheles gambiae. Science 2002; 298:176-178.
29. Kudo M, Chen T, Nakabayashi K et al. The nematode leucine-rich repeat-containing, G protein-coupled receptor (LGR) protein homologous to vertebrate gonadotropin and thyrotropin receptors is constitutively active in mammalian cells. Mol Endocrinol 2000; 14:272-284.
30. Park JI, Semyonov J, Chang CL et al. Conservation of the heterodimeric glycoprotein hormone subunit family proteins and the LGR signaling system from nematodes to humans. Endocrine 2005; 26:267-276.
31. Eriksen KK, Hauser F, Schiott M et al. Molecular cloning, genomic organization, developmental regulation and a knock-out mutant of a novel leu-rich repeats-containing G protein-coupled receptor (DLGR-2) from Drosophila melanogaster. Genome Res 2000; 10:924-938.
32. Mendive FM, Van Loy T, Claeysen S et al. Drosophila molting neurohormone bursicon is a heterodimer and the natural agonist of the orphan receptor DLGR2. FEBS Lett 2005; 579:2171-2176.
33. Hauser F, Cazzamali G, Williamson M et al. A review of neurohormone GPCRs present in the fruitfly Drosophila melanogaster and the honey bee Apis mellifera. Prog Neurobiol 2006; 80:1-19.
34. Dewey EM, Mcnabb SL, Ewer J et al. Identification of the gene encoding bursicon, an insect neuropeptide responsible for cuticle sclerotization and wing spreading. Curr Biol 2004; 14:1208-1213.
35. Kimura KI, Kodama A, Hayasaka Y et al. Activation of the cAMP/PKA signaling pathway is required for post-ecdysial cell death in wing epidermal cells of Drosophila melanogaster. Development 2004; 131:1597-1606.
36. Kumagai J, Hsu SY, Matsumi H et al. INSL3/Leydig insulin-like peptide activates the LGR8 receptor important in testis descent. J Biol Chem 2002; 277:31283-31286.
37. Zimmermann S, Steding G, Emmen JM et al. Targeted disruption of the Insl3 gene causes bilateral cryptorchidism. Mol Endocrinol 1999; 13:681-691.
38. Ruan Y, Chen C, Cao Y et al. The Drosophila insulin receptor contains a novel carboxyl-terminal extension likely to play an important role in signal transduction. J Biol Chem 1995; 270:4236-4243.
39. Graf R, Neuenschwander S, Brown MR et al. Insulin-mediated secretion of ecdysteroids from mosquito ovaries and molecular cloning of the insulin receptor homologue from ovaries of bloodfed Aedes aegypti. Insect Mol Biol 1997; 6:151-163.
40. Yenush L, Fernandez R, Myers MG et al. The Drosophila insulin receptor activates multiple signaling pathways but requires insulin receptor substrate proteins for DNA synthesis. Mol Cell Biol 1996; 16:2509-2517.
41. White MF. The IRS-signalling system: A network of docking proteins that mediate insulin action. Mol Cell Biochem 1998; 182:3-11.

42. Blenis J. Signal transduction via the map kinases—Proceed at your own Rsk. Proc Natl Acad Sci USA 1993; 90:5889-5892.
43. Shepherd PR, Withers DJ, Siddle K. Phosphoinositide 3-kinase: the key switch in insulin signalling. The Biochemical Journal 1998; 333:471-490.
44. Claeys I, Simonet G, Poels J et al. Insulin-related peptides and their conserved signal transduction pathway. Peptides 2002; 23:807-816.
45. Edgar BA. How flies get their size: genetics meets physiology. Nat Rev Genet 2006; 7:907-916.
46. Goberdhan DCI, Wilson C. The functions of insulin signaling: size isn't everything, even in Drosophila. Differentiation 2003; 71:375-397.
47. Mirth CK, Riddiford LM. Size assessment and growth control: how adult size is determined in insects. BioEssays 2007; 29:344-355.
48. Satake S, Masumura M, Ishizaki H et al. Bombyxin, an insulin-related peptide of insects, reduces the major storage carbohydrates in the silkworm Bombyx mori. Comp Biochem Physiol BBiochem Mol Biol 1997; 118:349-357.
49. Masumura M, Satake SI, Saegusa H et al. Glucose stimulates the release of bombyxin, an insulin-related peptide of the silkworm Bombyx mori. Gen Comp Endocrinol 2000; 118:393-399.
50. Ikeya T, Galic M, Belawat P et al. Nutrient-dependent expression of insulin-like peptides from neuroendocrine cells in the CNS contributes to growth regulation in Drosophila. Curr Biol 2002; 12:1293-1300.
51. Britton JS, Lockwood WK, Li L et al. Drosophila's insulin/P13-kinase pathway coordinates cellular metabolism with nutritional conditions. Developmental Cell 2002; 2:239-249.
52. Britton JS, Edgar BA. Environmental control of the cell cycle in Drosophila: nutrition activates mitotic and endoreplicative cells by distinct mechanisms. Development 1998; 125:2149-2158.
53. Edgar BA. From small flies come big discoveries about size control. Nat Cell Biol 1999; 1:E191-E193.
54. Fullbright G, Lacy ER, Bullesbach EE. The prothoracicotropic hormone bombyxin has specific receptors on insect ovarian cells. Eur J Biochem 1997; 245:774-780.
55. Riehle MA, Brown MR. Insulin stimulates ecdysteroid production through a conserved signaling cascade in the mosquito Aedes aegypti. Insect Biochem 1999; 29:855-860.
56. Fallon AM, Hagedorn HH. Synthesis of vitellogenin by fat body in Aedes aegypti: effect of injected ecdysone. American Zoology 1972; 12:697-697.
57. Huybrechts R, De Loof A. Induction of vitellogenin synthesis in male Sarcophaga bullata by ecdysterone. J Insect Physiol 1977; 23:1359-1362.
58. Riehle MA, Brown MR. Insulin receptor expression during development and a reproductive cycle in the ovary of the mosquito Aedes aegypti. Cell Tissue Res 2002; 308:409-420.
59. Drummond-Barbosa D, Spradling AC. Stem cells and their progeny respond to nutritional changes during Drosophila oogenesis. Dev Biol 2001; 231:265-278.
60. Reagan JD. Expression cloning of an insect diuretic hormone receptor. A member of the calcitonin/secretin receptor family. J Biol Chem 1994; 269:9-12.
61. Reagan JD. Molecular cloning and function expression of a diuretic hormone receptor from the house cricket, Acheta domesticus. Insect Biochem Mol Biol 1996; 26:1-6.
62. Ha SD, Kataoka H, Suzuki A et al. Cloning and sequence analysis of cDNA for diuretic hormone receptor from the Bombyx mori. Mol Cells 2000; 10:13-17.
63. Johnson EC, Bohn LM, Taghert PH. Drosophila CG8422 encodes a functional diuretic hormone receptor. J Exp Biol 2004; 207:743-748.
64. Johnson EC, Shafer OT, Trigg JS et al. A novel diuretic hormone receptor in Drosophila: evidence for conservation of CGRP signaling. J Exp Biol 2005; 208:1239-1246.
65. Lin YJ, Seroude L, Benzer S. Extended life-span and stress resistance in the Drosophila mutant methuselah. Science 1998; 282:943-946.
66. Cvejic S, Zhu Z, Felice SJ et al. The endogenous ligand Stunted of the GPCR Methuselah extends lifespan in Drosophila. Nat Cell Biol 2004; 6:540-546.
67. Ja WW, West AP Jr, Delker SL et al. Extension of Drosophila melanogaster life span with a GPCR peptide inhibitor. Nat Chem Biol 2007; 3:415-419.
68. Gade G. Regulation of intermediary metabolism and water balance of insects by neuropeptides. Annu Rev Entomol 2004; 49:93-113.
69. Kaufmann C, Brown MR. Adipokinetic hormones in the African malaria mosquito, Anopheles gambiae: identification and expression of genes for two peptides and a putative receptor. Insect Biochem Mol Biol 2006; 36:466-481.
70. Ziegler R, Jasensky RD, Morimoto H. Characterization of the adipokinetic hormone receptor form the fat body of Manduca sexta. Regul Pept 1995; 57:329-338.

71. Park Y, Kim YJ, Adams ME. Identification of G protein-coupled receptors for Drosophila PRXamide peptides, CCAP, corazonin and AKH supports a theory of ligand-receptor coevolution. Proc Natl Acad Sci USA 2002; 99:11423-11428.
72. Staubli F, Jorgensen TJ, Cazzamali G et al. Molecular identification of the insect adipokinetic hormone receptors. Proc Natl Acad Sci USA 2002; 99:3446-3451.
73. Wicher D, Agricola HJ, Sohler S et al. Differential receptor activation by cockroach adipokinetic hormones produces differential effects on ion currents, neuronal activity and locomotion. J Neurophysiol 2006; 95:2314-2325.
74. Hansen KK, Hauser F, Cazzamali G et al. Cloning and characterization of the adipokinetic hormone receptor from the cockroach Periplaneta americana. Biochem Biophys Res Commun 2006; 343:638-643.
75. Belmont M, Cazzamali G, Williamson M et al. Identification of four evolutionarily related G protein-coupled receptors from the malaria mosquito Anopheles gambiae. Biochem Biophys Res Commun 2006; 344:160-165.
76. Pennefather JN, Lecci A, Candenas ML et al. Tachykinins and tachykinin receptors: a growing family. Life Sci 2004; 74:1445-1463.
77. Palma C. Tachykinins and their receptors in human malignancies. Current Drug Targets 2006; 7:1043-1052.
78. Nässel DR. Neuropeptides in the nervous system of Drosophila and other insects: multiple roles as neuromodulators and neurohormones. Prog Neurobiol 2002; 68:1-84.
79. Li XJ, Wolfgang W, Wu YN et al. Cloning, heterologous expression and developmental regulation of a Drosophila receptor for tachykinin-like peptides. EMBO J 1991; 10:3221-3229.
80. Monnier D, Colas JF, Rosay P et al. NKD, a developmentally regulated tachykinin receptor in Drosophila. J Biol Chem 1992; 267:1298-1302.
81. Guerrero FD. Cloning of a cDNA from stable fly which encodes a protein with homology to a Drosophila receptor for tachykinin-like peptides. Ann NY Acad Sci 1997; 814:310-311.
82. Kawada T, Furukawa Y, Shimizu Y et al. A novel tachykinin-related peptide receptor. Sequence, genomic organization and functional analysis. Eur J Biochem 2002; 269:4238-4246.
83. Birse RT Johnson EC, Taghert PH et al. Widely distributed Drosophila G-protein-coupled receptor (CG7887) is activated by endogenous tachykinin-related peptides. J Neurobiol 2006; 66:33-46.
84. Poels J, Verlinden H, Fichna J et al. Functional comparison of two evolutionary conserved insect neurokinin-like receptors. Peptides 2007; 28:103-108.
85. Kanda A, Takuwa-Kuroda K, Aoyama M et al. A novel tachykinin-related peptide receptor of Octopus vulgaris—evolutionary aspects of invertebrate tachykinin and tachykinin-related peptide. FEBS J 2007; 274:2229-2239.
86. Siviter RJ, Coast GM, Winther AM et al. Expression and functional characterization of a Drosophila neuropeptide precursor with homology to mammalian preprotachykinin A. J Biol Chem 2000; 275:23273-23280.
87. Winther AM, Siviter RJ, Isaac RE et al. Neuronal expression of tachykinin-related peptides and gene transcript during postembryonic development of Drosophila. J Comp Neurol 2003; 464:180-196.
88. Birse RT, Johnson EC, Taghert PH et al. Widely distributed Drosophila G-protein-coupled receptor (CG7887) is activated by endogenous tachykinin-related peptides. J Neurobiol 2006; 66:33-46.
89. Poels J, Van Loy T, Franssens V et al. Substitution of conserved glycine residue by alanine in natural and synthetic neuropeptide ligands causes partial agonism at the stomoxytachykinin receptor. J Neurochem 2004; 90:472-478.
90. Nässel DR. Neuropeptides in the nervous system of Drosophila and other insects: multiple roles as neuromodulators and neurohormones. Prog Neurobiol 2002; 68:1-84.
91. Winther AM, Acebes A, Ferrus A. Tachykinin-related peptides modulate odor perception and locomotor activity in Drosophila. Mol Cell Neurosci 2006; 31:399-406.
92. Takeuchi H, Yasuda A, Yasuda-Kamatani Y et al. Identification of a tachykinin-related neuropeptide from the honeybee brain using direct MALDI-TOF MS and its gene expression in worker, queen and drone heads. Insect Mol Biol 2003; 12:291-298.
93. Takeuchi H, Yasuda A, Yasuda-Kamatani Y et al. Prepro-tachykinin gene expression in the brain of the honeybee Apis mellifera. Cell Tissue Res 2004; 316:281-293.
94. Nachman RJ, Moyna G, Williams HJ et al. Comparison of active conformations of the insectatachykinin/tachykinin and insect kinin/Tyr-W-MIF-1 neuropeptide family pairs. Ann NY Acad Sci 1999; 897:388-400.
95. Torfs H, Shariatmadari R, Guerrero F et al. Characterization of a receptor for insect tachykinin-like peptide agonists by functional expression in a stable Drosophila Schneider 2 cell line. J Neurochem 2000; 74:2182-2189.

96. Lundquist CT, Nässel DR. Peptidergic activation of locust dorsal unpaired median neurons: Depolarization induced by locustatachykinins may be mediated by cyclic AMP. J Neurobiol 1997; 33:297-315.
97. Jong-Brink M, ter Maat A, Tensen CP. NPY in invertebrates: molecular answers to altered functions during evolution. Peptides 2001; 22:309-315.
98. Brown MR, Crim JW, Arata RC et al. Identification of a Drosophila brain-gut peptide related to the neuropeptide Y family. Peptides 1999; 20:1035-1042.
99. Stanek DM, Pohl J, Crim JW et al. Neuropeptide F and its expression in the yellow fever mosquito, Aedes aegypti. Peptides 2002; 23:1367-1378.
100. Riehle MA, Garczynski SF, Crim JW et al. Neuropeptides and peptide hormones in Anopheles gambiae. Science 2002; 298:172-175.
101. Berglund MM, Hipskind PA, Gehlert DR. Recent developments in our understanding of the physiological role of PP-fold peptide receptor subtypes. Exp Biol Med (Maywood) 2003; 228:217-244.
102. Garczynski SF, Brown MR, Shen P et al. Characterization of a functional neuropeptide F receptor from Drosophila melanogaster. Peptides 2002; 23:773-780.
103. Garczynski SF, Crim JW, Brown MR. Characterization of neuropeptide F and its receptor from the African malaria mosquito, Anopheles gambiae. Peptides 2005; 26:99-107.
104. Shen P, Cai HN. Drosophila neuropeptide F mediates integration of chemosensory stimulation and conditioning of the nervous system by food. Journal of Neurobiology 2001; 47:16-25.
105. Wu Q, Wen TQ, Lee G et al. Developmental control of foraging and social behavior by the Drosophila neuropeptide Y-like system. Neuron 2003; 39:147-161.
106. Lingo PR, Zhao Z, Shen P. Coregulation of cold-resistant food acquisition by insulin- and neuropeptide Y-like systems in Drosophila melanogaster. Neuroscience 2007; 148:371-374.
107. Spittaels K, Verhaert P, Shaw C et al. Insect neuropeptide F (NPF)-related peptides: isolation from Colorado potato beetle (Leptinotarsa decemlineata) brain. Insect Biochem Mol Biol 1996; 26:375-382.
108. Mertens I, Meeusen T, Huybrechts R et al. Characterization of the short neuropeptide F receptor from Drosophila melanogaster. Biochem Biophys Res Commun 2002; 297:1140-1148.
109. Feng G, Reale V, Chatwin H et al. Functional characterization of a neuropeptide F-like receptor from Drosophila melanogaster. Eur J Neurosci 2003; 18:227-238.
110. Reale V, Chatwin HM, Evans PD. The activation of G-protein gated inwardly rectifying K+ channels by a cloned Drosophila melanogaster neuropeptide F-like receptor. Eur J Neurosci 2004; 19:570-576.
111. Chen ME, Pietrantonio PV. The short neuropeptide F-like receptor from the red imported fire ant, Solenopsis invicta Buren (Hymenoptera: Formicidae). Arch Insect Biochem Physiol 2006; 61:195-208.
112. Garczynski SF, Brown MR, Crim JW. Structural studies of Drosophila short neuropeptide F: Occurrence and receptor binding activity. Peptides 2006; 27:575-582.
113. Cerstiaens A, Benfekih L, Zouiten H et al. Led-NPF-1 stimulates ovarian development in locusts. Peptides 1999; 20:39-44.
114. Zitnan D, Kingan TG, Hermesman JL et al. Identification of ecdysis-triggering hormone from an epitracheal endocrine system. Science 1996; 271:88-91.
115. Zitnan D, Ross LS, Zitnanova I et al. Steroid induction of a peptide hormone gene leads to orchestration of a defined behavioral sequence. Neuron 1999; 23:523-535.
116. Ewer J, Gammie SC, Truman JW. Control of insect ecdysis by a positive-feedback endocrine system: roles of eclosion hormone and ecdysis triggering hormone. J Exp Biol 1997; 200:869-881.
117. Truman JW, Fallon AM, Wyatt GR. Hormonal release of programmed behavior in silk moths: probable mediation by cyclic AMP. Science 1976; 194:1432-1434.
118. Kim YJ, Spalovska-Valachova I, Cho KH et al. Corazonin receptor signaling in ecdysis initiation. Proc Natl Acad Sci USA 2004; 101:6704-6709.
119. Ewer J, Truman JW. Increases in cyclic 3′, 5′-guanosine monophosphate (cGMP) occur at ecdysis in an evolutionarily conserved crustacean cardioactive peptide-immunoreactive insect neuronal network. J Comp Neurol 1996; 370:330-341.
120. Zitnan D, Kingan TG, Hermesman JL et al. Identification of ecdysis-triggering hormone from an epitracheal endocrine system. Science 1996; 271:88-91.
121. Stangier J, Hilbich C, Beyreuther K et al. Unusual cardioactive peptide (CCAP) from pericardial organs of the shore crab Carcinus maenas. Proc Natl Acad Sci USA 1987; 84:575-579.
122. Ewer J, Truman JW. Increases in cyclic 3′, 5′-guanosine monophosphate (cGMP) occur at ecdysis in an evolutionarily conserved crustacean cardioactive peptide-immunoreactive insect neuronal network. J Comp Neurol 1996; 370:330-341.
123. Gammie SC, Truman JW. Neuropeptide hierarchies and the activation of sequential motor behaviors in the hawkmoth, Manduca sexta. J Neurosci 1997; 17:4389-4397.

124. Park Y, Kim YJ, Dupriez V et al. Two subtypes of ecdysis-triggering hormone receptor in Drosophila melanogaster. J Biol Chem 2003; 278:17710-17715.
125. Iversen A, Cazzamali G, Williamson M et al. Molecular identification of the first insect ecdysis triggering hormone receptors. Biochem Biophys Res Commun 2002; 299:924-931.
126. Cazzamali G, Hauser F, Kobberup S et al. Molecular identification of a Drosophila G protein-coupled receptor specific for crustacean cardioactive peptide. Biochem Biophys Res Commun 2003; 303:146-152.
127. Sakai T, Satake H, Takeda M. Nutrient-induced alpha-amylase and protease activity is regulated by crustacean cardioactive peptide (CCAP) in the cockroach midgut. Peptides 2006; 27:2157-2164.
128. Veelaert D, Passier P, Devreese B et al. Isolation and characterization of an adipokinetic hormone release-inducing factor in locusts: the crustacean cardioactive peptide. Endocrinology 1997; 138:138-142.
129. Kim YJ, Spalovska-Valachova I, Cho KH et al. Corazonin receptor signaling in ecdysis initiation. Proc Natl Acad Sci USA 2004; 101:6704-6709.
130. Cazzamali G, Saxild N, Grimmelikhuijzen C. Molecular cloning and functional expression of a Drosophila corazonin receptor. Biochem Biophys Res Commun 2002; 298:31-36.
131. Badisco L, Claeys I, Van Loy T et al. Neuroparsins, a family of conserved arthropod neuropeptides. Gen Comp Endocrinol 2007; 153:64-71.
132. Altstein, M. Role of neuropeptides in sex pheromone production in moths. Peptides 2004; 25:1491-1501.
133. Rosenkilde C, Cazzamali G, Williamson M et al. Molecular cloning, functional expression and gene silencing of two Drosophila receptors for the Drosophila neuropeptide pyrokinin-2. Biochem Biophys Res Commun 2003; 309:485-494.
134. Cazzamali G, Torp M, Hauser F et al. The Drosophila gene CG9918 codes for a pyrokinin-1 receptor. Biochem Biophys Res Commun 2005; 335:14-19.
135. Hull JJ, Ohnishi A, Moto K et al. Cloning and characterization of the pheromone biosynthesis activating neuropeptide receptor from the silkmoth, Bombyx mori. Significance of the carboxyl terminus in receptor internalization. J Biol Chem 2004; 279:51500-51507.
136. Choi MY, Fuerst EJ, Rafaeli A et al. Identification of a G protein-coupled receptor for pheromone biosynthesis activating neuropeptide from pheromone glands of the moth Helicoverpa zea. Proc Natl Acad Sci USA 2003; 100:9721-9726.
137. Zheng L, Lytle C, Njauw CN et al. Cloning and characterization of the pheromone biosynthesis activating neuropeptide receptor gene in Spodoptera littoralis larvae. Gene 2007; 393:20-30.
138. Monnier D, Colas JF, Rosay P et al. NKD, a developmentally regulated tachykinin receptor in Drosophila. J Biol Chem 1992; 267:1298-1302.
139. Li XJ, Wolfgang W, Wu YN et al. Cloning, heterologous expression and developmental regulation of a Drosophila receptor for tachykinin-like peptides. EMBO J 1991; 10:3221-3229.
140. Torfs H, Shariatmadari R, Guerrero F et al. Characterization of a receptor for insect tachykinin-like peptide agonists by functional expression in a stable Drosophila Schneider 2 cell line. J Neurochem 2000; 74:2182-2189.
141. Poels J, Van Loy T, Franssens V et al. Substitution of conserved glycine residue by alanine in natural and synthetic neuropeptide ligands causes partial agonism at the stomoxytachykinin receptor. J Neurochem 2004; 90:472-478.
142. Luo CW, Dewey EM, Sudo S et al. Bursicon, the insect cuticle-hardening hormone, is a heterodimeric cystine knot protein that activates G protein-coupled receptor LGR2. Proc Natl Acad Sci USA 2005; 102:2820-2825.
143. Cazzamali G, Hauser F, Kobberup S et al. Molecular identification of a Drosophila G protein-coupled receptor specific for crustacean cardioactive peptide. Biochem Biophys Res Commun 2003; 303:146-152.

INDEX

E

F

G

H

I

K

L

M

N

O

P

R

S

T

U

V

Y